Lecture Notes in Mathematics 2257

More information about this series at http://www.springer.com/series/304

Maria Fragoulopoulou • Camillo Trapani

Locally Convex Quasi *-Algebras and their Representations

Springer

Maria Fragoulopoulou
Department of Mathematics
National and Kapodistrian University
of Athens
Athens, Greece

Camillo Trapani
Matematica e Informatica
Università degli Studi di Palermo
Palermo, Italy

ISSN 0075-8434 ISSN 1617-9692 (electronic)
Lecture Notes in Mathematics
ISBN 978-3-030-37704-5 ISBN 978-3-030-37705-2 (eBook)
https://doi.org/10.1007/978-3-030-37705-2

Mathematics Subject Classification (2010): Primary: 46H05, 46H15, 46H20; Secondary: 47L60

This Springer imprint is published by the registered company Springer Nature Switzerland AG.
The registered company address is: Gewerbestrasse 11, 6330 Cham, Switzerland

Contents

Chapter 1
Introduction

A locally convex quasi *-algebra is a pair consisting of a locally convex space $\mathfrak{A}[\tau]$ containing densely a *-algebra $\mathfrak{A}_0$, whose multiplication and involution extend to $\mathfrak{A}$, in the sense that for any $a \in \mathfrak{A}$ and $x \in \mathfrak{A}_0$, the elements ax, xa belong to $\mathfrak{A}$, in such a way that $(ax)^* = x^*a^*$, $(xa)^* = a^*x^*$ and moreover the left and right multiplications of the elements of $\mathfrak{A}$ by a fixed $x \in \mathfrak{A}_0$, as well as the involution on $\mathfrak{A}[\tau]$ are continuous. A typical example is obtained by taking as $\mathfrak{A}$ the completion of a locally convex *-algebra $\mathfrak{A}_0[\tau]$ with continuous involution and separately continuous multiplication.

The book in hands aims to present the essential aspects of the theory of locally convex quasi *-algebras as it has been developed so far. We begin with giving a brief account of their history and of the motivations behind their initiation.

Even though there are several familiar instances, where the preceding structure appears in a natural way (consider, for instance, the L^p spaces on an interval or the spaces of distributions), until the 1980s only a little attention had been paid, in general, to the interplay between the 'partially defined multiplication' and the topological structure they possess.

The theory of Banach and C*-algebras was recent enough (Banach algebras appeared in 1938 with Gelfand's thesis and C*-algebras first appeared in 1943 in the famous paper of Gelfand and Naimark [60]) to let researchers take into account a structure with a partial multiplication, looking as an exotic feature with very few or not so exciting consequences. Actually, the beauty of the theory of C*-algebras consists in the fact that an apparently simple condition, the well-known C*-condition, has such a large number of nontrivial consequences. The noncommutative Gelfand–Naimark theorem crowns the theory by showing that any C*-algebra can be realized as a C*-algebra of bounded linear operators on a Hilbert space. The main outcome of this is a unification of an abstract set-up with the theory of operator algebras, mostly developed by von Neumann. This deep result made of C*-algebras a very popular research subject evolving in several different and sometimes unexpected directions. Today, when thousands of papers on this subject

M. Fragoulopoulou, C. Trapani, *Locally Convex Quasi *-Algebras and their Representations*, Lecture Notes in Mathematics 2257,
https://doi.org/10.1007/978-3-030-37705-2_1

are at our disposal, we have a strong awareness of the wide range of applications that C*-algebras have in many fields of mathematics (noncommutative geometry, noncommutative analysis, harmonic analysis, group representations) and in physics (quantum statistical mechanics, quantum field theory).

The first suggestions to go towards partial algebraic structures came from quantum physics and from the theory of representations of Lie algebras. However, apart from the short discussion below, the reader will not find any reference to these theories in this book. Nevertheless, many books and papers on these subjects are listed in the bibliography. But, let us now come back to our brief historic overview.

In 1964 Haag and Kastler [61] proposed the so-called algebraic approach to quantum theories, whose cornerstone is the assumption that the local observables of a quantum system (that is physical quantities that can be measured within a region of finite measure) constitute a C*-algebra and global quantities are intended to be limits in the C*-norm of nets of local observables. Representing the C*-algebra of local observables with (necessarily) bounded operators, this approach provides the recipe for theoretical physicists to develop their computations on physical systems. So, after the publication of Haag and Kastler's paper, the research of physical models fitting into their set-up, has been on the stage and for a large number of models, mostly taken from quantum field theory and quantum statistical mechanics, this attempt has been successful. On the other hand, several models, like the Bose gas [40], go beyond this algebraic framework. This happens, in general, for models, where unbounded operators play a relevant role, and the representation of observables by means of bounded operators determines, in a sense, loss of information.

One possible way to include a larger number of models in the formulation of Haag and Kastler is to enlarge the algebraic set-up, considering also families of unbounded operators (and not only von Neumann algebras) for the representation of local observable algebras. For these reasons, algebras of unbounded operators, the so-called O*-algebras (see [42, 65, 72, 91] and [23], for a complete overview), partial *-algebras [2, 36] and quasi *-algebras [67, 68] have been studied, mainly from the mathematical point of view, deeply enough as to achieve the status of complete theories. In particular, the appearance of partial *-algebras and quasi *-algebras found its impetus in the non-neglectable number of instances in quantum theories, where one has to deal with sets of operators, whose multiplication is not defined for arbitrary pairs of elements. This unpleasant feature depends essentially on two facts: first, the operators, being in general, unbounded cannot be defined on the whole Hilbert space of states; second, they may not have a common invariant dense domain of definition (the existence of local von Neumann algebras, which do not leave the natural domain of Wightman fields invariant, has been shown by Horuzhy and Voronin in [62]). Following an idea of Borchers [56], Antoine and Karwowski [36] began an analysis of *partial *-algebras*, which was then continued in a long series of papers with Inoue, Mathot and one of us (CT) (see, e.g., [34, 35, 37, 38]). In 2002 a monograph [2] on this subject appeared, where the results obtained so far were systematically collected.

The problem of a rigorous mathematical description of the thermodynamical limit of the local Heisenberg dynamics for certain quantum statistical systems led Lassner, more or less in the same period, to introduce the notion of a *quasi *-algebra* [67, 68], which from the algebraic point of view, is a structure simpler than that of a general partial *-algebra and can be more easily cast into a topological framework.

As mentioned before, the most basic example of a locally convex quasi *-algebra is the completion of a locally convex *-algebra, whose involution is continuous and multiplication is separately continuous. Completions of this sort may really occur in quantum statistics. In fact, in this case, the observable algebra $\mathfrak{A}$, which is supposed to be a C*-algebra, does not contain, in general, the thermodynamical limit of the local Heisenberg dynamics. Then, the procedure to circumvent this difficulty, is to define on $\mathfrak{A}$ a locally convex topology τ, (which, in general, depends on the model under consideration) in such a way that the dynamics at the thermodynamical limit belongs to the completion of $\mathfrak{A}$ under τ.

More general structures than C*-algebras are also needed if one wants to represent Lie algebras by operators in infinite dimensional Hilbert spaces. In fact, the Heisenberg Lie algebra $\mathfrak{h}$ generated by three elements $a, b, c \in \mathfrak{h}$, whose Lie brackets are defined by

$$[a, b] = c, \quad [a, c] = [b, c] = 0,$$

does not possess any bounded representation. Actually, the Wiener–Wielandt theorem states that there exists no bounded representation π (with carrier space a Hilbert space $\mathcal{H}$) of the Heisenberg Lie algebra $\mathfrak{h}$, satisfying the equality $\pi([a, b]) = I$, where I is the identity operator on $\mathcal{H}$.

A familiar representation of $\mathfrak{h}$ is provided by a pair of operators P, Q satisfying the *canonical commutation relation* (CCR) $PQ - QP = -iI$, which in physics expresses the Heisenberg uncertainty principle. This relation is nothing, but a special representation π of pairs of operators satisfying the CCR condition, that allow the construction of a locally convex quasi *-algebra of pseudo differential operators (whose coefficients are distributions!) [69].

Operators acting on distribution spaces provide another interesting field of applications of locally convex quasi *-algebras, which is fairly important also from the mathematical point of view. Indeed, these arise in a natural way, when one considers the space $\mathfrak{L}(\mathcal{D}, \mathcal{D}^{\times})$ of continuous operators acting on *rigged Hilbert spaces* $\mathcal{D} \subset \mathcal{H} \subset \mathcal{D}'$ (under certain assumptions on the topology of $\mathcal{D}$). The most frequent situation in the applications occurs, when $\mathcal{D}$ is the space of C^{∞}-vectors of a self-adjoint operator H. The quasi *-algebra $\mathfrak{L}(\mathcal{D}, \mathcal{D}^{\times})$ plays also a role in Wightman quantum field theory: point-like fields satisfying a certain *regularity condition* are indeed elements of $\mathfrak{L}(\mathcal{D}, \mathcal{D}^{\times})$, for an appropriate domain $\mathcal{D}$. On the other hand, a smeared field satisfying the usual Wightman axioms on $\mathcal{D}$, gives rise to a point-like field with values in $\mathfrak{L}(\mathcal{D}, \mathcal{D}^{\times})$ [2, Chapter 11], [58].

These considerations motivate a study of locally convex quasi *-algebras, to which this book is devoted. Even though many aspects of the theory are yet unexplored, the amount of results that can be found in the literature is quite large and in authors' opinion it is time for a monographic synthesis on the subject, so that any interested researchers could have a reference guide to the theory developed so far.

This book is, in a sense, a cadet child of the monograph [2] coauthored by one of us (CT) and appeared in 2002. Like every child, it inherited something from his parents, but not so much as to be indistinguishable and, after all, seventeen years have not passed in vain. Thc book [2] was mainly concerned, as the title itself announced, with representations by means of families of unbounded operators. Only in Chap. 7 locally convex topologies on partial *-algebras were considered. Here, on the contrary the aspects of representation theory, which do not need topology, cover only Chap. 2. We do not claim originality of the contents of this book: they have mostly appeared in journal articles, with the possible exception of some parts of spectral theory, which are published here first time (a subsection in Chap. 4).

Chapter 2 is entirely dedicated to the algebraic aspects of the theory. After giving the basic definitions and properties of quasi *-algebras, we present a study of their *-representations. A *-representation of a given quasi *-algebra $(\mathfrak{A}, \mathfrak{A}_0)$ is a *-homomorphism π from $(\mathfrak{A}, \mathfrak{A}_0)$ into the *partial* *-algebra $\mathcal{L}^\dagger(\mathcal{D}, \mathcal{H})$ of closable linear operators defined on a dense domain $\mathcal{D}$ of a Hilbert space $\mathcal{H}$. These representations may be *unbounded,* in the sense that for some $a \in \mathfrak{A}$, $\pi(a)$ can be a true unbounded operator in $\mathcal{H}$. A standard way for studying *-representations is the use of the Gelfand–Naimark–Segal (GNS) construction, which allows to retrieve *-representations from certain positive linear functionals. The lack of an everywhere defined multiplication makes the situation more convenient, by replacing linear functionals with a family of positive sesquilinear forms on $\mathfrak{A}$ satisfying certain *invariance* properties. The possibility of performing a Hilbert space *-representation, starting either from an element of this class of positive sesquilinear forms, or from certain positive linear functionals on $\mathfrak{A}$, called representable, the classical GNS procedure is discussed, without having to assume, a priori, the existence of a topology on $\mathfrak{A}$.

In Chap. 3 we begin the analysis of locally convex quasi *-algebras starting from the simplest case: normed and Banach quasi *-algebras. This situation covers very familiar examples such as L^p-spaces (both commutative and noncommutative). In this chapter, we will mainly consider the case, where $(\mathfrak{A}[\|\cdot\|], \mathfrak{A}_0)$ is a *Banach quasi *-algebra.* This means, in a few words, that $\mathfrak{A}$ is a Banach space, whose norm $\|\cdot\|$ satisfies certain coupling properties related to the *partial* multiplication of $(\mathfrak{A}[\|\cdot\|], \mathfrak{A}_0)$. The main goal is the study of structure of normed quasi *-algebras and particular attention is paid to the role of certain families of positive sesquilinear forms and to the *-representations they define. Several examples are also discussed. In particular, since the range $\mathcal{L}^\dagger(\mathcal{D}, \mathcal{H})$ of *-representations carries a number of topologies, making of it a *locally convex* partial *-algebra, it makes sense to consider the problem of continuity of *-representations on topological quasi *-algebras. It turns out that in the case of a normed quasi *-algebra, this problem is closely linked with the normed structure of the quasi *-algebra under consideration: thus, as in the case of *-representations of Banach *-algebras, a certain amount of information on the structure of a Banach quasi *-algebra can be obtained from the knowledge of the properties of the family of its *-representations.

Chapter 4 is devoted to the special role the set $\mathfrak{A}_b$ of *bounded* elements of a normed quasi *-algebra $(\mathfrak{A}[\|\cdot\|], \mathfrak{A}_0)$ plays in the study of its structure. Bounded elements are characterized by the fact that the corresponding multiplication operators are bounded linear maps. More precisely, an element $a \in \mathfrak{A}$ is said to be bounded if both $L_a : x \in \mathfrak{A}_0[\|\cdot\|] \to ax \in \mathfrak{A}[\|\cdot\|]$ and $R_a : x \in \mathfrak{A}_0[\|\cdot\|] \to xa \in \mathfrak{A}[\|\cdot\|]$ are bounded linear maps. Then, we focus our attention to the class of normal Banach quasi *-algebras: they are characterized by the fact that $\mathfrak{A}_b$ is a Banach *-algebra. If $(\mathfrak{A}[\|\cdot\|], \mathfrak{A}_0)$ is normal, the Banach *-algebra $\mathfrak{A}_b$ allows the notion of *spectrum* of an element $a \in \mathfrak{A}$, which enjoys properties analogous to the spectrum of an element in a Banach *-algebra. An important role in this study, in particular for the analysis of *-representations, is played by two seminorms $\mathfrak{p}$, $\mathfrak{q}$ (defined in Chap. 3) that emulate the Gelfand–Naimark seminorm on a Banach *-algebra (but $\mathfrak{q}$ is only defined on a domain $D(\mathfrak{q}) \subseteq \mathfrak{A}$: it is actually an *unbounded* C*-seminorm, in the sense of [46, 55, 63, 78]). This approach, already extensively used in the study of locally convex *-algebras [54, 55, 59] has given a quite deep insight into their structure, in particular for the existence of *well-behaved* *-representations (see, also [54, 74]).

Chapter 5 deals with *CQ*-algebras*. From the historical point of view, the first results on the Banach case were obtained by F. Bagarello and CT, in 1990s [48, 49], for the so called CQ*-algebras, which roughly speaking, are completions of C*-algebras with respect to a second norm defining a topology coarser than that induced by the C*-norm. At the beginning, a quite general construction was performed, giving large room to the case, where the involution of the C*-algebra was not necessarily the same like that of the Banach space enclosing the whole structure. This may appear to be a little artificial, but in a cooperation of F. Bagarello, A. Inoue and CT was realized that this was the right environment, where some aspects of the Tomita–Takesaki theory could be naturally cast [47]. On the other hand, the family of *proper* CQ*-algebras enters more directly in the framework of locally convex quasi *-algebras and a lot of results were demonstrated. Here, it is worth mentioning the characterization of *-semisimple proper CQ*-algebras in terms of function spaces [50] in the commutative case, or in terms of measurable operators in the noncommutative case [52].

With Chap. 6, we finally go to a more general set-up. The main topic is, in fact, to give conditions under which a locally convex quasi *-algebra $(\mathfrak{A}[\tau], \mathfrak{A}_0)$ attains sufficiently many (τ, t_w)-continuous *-representations in $\mathcal{L}^\dagger(\mathcal{D}, \mathcal{H})$, to separate its points. This leads to the notion of fully-representable quasi *-algebras, which are explored in detail. Once we have at our disposal a sufficiently large family of *-representations, a usual notion of bounded elements on $\mathfrak{A}[\tau]$ rises. Of course, this generalizes what was done in Chap. 5, for the Banach case. On the other hand, a natural order exists on $(\mathfrak{A}[\tau], \mathfrak{A}_0)$ related to the topology τ, which also leads to a different kind of bounded elements, which we call *order bounded*. Under certain conditions, the latter notion of boundedness coincides with the usual one. In a fully representable quasi *-algebra $(\mathfrak{A}[\tau], \mathfrak{A}_0)$ a *weak* partial multiplication $\square$ can be introduced and an *unbounded* C*-seminorm $\|\cdot\|_b$ is defined on $\mathfrak{A}$ (by means of the order boundedness), with domain the partial *-subalgebra $\mathfrak{A}_b$ of $\mathfrak{A}[\tau]$ consisting of all order bounded elements. In this way, under certain conditions, $\mathfrak{A}_b$ becomes a C*-algebra, with respect to the weak multiplication $\Box$ and the C*-norm $\|\cdot\|_b$.

Chapter 7 is closely connected with Chaps. 5 and 6. In fact, the attention is again focused on the completion of a given C*-algebra $\mathfrak{A}_0$, but this time with respect to a second topology that is not necessarily a normed one. Clearly, the techniques to be used become more and more complicated and also the understanding becomes more difficult. The study of the structure and representation theory of the completion of a (normed) C*-algebra $\mathfrak{A}_0[\|\cdot\|_0]$, under a locally convex *-algebra topology τ on $\mathfrak{A}_0$, was started in [44] and continued in [59]. When the multiplication of $\mathfrak{A}_0$, with respect to τ is jointly continuous, the completion $\widetilde{\mathfrak{A}}_0[\tau]$ of $\mathfrak{A}_0[\tau]$ is a GB*-algebra (of unbounded operators) over the unit ball $\mathcal{U}(\mathfrak{A}_0) = \{x \in \mathfrak{A}_0 : \|x\|_0 \leq 1\}$ of $\mathfrak{A}_0[\|\cdot\|_0]$, if and only if, $\mathcal{U}(\mathfrak{A}_0)$ is τ-closed [59, Corollary 2.2]. When the multiplication of $\mathfrak{A}_0$, with respect to τ is just separately continuous, $\widetilde{\mathfrak{A}}_0[\tau]$ may fail to be a locally convex $*$-algebra, but may well carry the structure of a quasi $*$-algebra. First studies on the properties and the $*$-representation theory of $\widetilde{\mathfrak{A}}_0[\tau]$, in this case, were done in [44, Section 3] and [59, Section 3]. These studies have led to the introduction of *locally convex quasi C*-algebras*, which are discussed in this chapter. As usually, a more general set-up requires a strengthening of several notions and adaptation of many others. This is the reason why two notions of positivity are introduced in the locally convex quasi *-algebra $\widetilde{\mathfrak{A}}_0[\tau]$. One is called just "positivity" and the second one "commutatively positivity" (see Definition 7.1.1 and after that discussion before (T_3)). Locally convex quasi C*-algebras are then defined and some examples from various classes of topological algebras are discussed. The chapter continues with a study of locally convex quasi C*-algebras of operators and with the analysis of the structure of commutative locally convex quasi C*-algebras, which is discussed by taking into account [32, Section 6] and [44, 59]. By applying the results of the previous sections and also some ideas developed in [57, Section 4] and [59] one obtains a functional calculus for the positive elements of a commutative locally convex quasi C*-algebra. Furthermore, if $\mathfrak{A}[\tau]$ is a noncommutative locally convex quasi C*-algebra, necessary and sufficient conditions are given for $\mathfrak{A}[\tau]$ to be continuously embedded in a locally convex quasi C*-algebra of operators, obtaining thus Gelfand–Naimark type theorems for the aforementioned topological quasi *-algebras. In the final part of this chapter, some results concerning the possibility of representing a noncommutative locally convex quasi C*-algebra as locally convex quasi *-algebra of operators, measurable in the sense of Segal and Nelson, are presented.

Two appendices close this book: in the first one we collect results on *-algebras and their representations, so that the reader can find a handy reference to classical results of the theory of Banach (*-)algebras, as well as of C*-algebras that are systematically used throughout this volume. The second appendix is a short exhibition of the basic theorems and facts of (bounded and unbounded) operator theory. We believe they can be useful, not only for a better comprehension of this book, but also as a guide text for mini courses on these topics. Unfortunately the contents of the two appendices do not exhaust what is needed for reading this book:

familiarity is also presumed with general functional analysis, harmonic analysis, operator theory and von Neumann algebras theory.

Finally, at the end of the book, we add a list of further references, under the name "Suggested Readings", in order to provide the young researchers with an adequate literature that will possibly help them to explore further, topological (quasi) *-algebras and their applications. We also remark that the bibliography of [2] provides the reader with an extensive literature covering the representation theory of unbounded operator algebras and not only.

Last but not least, we want to extend our heartfelt thanks to all those, who have cooperated with us in these years in writing the papers from which this book originated and also to colleagues and students involved with us into discussions on the topics of the present work.

In the end, we want to express our gratitude to the *Gruppo Nazionale per l'Analisi Matematica, la Probabilità e le loro Applicazioni* of the *Istituto Nazionale di Alta Matematica*, whose financial support has allowed us several mutual visits during the preparation of this book.

Chapter 2
Algebraic Aspects

The notion of a *quasi *-algebra* (see Definition 2.1.1 below) was first introduced by G. Lassner in the early 80s of last century ([67, 68], see also [23]). In this chapter, we discuss some algebraic aspects of the theory of these *-algebras. All vector spaces or algebras considered throughout this book are over the field $\mathbb{C}$ of complexes and all locally convex spaces are supposed to be Hausdorff. Our basic definitions and notation mainly come from [2].

2.1 Definitions and Examples

2.1.1 Quasi *-Algebras and Partial *-Algebras

Definition 2.1.1 A *quasi *-algebra* $(\mathfrak{A}, \mathfrak{A}_0)$ is a pair consisting of a vector space $\mathfrak{A}$ and a *-algebra $\mathfrak{A}_0$ contained in $\mathfrak{A}$ as a subspace, such that

(i) $\mathfrak{A}$ carries an involution $a \mapsto a^*$ extending the involution of $\mathfrak{A}_0$;
(ii) $\mathfrak{A}$ is a bimodule over $\mathfrak{A}_0$ and the module multiplications extend the multiplication of $\mathfrak{A}_0$; In particular, the following associative laws hold:

$$(xa)y = x(ay); \quad a(xy) = (ax)y, \quad \forall a \in \mathfrak{A},\ x, y \in \mathfrak{A}_0; \tag{2.1.1}$$

(iii) $(ax)^* = x^*a^*$, for every $a \in \mathfrak{A}$ and $x \in \mathfrak{A}_0$.

We say that a quasi *-algebra $(\mathfrak{A}, \mathfrak{A}_0)$ is *unital*, if there is an element $e \in \mathfrak{A}_0$, such that $ae = a = ea$, for all $a \in \mathfrak{A}$; e is unique and called *unit* of $(\mathfrak{A}, \mathfrak{A}_0)$.

We say that $(\mathfrak{A}, \mathfrak{A}_0)$ has a *quasi-unit* if there exists an element $q \in \mathfrak{A}$, such that $qx = xq = x$, for every $x \in \mathfrak{A}_0$. It is clear that the unit e, if any, is a quasi-unit but the converse is false, in general.

M. Fragoulopoulou, C. Trapani, *Locally Convex Quasi *-Algebras and their Representations*, Lecture Notes in Mathematics 2257,
https://doi.org/10.1007/978-3-030-37705-2_2

A quasi *-algebra is a particular type of a *partial *-algebra*, in the sense of the following definition.

Definition 2.1.2 A *partial *-algebra* is a complex vector space $\mathfrak{A}$, endowed with an involution $a \mapsto a^*$ (that is, a bijection, such that $a^{**} = a$, for all $a \in \mathfrak{A}$) and a partial multiplication defined by a set $\Gamma \subset \mathfrak{A} \times \mathfrak{A}$ (a binary relation), with the following properties

(i) $(a, b) \in \Gamma$ implies $(b^*, a^*) \in \Gamma$;
(ii) $(a, b_1), (a, b_2) \in \Gamma$ implies $(a, \lambda b_1 + \mu b_2) \in \Gamma$, $\forall\, \lambda, \mu \in \mathbb{C}$;
(iii) for any $(a, b) \in \Gamma$, a product $a \cdot b \in \mathfrak{A}$ is defined, which is distributive with respect to the addition and satisfies the relation $(a \cdot b)^* = b^* \cdot a^*$.

We shall assume that the partial *-algebra $\mathfrak{A}$ contains a unit e, if

$$e^* = e,\ (e, a) \in \Gamma,\ \forall\, a \in \mathfrak{A} \text{ and } e \cdot a = a \cdot e = a, \quad \forall\, a \in \mathfrak{A}.$$

If $\mathfrak{A}$ has no unit, it may always be embedded into a larger partial *-algebra with unit (the so-called *unitization of* $\mathfrak{A}$), in the standard fashion [37].

Given the defining set Γ, spaces of multipliers are defined in obvious way. More precisely,

$$(a, b) \in \Gamma \iff a \in L(b) \text{ i.e., } a \text{ is a left multiplier of } b$$
$$\text{and } b \in R(a) \text{ i.e., } b \text{ is a right multiplier of } a.$$

By Definition 2.1.2, (ii), $L(a)$, $L(b)$ are clearly vector subspaces of $\mathfrak{A}$.
For any subset $\mathfrak{N} \subset \mathfrak{A}$, we put

$$L\mathfrak{N} = \bigcap_{a \in \mathfrak{N}} L(a), \quad R\mathfrak{N} = \bigcap_{a \in \mathfrak{N}} R(a)$$

and, of course, the involution exchanges 'L and R', as follows

$$(L\mathfrak{N})^* = R\mathfrak{N}^*, \quad (R\mathfrak{N})^* = L\mathfrak{N}^*.$$

Clearly all these multiplier spaces are vector subspaces of $\mathfrak{A}$, containing e.

A partial *-algebra $\mathfrak{A}$ is called *abelian* if $L(a) = R(a)$, for all $a \in \mathfrak{A}$ and

$$a \cdot b = b \cdot a, \quad \forall\, b \in R(a).$$

In that case, we simply write for the multiplier spaces

$$L(a) = R(a) \equiv M(a),\ L\mathfrak{N} = R\mathfrak{N} \equiv M\mathfrak{N},\ (\text{where } \mathfrak{N} \subset \mathfrak{A}).$$

Notice that the partial multiplication is *not* required to be associative (and often it is not). A partial *-algebra $\mathfrak{A}$ is said to be *associative* if the following condition holds for any $a, b, c \in \mathfrak{A}$: whenever $a \in L(b)$, $b \in L(c)$ and $a \cdot b \in L(c)$, then $b \cdot c \in R(a)$ and

$$(a \cdot b) \cdot c = a \cdot (b \cdot c). \tag{2.1.2}$$

This condition is rather strong and rarely realized in practice. However, a weaker notion is sometimes useful.

A partial *-algebra $\mathfrak{A}$ is said to be *semi-associative* if $b \in R(a)$ implies $b \cdot c \in R(a)$, for every $c \in R\mathfrak{A}$; then (2.1.2) holds. Of course, if the partial *-algebra $\mathfrak{A}$ is semi-associative, both $R\mathfrak{A}$ and $L\mathfrak{A}$ are algebras.

➽ From here on, *we shall simply write ab for the product* $a \cdot b$.

A quasi *-algebra is a partial *-algebra for which $L(a) = R(a) = \mathfrak{A}_0$, if $a \in \mathfrak{A}$, and $L(x) = R(x) = \mathfrak{A}$, if $x \in \mathfrak{A}_0$. In other words, if $(\mathfrak{A}, \mathfrak{A}_0)$ is a quasi *-algebra, then $\mathfrak{A}$ is a partial *-algebra, where the product ab of $a, b \in \mathfrak{A}$ is well-defined if either $a \in \mathfrak{A}_0$ or $b \in \mathfrak{A}_0$. It is clear that a quasi *-algebra is semi-associative, because of (2.1.1).

2.1.2 Basic Examples

Example 2.1.3 Let $\mathfrak{A}_0$ be a locally convex *-algebra with topology τ (the involution * is supposed to be τ-continuous). If the multiplication in $\mathfrak{A}_0$ is jointly continuous with respect to τ, the completion $\mathfrak{A} := \widetilde{\mathfrak{A}_0}^{\tau}$ of $\mathfrak{A}_0[\tau]$ is a complete locally convex *-algebra. The operations in $\mathfrak{A}$ are defined by a limiting process. In particular, the multiplication of two elements $a, b \in \mathfrak{A}$ is defined as the limit of the products of elements from two nets $\{x_\alpha\}$, $\{y_\beta\}$ in $\mathfrak{A}_0$ approximating respectively, a and b. If the multiplication is only separately continuous, an analogous procedure can be used only if one takes an arbitrary fixed element in $\mathfrak{A}_0$. More precisely, if $a = \lim_\alpha x_\alpha$, $\{x_\alpha\} \subset \mathfrak{A}_0$, then one can define, for each $y \in \mathfrak{A}_0$,

$$ay = \lim_\alpha x_\alpha y \quad \text{and} \quad ya = \lim_\alpha y x_\alpha.$$

Thus, for $\mathfrak{A} := \widetilde{\mathfrak{A}_0}^{\tau}$, the pair $(\mathfrak{A}, \mathfrak{A}_0)$ is, in general, only a quasi *-algebra. This example is quite important because it was the starting point of the theory [67, 68].

Example 2.1.4 Let $I = [0, 1]$. Then the pair of Lebesgue spaces $(L^p(I), L^\infty(I))$ (with the Lebesgue measure dt on I), $1 \leq p < \infty$, is a quasi *-algebra, with respect to the usual operations. The involution is the complex conjugation.

Example 2.1.5 Let $\mathcal{S}(\mathbb{R})$ be the Schwartz space of rapidly decreasing C^∞-functions on the real line and $\mathcal{S}'(\mathbb{R})$ the space of tempered distributions. As it is well known, $\mathcal{S}(\mathbb{R})$, which is a *-algebra with respect to the usual operations and complex conjugation, can be identified with a subspace of $\mathcal{S}'(\mathbb{R})$. The (commutative) multiplication of an element of $\mathcal{S}'(\mathbb{R})$ and one of $\mathcal{S}(\mathbb{R})$ is defined as follows

$$(\phi F)(\psi) = (F\phi)(\psi) := F(\phi\psi), \quad \phi, \psi \in \mathcal{S}(\mathbb{R}),\ F \in \mathcal{S}'(\mathbb{R}).$$

It is readily checked that $(\mathcal{S}'(\mathbb{R}), \mathcal{S}(\mathbb{R}))$ is a quasi *-algebra. For a more detailed discussion see [73, 88].

Example 2.1.6 Let $\mathcal{D}$ be a dense linear subspace of a Hilbert space $\mathcal{H}$ and t a locally convex topology on $\mathcal{D}$, finer than the topology induced by the Hilbert norm. Then, the space $\mathcal{D}^\times$ of all continuous conjugate linear functionals on $\mathcal{D}[t]$, i.e., the conjugate dual of $\mathcal{D}[t]$, is a linear vector space that *contains* $\mathcal{H}$, in the sense that $\mathcal{H}$ can be identified with a subspace of $\mathcal{D}^\times$. These identifications imply that the sesquilinear form $B(\cdot,\cdot)$ that puts $\mathcal{D}$ and $\mathcal{D}^\times$ in duality is an extension of the inner product of $\mathcal{D}$; i.e., $B(\xi, \eta) = \langle \xi | \eta \rangle$, for every $\xi, \eta \in \mathcal{D}$ (to simplify notation we adopt the symbol $\langle \cdot | \cdot \rangle$ for both of them). The space $\mathcal{D}^\times$ will always be considered as endowed with the *strong dual topology* $t^\times = \beta(\mathcal{D}^\times, \mathcal{D})$. The Hilbert space $\mathcal{H}$ is dense in $\mathcal{D}^\times[t^\times]$.

We get in this way a *Gelfand triplet* or *rigged Hilbert space* (RHS) [2, 3]

$$\mathcal{D}[t] \hookrightarrow \mathcal{H} \hookrightarrow \mathcal{D}^\times[t^\times], \tag{2.1.3}$$

where $\hookrightarrow$ denotes a continuous embedding with dense range. As it is usual, we will systematically read (2.1.3) as a chain of inclusions and we will write $\mathcal{D}[t] \subset \mathcal{H} \subset \mathcal{D}^\times[t^\times]$ or $(\mathcal{D}[t], \mathcal{H}, \mathcal{D}^\times[t^\times])$ for denoting a RHS.

Let $\mathfrak{L}(\mathcal{D}, \mathcal{D}^\times)$ denote the vector space of all continuous linear maps from $\mathcal{D}[t]$ into $\mathcal{D}^\times[t^\times]$.

Let $\mathcal{D}[t] \hookrightarrow \mathcal{H} \hookrightarrow \mathcal{D}^\times[t^\times]$ be a rigged Hilbert space. We consider the following spaces of continuous linear maps

- $\mathfrak{L}(\mathcal{D}, \mathcal{D}^\times)$: the linear space of all continuous linear maps from $\mathcal{D}[t]$ into $\mathcal{D}^\times[t^\times]$;
- $\mathfrak{L}(\mathcal{D})$: the algebra of all continuous linear maps from $\mathcal{D}[t]$ into itself;
- $\mathfrak{L}(\mathcal{D}^\times)$: the algebra of all continuous linear maps from $\mathcal{D}^\times[t^\times]$ into itself.

Both $\mathfrak{L}(\mathcal{D})$ and $\mathfrak{L}(\mathcal{D}^\times)$ can be regarded as subspaces of $\mathfrak{L}(\mathcal{D}, \mathcal{D}^\times)$, in the sense that $\mathfrak{L}(\mathcal{D}, \mathcal{D}^\times)$ contains subspaces isomorphic to $\mathfrak{L}(\mathcal{D})$ and $\mathfrak{L}(\mathcal{D}^\times)$.

If $X \in \mathfrak{L}(\mathcal{D}, \mathcal{D}^\times)$ we can define the *adjoint* $X^\dagger$ of X by the equality

$$\langle X\xi | \eta \rangle = \overline{\langle X^\dagger \eta | \xi \rangle}, \quad \forall\, \xi, \eta \in \mathcal{D}.$$

Let us now assume that $\mathcal{D}[t]$ is a reflexive space so that both the topologies t and $t^\times$ coincide with the corresponding Mackey topologies of the conjugate dual pair $(\mathcal{D}, \mathcal{D}^\times)$. In this case, we have

1. The map $X \mapsto X^\dagger$ is an involution of $\mathfrak{L}(\mathcal{D}, \mathcal{D}^\times)$, i.e., $X^{\dagger\dagger} = X$, for each $X \in \mathfrak{L}(\mathcal{D}, \mathcal{D}^\times)$.
2. $\mathfrak{L}(\mathcal{D})^\dagger = \mathfrak{L}(\mathcal{D}^\times)$ and $\mathfrak{L}(\mathcal{D}) \cap \mathfrak{L}(\mathcal{D}^\times)$ is a *-algebra.

3. For each $X \in \mathfrak{L}(\mathcal{D}, \mathcal{D}^\times)$ and $Y \in \mathfrak{L}(\mathcal{D})$ one has $(XY)^\dagger = Y^\dagger X^\dagger$.
4. For each $X \in \mathfrak{L}(\mathcal{D}, \mathcal{D}^\times)$ and $Z \in \mathfrak{L}(\mathcal{D}^\times)$ one has $(ZX)^\dagger = X^\dagger Z^\dagger$.

Let us denote by $\mathcal{L}^\dagger(\mathcal{D})$ the space of all linear operators $X : \mathcal{D} \to \mathcal{D}$, having an adjoint $X^\dagger : \mathcal{D} \to \mathcal{D}$, by which we simply mean that

$$\langle X\xi|\eta\rangle = \langle \xi|X^\dagger \eta\rangle, \quad \forall\, \xi, \eta \in \mathcal{D}.$$

The space $\mathfrak{L}^\dagger(\mathcal{D})$ of continuous elements of $\mathcal{L}^\dagger(\mathcal{D})$, i.e.,

$$\mathfrak{L}^\dagger(\mathcal{D}) \equiv \left\{X \in \mathcal{L}^\dagger(\mathcal{D}) : X \in \mathfrak{L}(\mathcal{D}),\ X^\dagger \in \mathfrak{L}(\mathcal{D})\right\} \tag{2.1.4}$$

is a *-algebra and $\mathfrak{L}^\dagger(\mathcal{D}) \subset \mathfrak{L}(\mathcal{D}, \mathcal{D}^\times)$.

Moreover, $\mathfrak{L}^\dagger(\mathcal{D}) = \mathfrak{L}(\mathcal{D}) \cap \mathfrak{L}(\mathcal{D}^\times)$. Thus, from the previous considerations it follows that $(\mathfrak{L}(\mathcal{D}, \mathcal{D}^\times), \mathfrak{L}^\dagger(\mathcal{D}))$ is a quasi *-algebra.

2.1.3 Quasi *-Algebras and Partial *-Algebras of Operators

Some families of unbounded operators in Hilbert spaces can be cast into the framework of quasi or partial *-algebras. This fact is important and will be used throughout this book since the partial *-algebra of operators (or partial O*-algebras [2]), we are going to introduce, is the space, where representations will take values. Let $\mathcal{D}$ be a dense subspace of a Hilbert space $\mathcal{H}$. We denote by $\mathcal{L}^\dagger(\mathcal{D}, \mathcal{H})$ the set of all (closable) linear operators X in $\mathcal{H}$, such that $D(X) = \mathcal{D}$, $D(X^*) \supseteq \mathcal{D}$, where $D(X)$ denotes the domain of X.

The set $\mathcal{L}^\dagger(\mathcal{D}, \mathcal{H})$ is a partial *-algebra with respect to the following operations: the usual sum $X_1 + X_2$, the scalar multiplication λX, the involution $X \mapsto X^\dagger = X^* \upharpoonright \mathcal{D}$ and the (*weak*) partial multiplication $X_1 \Box X_2 = {X_1}^{\dagger *} X_2$ (where ${X_1}^{\dagger *} \equiv ({X_1}^\dagger)^*$). The latter is defined whenever X_2 is a *weak right multiplier* of X_1 (for this, we shall write $X_2 \in R^{\mathrm{w}}(X_1)$ or $X_1 \in L^{\mathrm{w}}(X_2)$, that is, if and only if, $X_2\mathcal{D} \subset D({X_1}^{\dagger *})$ and $X_1^\dagger \mathcal{D} \subset D({X_2}^*)$. The operator $\mathbb{I}_\mathcal{D}$, restriction to $\mathcal{D}$ of the identity operator $\mathbb{I}$ on $\mathcal{H}$, is the unit of the partial *-algebra $\mathcal{L}^\dagger(\mathcal{D}, \mathcal{H})$. By $\mathcal{L}^\dagger(\mathcal{D}, \mathcal{H})_b$ we shall denote the *bounded part* of $\mathcal{L}^\dagger(\mathcal{D}, \mathcal{H})$; i.e.,

$$\mathcal{L}^\dagger(\mathcal{D}, \mathcal{H})_b = \left\{X \in \mathcal{L}^\dagger(\mathcal{D}, \mathcal{H}) \colon \overline{X} \in \mathcal{B}(\mathcal{H})\right\},$$

where $\overline{X}$ is the closure of X, i.e., a minimal closed extension of X. Recall that when an operator X admits a closed extension is called *closable* and in this case, there exists a minimal closed extension of it, denoted by $\overline{X}$ and called *closure* of X [23, p. 28]. The space $\mathcal{L}^\dagger(\mathcal{D})$, already introduced in Example 2.1.6 can be identified with the subspace of $\mathcal{L}^\dagger(\mathcal{D}, \mathcal{H})$ consisting of all elements in $\mathcal{L}^\dagger(\mathcal{D}, \mathcal{H})$ that leave, together with their adjoints, the domain $\mathcal{D}$ invariant. It is clear that if $X_1, X_2 \in$

$\mathcal{L}^\dagger(\mathcal{D})$, then $(X_1 \square X_2)\xi = X_1{}^{\dagger *} X_2 \xi = X_1 X_2 \xi$, for every $\xi \in \mathcal{D}$. Hence, $\mathcal{L}^\dagger(\mathcal{D})$ is a *-algebra with respect to the usual algebraic operations.

Let $\mathcal{L}^\dagger(\mathcal{D})_b$ denote the bounded part of $\mathcal{L}^\dagger(\mathcal{D})$; i.e., $\mathcal{L}^\dagger(\mathcal{D})_b = \mathcal{L}^\dagger(\mathcal{D}, \mathcal{H})_b \cap \mathcal{L}^\dagger(\mathcal{D})$. Then, $(\mathcal{L}^\dagger(\mathcal{D}, \mathcal{H}), \mathcal{L}^\dagger(\mathcal{D})_b)$ is a quasi *-algebra.

Remark 2.1.7 In general, $(\mathcal{L}^\dagger(\mathcal{D}, \mathcal{H}), \mathcal{L}^\dagger(\mathcal{D}))$ is not a quasi *-algebra; indeed, if $X \in \mathcal{L}^\dagger(\mathcal{D}, \mathcal{H})$ and $Y \in \mathcal{L}^\dagger(\mathcal{D})$, we cannot say that $X\mathcal{D} \subset D(Y^{\dagger *})$. It is clear that the latter is obvious, when Y is bounded.

For later use, we remind the definitions of the main topologies usually considered on $\mathcal{L}^\dagger(\mathcal{D}, \mathcal{H})$. All these topologies are locally convex, making $\mathcal{L}^\dagger(\mathcal{D}, \mathcal{H})$ into a locally convex space. They are defined as follows:

- the *weak topology* t_w, defined by the family of seminorms $\{p_{\xi,\eta}\}$, with $p_{\xi,\eta}(X) = |\langle X\xi | \eta \rangle|$, $X \in \mathcal{L}^\dagger(\mathcal{D}, \mathcal{H})$, $\xi, \eta \in \mathcal{D}$;
- the *strong topology* t_s, defined by the family of seminorms $\{p_\xi\}$, with $p_\xi(X) = \|X\xi\|$, $X \in \mathcal{L}^\dagger(\mathcal{D}, \mathcal{H})$, $\xi \in \mathcal{D}$;
- the *strong* topology* t_{s^*}, defined by the family of seminorms $\{p^*_\xi\}$, with $p^*_\xi(X) = \max\left\{\|X\xi\|, \|X^\dagger\xi\|\right\}$, $X \in \mathcal{L}^\dagger(\mathcal{D}, \mathcal{H})$, $\xi \in \mathcal{D}$.

Clearly, t_{s^*} is finer than t_s, which in turn is finer than t_w, in general. The involution * is continuous with respect to t_w and t_{s^*}, while none of these topologies makes the weak multiplication continuous.

2.2 Homomorphisms, Ideals and Representations

Definition 2.2.1 Let $(\mathfrak{A}, \mathfrak{A}_0)$ be a quasi *-algebra and $\mathfrak{M}$ a subspace of $\mathfrak{A}$. Consider the conditions:

(i) $a \in \mathfrak{M} \Leftrightarrow a^* \in \mathfrak{M}$;
(ii) $xa \in \mathfrak{M}$, for every $a \in \mathfrak{M}$, $x \in \mathfrak{A}_0$;

if (i) and (ii) hold, then $\mathfrak{M}$ is called an $\mathfrak{A}_0$-**-submodule of* $\mathfrak{A}$. If, in addition, $\mathfrak{A}_0 \subset \mathfrak{M}$, then $(\mathfrak{M}, \mathfrak{A}_0)$ is called a *quasi *-subalgebra of* $(\mathfrak{A}, \mathfrak{A}_0)$.

Remark 2.2.2 Clearly (i) and (ii) of the previous definition imply that $ax \in \mathfrak{M}$, for every $a \in \mathfrak{M}$, $x \in \mathfrak{A}_0$. Moreover, if $(\mathfrak{A}, \mathfrak{A}_0)$ has a quasi-unit q, then for every $\mathfrak{A}_0$-*-submodule $\mathfrak{M}$ of $\mathfrak{A}$, such that $q \in \mathfrak{M}$, $(\mathfrak{M}, \mathfrak{A}_0)$ is a quasi *-subalgebra of $(\mathfrak{A}, \mathfrak{A}_0)$.

Definition 2.2.3 Let $(\mathfrak{A}, \mathfrak{A}_0)$ be a quasi *-algebra and $\mathfrak{B}$ a partial *-algebra. A linear map Φ from $\mathfrak{A}$ into $\mathfrak{B}$ is called a *homomorphism* of $(\mathfrak{A}, \mathfrak{A}_0)$ into $\mathfrak{B}$ if, for $a \in \mathfrak{A}$ and $x \in \mathfrak{A}_0$, $\Phi(a)\Phi(x)$ and $\Phi(x)\Phi(a)$ are well defined in $\mathfrak{B}$ and $\Phi(a)\Phi(x) = \Phi(ax)$, $\Phi(x)\Phi(a) = \Phi(xa)$, respectively. The homomorphism Φ is called a **-homomorphism* if $\Phi(a^*) = \Phi(a)^*$, for every $a \in \mathfrak{A}$.

If $(\mathfrak{B}, \mathfrak{B}_0)$ is a quasi *-algebra and Φ is a homomorphism (resp., *-homomorphism) of $(\mathfrak{A}, \mathfrak{A}_0)$ into $\mathfrak{B}$, then the above definition implies that $(\Phi(a), \Phi(b)) \in \mathfrak{B} \times \mathfrak{B}_0 \cup \mathfrak{B}_0 \times \mathfrak{B}$, whenever $a, b \in \mathfrak{A}$ and either $a \in \mathfrak{A}_0$ or $b \in \mathfrak{A}_0$. This in turn implies that $\Phi(\mathfrak{A}_0) \subseteq \mathfrak{B}_0$. For this reason, we call Φ a *qu-homomorphism* (resp., *qu*-homomorphism*). If Φ is a qu*-homomorphism, $(\Phi(\mathfrak{A}), \Phi(\mathfrak{A}_0))$ is a quasi *-subalgebra of $(\mathfrak{B}, \mathfrak{B}_0)$.

Definition 2.2.4 Let $(\mathfrak{A}, \mathfrak{A}_0)$ be a quasi *-algebra and $\mathfrak{I}$ a subspace of $\mathfrak{A}$. We say that $\mathfrak{I}$ is a *left qu-ideal* if $ax \in \mathfrak{I}$, for every $x \in \mathfrak{I}$ and $a \in L(x)$. Similarly, $\mathfrak{I}$ is a *right qu-ideal* if $xa \in \mathfrak{I}$, for every $x \in \mathfrak{I}$ and $a \in R(x)$. Finally, $\mathfrak{I}$ is a (two-sided) *qu-ideal* if it is both a left and right qu-ideal. A left or right qu-ideal $\mathfrak{I}$ is a qu*-ideal if $a \in \mathfrak{I}$ implies $a^* \in \mathfrak{I}$. A qu*-ideal is necessarily a two-sided qu-ideal.

The kernel of a *-homomorphism is clearly a qu*-ideal.

If $\mathfrak{I}$ is a left qu-ideal, then $\mathfrak{I}$ is a left $\mathfrak{A}_0$-module, since $\mathfrak{A}_0 \subset L\mathfrak{I}$.

Let $\mathfrak{I}$ be a qu*-ideal of $\mathfrak{A}$, such that $\mathfrak{I} \subset R\mathfrak{I}$. Then, the quotient $\mathfrak{A}/\mathfrak{I}$, which is a vector space with the usual sum, can be made into a partial *-algebra by setting

$$\Gamma = \big\{ (a + \mathfrak{I}, b + \mathfrak{I}) : a \in L(b), a \in L\mathfrak{I}, b \in R\mathfrak{I} \big\},$$

$$(a + \mathfrak{I}) \cdot (b + \mathfrak{I}) = ab + \mathfrak{I},$$

$$(a + \mathfrak{I})^* = a^* + \mathfrak{I}.$$

Definition 2.2.5 Let $(\mathfrak{A}, \mathfrak{A}_0)$ be a quasi *-algebra and $\mathcal{D}_\pi$ a dense domain in a certain Hilbert space $\mathcal{H}_\pi$. A linear map π from $\mathfrak{A}$ into $\mathcal{L}^\dagger(\mathcal{D}_\pi, \mathcal{H}_\pi)$ is called a *-representation* of $(\mathfrak{A}, \mathfrak{A}_0)$, if the following properties are fulfilled:

(i) $\pi(a^*) = \pi(a)^\dagger$, $\forall a \in \mathfrak{A}$;
(ii) for $a \in \mathfrak{A}$ and $x \in \mathfrak{A}_0$, $\pi(a)\square\pi(x)$ is well defined and $\pi(a)\square\pi(x) = \pi(ax)$.

In other words, π is a *-homomorphism of $(\mathfrak{A}, \mathfrak{A}_0)$ into the partial *-algebra $\mathcal{L}^\dagger(\mathcal{D}_\pi, \mathcal{H}_\pi)$.

If $(\mathfrak{A}, \mathfrak{A}_0)$ has a unit $e \in \mathfrak{A}_0$, we assume $\pi(e) = \mathbb{I}_{\mathcal{D}_\pi}$, where $\mathbb{I}_{\mathcal{D}_\pi}$ is the identity operator on the space $\mathcal{D}_\pi$.

If $\pi_o := \pi \upharpoonright_{\mathfrak{A}_0}$ is a *-representation of the *-algebra $\mathfrak{A}_0$ into $\mathcal{L}^\dagger(\mathcal{D}_\pi)$ we say that π is a *qu*-representation*.

If π is a *-representation of $(\mathfrak{A}, \mathfrak{A}_0)$, then the *closure* $\tilde{\pi}$ of π is defined, for each $a \in \mathfrak{A}$, as the restriction of $\overline{\pi(a)}$ to the domain $\widetilde{\mathcal{D}}_\pi$, which is the completion of $\mathcal{D}_\pi$ under the *graph topology* t_π [12, p. 9] defined by the seminorms $\xi \in \mathcal{D}_\pi \to \|\xi\| + \|\pi(a)\xi\|$, $a \in \mathfrak{A}$, where $\| \cdot \|$ is the norm induced by the inner product on $\mathcal{D}_\pi$. If $\pi = \tilde{\pi}$, the *-representation is said to be *closed*.

The *adjoint* of a *-representation π of a quasi *-algebra $(\mathfrak{A}, \mathfrak{A}_0)$, denoted by π^*, is defined as follows; see [2, 23]

$$\mathcal{D}_{\pi^*} \equiv \bigcap_{a \in \mathfrak{A}} D(\pi(a)^*) \quad \text{and} \quad \pi^*(a) = \pi(a^*)^* \upharpoonright_{\mathcal{D}_{\pi^*}}, \quad a \in \mathfrak{A}.$$

The *-representation π is said to be *selfadjoint* if $\pi = \pi^*$.

Finally, a *-representation π is called *bounded* if $\pi(a)$ is a bounded operator in $\mathcal{D}_\pi$, for every $a \in \mathfrak{A}$.

The *-representation π is said to be

- *ultra-cyclic*, if there exists $\xi_0 \in \mathcal{D}_\pi$, such that $\mathcal{D}_\pi = \pi(\mathfrak{A}_0)\xi_0$;
- *strongly-cyclic*, if there exists $\xi_0 \in \mathcal{D}_\pi$, such that $\pi(\mathfrak{A}_0)\xi_0$ is dense in $\mathcal{D}_\pi$ with respect to the graph topology t_π;
- *cyclic*, if there exists $\xi_0 \in \mathcal{D}_\pi$, such that $\pi(\mathfrak{A}_0)\xi_0$ is dense in $\mathcal{H}$ in its norm topology.

Let $\{\pi_\alpha : \alpha \in I\}$ be a family of closed *-representations of $\mathfrak{A}$. The *algebraic direct sum* of the *-representations π_α's, denoted by π, is defined on the pre-Hilbert space

$$\mathcal{D}_\pi = \sum_{\alpha\in I} \mathcal{D}_{\pi_\alpha} := \Big\{(\xi_\alpha)_{\alpha\in I} \in \Pi_{\alpha\in I}\mathcal{D}_{\pi_\alpha}, \text{ with } \xi_\alpha \neq 0, \text{ for at most finitely many } \alpha \in I\Big\}.$$

Inner product on $\mathcal{D}_\pi$ and $\pi(x)$ on $\mathcal{D}_\pi$, $x \in \mathfrak{A}$, are given as follows

$$\big\langle(\xi_\alpha)_{\alpha\in I}), (\eta_\alpha)_{\alpha\in I})\big\rangle := \sum_{\alpha\in I}\langle \xi_\alpha, \eta_\alpha\rangle_\alpha, \ \forall\, (\xi_\alpha)_{\alpha\in I}, (\eta_\alpha)_{\alpha\in I} \in \mathcal{D} \text{ and}$$

$$\pi(x)\big((\xi_\alpha)_{\alpha\in I}\big) := \big(\pi_\alpha(x)\xi_\alpha\big)_{\alpha\in I}, \ \forall\, x \in \mathfrak{A} \text{ and } (\xi_\alpha)_{i\in I} \in \mathcal{D}_\pi.$$

Considering now the Hilbert spaces $\mathcal{H}_\alpha$'s, completions of the pre-Hilbert spaces $\mathcal{D}_\alpha$'s, $\alpha \in I$, we take the *Hilbert space direct sum* of $\mathcal{H}_\alpha$'s, given by

$$\mathcal{H}_\pi = \Big\{\xi = (\xi_\alpha)_{\alpha\in I} : \xi_\alpha \in \mathcal{D}_{\pi_\alpha}, \sum_{\alpha\in I} \|\pi_\alpha(a)\|^2 < \infty, \ \forall\, a \in \mathfrak{A}\Big\}.$$

We clearly have that $\mathcal{D}_\pi$ is dense in $\mathcal{H}_\pi$. Now, the closure $\widetilde{\pi}$, of the *-representation algebraic direct sum π, is called *direct sum* of the family of the closed *-representations $\{\pi_\alpha : \alpha \in I\}$, and it is denoted by $\oplus_{\alpha\in I}\pi_\alpha$.

Following [23], we say that a subspace $\mathcal{M}$ of $\mathcal{D}_\pi$ is *invariant* for a *-representation π of $(\mathfrak{A}, \mathfrak{A}_0)$ if $\pi(a)\xi \in \mathcal{M}$, for every $a \in \mathfrak{A}$ and $\xi \in \mathcal{M}$. A closed subspace $\mathcal{K}$ of $\mathcal{H}_\pi$ is called invariant for π if there exists a subspace $\mathcal{M}$ of $\mathcal{D}_\pi$, which is dense in $\mathcal{K}$ and invariant for π. If $\mathcal{M} \subset \mathcal{D}_\pi$ is invariant for π, then the mapping $a \mapsto \pi(a)\restriction_\mathcal{M}$ defines a *-representation $\pi_\mathcal{M}$ of $(\mathfrak{A}, \mathfrak{A}_0)$ and clearly $\overline{\mathcal{M}}$ is invariant for $\pi_\mathcal{M}$.

We say that a subspace $\mathcal{M}$ of $\mathcal{D}_\pi$ is *reducing* for π if there exist *-representations π_1, π_2 of $\mathfrak{A}$, with π_1 having $\mathcal{M}$ as domain, such that $\pi = \pi_1 \oplus \pi_2$.

As for bounded representations, reducing subspaces can be described in terms of projections in the commutant of the image of the *-representation involved.

Let $\mathcal{D}$ be a subspace of a Hilbert space $\mathcal{H}$. If $\mathcal{Y}$ is a †-invariant subset of $\mathcal{L}^\dagger(\mathcal{D}, \mathcal{H})$, the *weak bounded commutant* of $\mathcal{Y}$ is defined to be the set

$$(\mathcal{Y}, \mathcal{D})'_{\mathrm{w}} = \{B \in \mathcal{B}(\mathcal{H}) : \langle BY\xi|\eta\rangle = \langle B\xi|Y^*\eta\rangle, \quad \forall\, Y \in \mathcal{Y},\ \xi, \eta \in \mathcal{D}\}.$$

It is easily seen that $(\mathcal{Y}, \mathcal{D})'_{\mathrm{w}}$ is a weakly closed *-invariant subspace of $\mathcal{B}(\mathcal{H})$. One also defines the *strong bounded commutant* of $\mathcal{Y}$ as

$$(\mathcal{Y}, \mathcal{D})'_{\mathrm{s}} = \{B \in (\mathcal{Y}, \mathcal{D})'_{\mathrm{w}} : B\mathcal{D} \subset \mathcal{D}\}.$$

It is easily seen that $(\mathcal{Y}, \mathcal{D})'_{\mathrm{s}}$ is an algebra, but it is not *-invariant, in general.

Then, one has the following (for the relevant terminology, see [23, p. 214, Lemma 8.3.5]).

Proposition 2.2.6 *Let π be a *-representation of $\mathfrak{A}$ defined on $\mathcal{D}_\pi$ and let $\mathcal{M}$ be a subspace of $\mathcal{D}_\pi$. Denote by $P_{\overline{\mathcal{M}}}$ the projection onto the closure $\overline{\mathcal{M}}$ of $\mathcal{M}$ in $\mathcal{H}$. The following statements are equivalent:*

(i) *$\mathcal{M}$ is reducing for π.*
(ii) *$\mathcal{M}$ and $\mathcal{M}^\perp \cap \mathcal{D}_\pi$ are both invariant for π and $P_{\overline{\mathcal{M}}}\mathcal{D}_\pi \subset \mathcal{D}_\pi$.*
(iii) *$P_{\overline{\mathcal{M}}} \in (\pi(\mathfrak{A}), \mathcal{D}_\pi)'_{\mathrm{s}}$.*

A *-representation π of $\mathfrak{A}$, defined on $\mathcal{D}_\pi$, is said to be *irreducible* if there is no nontrivial subspace $\mathcal{M}$ of $\mathcal{D}_\pi$, which is reducing for π.

By Proposition 2.2.6 it follows immediately that π *is irreducible, if and only if,*

$$(\pi(\mathfrak{A}), \mathcal{D}_\pi)'_{\mathrm{s}} = \mathbb{C}\mathbb{I}.$$

2.3 Families of Sesquilinear Forms

Definition 2.3.1 Let $(\mathfrak{A}, \mathfrak{A}_0)$ be a quasi *-algebra. We denote with $\mathcal{Q}_{\mathfrak{A}_0}(\mathfrak{A})$ the set of all sesquilinear forms on $\mathfrak{A} \times \mathfrak{A}$, such that

(i) φ is positive, i.e., $\varphi(a, a) \geq 0, \quad \forall\, a \in \mathfrak{A}$;
(ii) $\varphi(ax, y) = \varphi(x, a^*y), \quad \forall\, a \in \mathfrak{A},\ x, y \in \mathfrak{A}_0$.

Let $\varphi \in \mathcal{Q}_{\mathfrak{A}_0}(\mathfrak{A})$. Then, the positivity of φ implies that

$$\varphi(a, b) = \overline{\varphi(b, a)}, \quad \forall\, a, b \in \mathfrak{A};$$
$$|\varphi(a, b)|^2 \leq \varphi(a, a)\varphi(b, b), \quad \forall\, a, b \in \mathfrak{A}.$$

Hence,

$$N_\varphi := \{a \in \mathfrak{A} : \varphi(a, a) = 0\} = \{a \in \mathfrak{A} : \varphi(a, b) = 0,\ \forall\, b \in \mathfrak{A}\},$$

so N_φ is a subspace with the property that, if

$$x \in N_\varphi \cap \mathfrak{A}_0 \text{ then } \varphi(ax, y) = 0, \quad \forall\, y \in \mathfrak{A}_0,$$

but it is not necessarily a left qu-ideal of $\mathfrak{A}$. Let $\lambda_\varphi : \mathfrak{A} \to \mathfrak{A}/N_\varphi$ be the usual quotient map and for each $a \in \mathfrak{A}$, let $\lambda_\varphi(a)$ be the corresponding coset of $\mathfrak{A}/N_\varphi$, which contains a. We define an inner product $\langle\cdot|\cdot\rangle$ on $\lambda_\varphi(\mathfrak{A}) = \mathfrak{A}/N_\varphi$ by

$$\langle \lambda_\varphi(a)|\lambda_\varphi(b)\rangle := \varphi(a, b), \quad \forall\, a, b \in \mathfrak{A}. \tag{2.3.5}$$

Denote by $\mathcal{H}_\varphi$ the Hilbert space obtained by the completion of the pre-Hilbert space $\lambda_\varphi(\mathfrak{A})$.

Proposition 2.3.2 *Let $\varphi \in \mathcal{Q}_{\mathfrak{A}_0}(\mathfrak{A})$. The following statements are equivalent:*

(i) *$\lambda_\varphi(\mathfrak{A}_0)$ is dense in $\mathcal{H}_\varphi$.*
(ii) *If $\{a_n\}$ is a sequence of elements of $\mathfrak{A}$ such that:*

 (ii.a) *$\varphi(a_n, x) \to 0$, as $n \to \infty$, for every $x \in \mathfrak{A}_0$;*
 (ii.b) *$\varphi(a_n - a_m, a_n - a_m) \to 0$, as $n, m \to \infty$;*

 then, $\lim_{n\to\infty} \varphi(a_n, a_n) = 0$.

Proof (i) $\Rightarrow$ (ii) Let $\{a_n\}$ be a sequence of elements of $\mathfrak{A}$ for which (ii.a) and (ii.b) hold. By (ii.b) the sequence $\{\lambda_\varphi(a_n)\}$ is Cauchy in $\mathcal{H}_\varphi$. Let ξ be its limit. By (ii.a) it follows that

$$\langle \xi|\lambda_\varphi(x)\rangle = \lim_{n\to\infty} \varphi(a_n, x) = 0, \quad \forall\, x \in \mathfrak{A}_0.$$

Hence, ξ is orthogonal to $\lambda_\varphi(\mathfrak{A}_0)$. This implies that $\xi = 0$ and, therefore,

$$\lim_{n\to\infty} \varphi(a_n, a_n) = \|\xi\|^2 = 0.$$

(ii) $\Rightarrow$ (i) Let $\xi \in \mathcal{H}_\varphi$ be a vector orthogonal to $\lambda_\varphi(\mathfrak{A}_0)$ and $\{a_n\}$ a sequence in $\mathfrak{A}$ such that $\lambda_\varphi(a_n) \to \xi$. Then, it is easily seen that $\{a_n\}$ satisfies (ii.a) and (ii.b). Moreover,

$$\|\xi\|^2 = \lim_{n\to\infty} \varphi(a_n, a_n) = 0.$$

This proves that $\lambda_\varphi(\mathfrak{A}_0)$ is dense in $\mathcal{H}_\varphi$. □

We denote by $\mathcal{I}_{\mathfrak{A}_0}(\mathfrak{A})$ the subset of all sesquilinear forms $\varphi \in \mathcal{Q}_{\mathfrak{A}_0}(\mathfrak{A})$ satisfying (i) or (ii) of Proposition 2.3.2.

Remark 2.3.3 Elements of $\mathcal{I}_{\mathfrak{A}_0}(\mathfrak{A})$ are instances of *invariant positive sesquilinear forms* (*ips-forms*, for short) as defined in [39, p. 5] for partial *-algebras. The elements $\varphi \in \mathcal{I}_{\mathfrak{A}_0}(\mathfrak{A})$ are characterized by the existence of a subspace $B(\varphi)$ of

$\mathfrak{A}$ (called a *core for* φ), whose properties are essentially the same as those that $\mathfrak{A}_0$ possesses, in view of Definition 2.3.1 and Proposition 2.3.2. In fact, in the definition of an ips-form, there is an additional condition, which takes into account the possible lack of associativity in a partial *-algebra. In particular, an ips-form is an *everywhere defined biweight* in the sense of [2]. Elements of $\mathcal{I}_{\mathfrak{A}_0}(\mathfrak{A})$ are nothing but ips-forms on $\mathfrak{A}$ with fixed core the *-algebra $\mathfrak{A}_0$.

The opposite of ips-forms are singular forms, which we define as follows [53, 82]:

Definition 2.3.4 Let $(\mathfrak{A}, \mathfrak{A}_0)$ be a quasi *-algebra and $\psi \in \mathcal{Q}_{\mathfrak{A}_0}(\mathfrak{A})$. We say that ψ is $\mathfrak{A}_0$-*singular* if there exists $a_0 \in \mathfrak{A}$ with $\psi(a_0, a_0) > 0$ and $\psi(a, x) = 0$, for every $a \in \mathfrak{A}$, $x \in \mathfrak{A}_0$.

Example 2.3.5 In general, $\mathcal{Q}_{\mathfrak{A}_0}(\mathfrak{A}) \supsetneq \mathcal{I}_{\mathfrak{A}_0}(\mathfrak{A})$ and therefore singular forms do really exist. We construct such an example as follows. Let $\mathfrak{B}$ be a C*-algebra with norm $\|\cdot\|$ and unit e. Let $\mathcal{J}$ be a proper dense *-ideal of $\mathfrak{B}$. Let $v_0 \in \mathfrak{B}$, $v_0 = v_0^*$, be an element of $\mathfrak{B}$ satisfying the following condition

$$\alpha e + \beta v_0 + \gamma v_0^2 \in \mathcal{J} \Leftrightarrow \alpha = \beta = \gamma = 0. \tag{2.3.6}$$

Let

$$\mathfrak{A} := \{\lambda e + \mu v_0 + u : \lambda, \mu \in \mathbb{C},\ u \in \mathcal{J}\}.$$

From (2.3.6) it follows that for every $a \in \mathfrak{A}$ the decomposition $a = \lambda_a e + \mu_a v_0 + u_a$ with $\lambda_a, \mu_a \in \mathbb{C}$ and $u_a \in \mathcal{J}$ is unique. Put

$$\mathfrak{A}_0 = \{\lambda e + u : \lambda \in \mathbb{C},\ u \in \mathcal{J}\}.$$

From (2.3.6) it also follows that there is no element $x \in \mathfrak{A}_0$, such that $v_0 x = v_0^2$, hence $(\mathfrak{A}, \mathfrak{A}_0)$ is a true quasi *-algebra with respect to the multiplication inherited on $\mathfrak{A}_0$ by the C*-algebra $\mathfrak{B}$. Now we define

$$\varphi(\lambda_a e + \mu_a v_0 + u_a, \lambda_b e + \mu_b v_0 + u_b) = \mu_a \overline{\mu_b}\, \omega(v_0^* v_0)$$

where ω is a positive linear functional on $\mathfrak{B}$, such that $\omega(v_0^* v_0) > 0$. Then, it is easily seen that $\varphi \in \mathcal{Q}_{\mathfrak{A}_0}(\mathfrak{A})$. But $\varphi \notin \mathcal{I}_{\mathfrak{A}_0}(\mathfrak{A})$, since $N_\varphi = \mathfrak{A}_0$ and so $\mathfrak{A}/N_\varphi = \mathbb{C}$ (up to an algebraic isomorphism), while $\lambda_\varphi(\mathfrak{A}_0) = \{0\}$.

2.4 Construction of *-Representations

The Gelfand–Naimark–Segal (GNS) construction for positive linear functionals is one of the most relevant tools for the study of (locally convex) *-algebras [9, 19, 23]. As customary, when a partial multiplication is involved [2], we consider, as starting point for the construction, a positive sesquilinear form enjoying certain *invariance* properties (see, also, [49]). For a quasi *-algebra $(\mathfrak{A}, \mathfrak{A}_0)$ this set of sesquilinear forms is exactly $\mathcal{I}_{\mathfrak{A}_0}(\mathfrak{A})$. An analogous GNS construction for positive *linear* functionals is also possible, under certain circumstances, and it will be discussed later (Theorem 2.4.8).

2.4.1 GNS-Like Construction with Sesquilinear Forms

Proposition 2.4.1 *Let $(\mathfrak{A}, \mathfrak{A}_0)$ be a quasi *-algebra with unit e and φ a sesquilinear form on $\mathfrak{A} \times \mathfrak{A}$. The following statements are equivalent:*

(i) $\varphi \in \mathcal{I}_{\mathfrak{A}_0}(\mathfrak{A})$.
(ii) *There exist a Hilbert space $\mathcal{H}_\varphi$, a dense domain $\mathcal{D}_\varphi$ of the Hilbert space $\mathcal{H}_\varphi$ and a closed cyclic *-representation π_φ in $\mathcal{L}^\dagger(\mathcal{D}_\varphi, \mathcal{H}_\varphi)$, with cyclic vector ξ_φ (in the sense that $\pi_\varphi(\mathfrak{A}_0)\xi_\varphi$ is dense in $\mathcal{H}_\varphi$), such that*

$$\varphi(a, b) = \langle \pi_\varphi(a)\xi_\varphi | \pi_\varphi(b)\xi_\varphi \rangle, \quad \forall\, a, b \in \mathfrak{A}. \tag{2.4.7}$$

Proof (i) $\Rightarrow$ (ii) Let $\varphi \in \mathcal{I}_{\mathfrak{A}_0}(\mathfrak{A})$. Put

$$\pi_\varphi^\circ(a)\lambda_\varphi(x) := \lambda_\varphi(ax), \quad a \in \mathfrak{A},\ x \in \mathfrak{A}_0. \tag{2.4.8}$$

First we prove that, for every $a \in \mathfrak{A}$, the map $\pi_\varphi^\circ(a)$ is well-defined. Assume that, for $x \in \mathfrak{A}_0$, $\lambda_\varphi(x) = 0$. If $a \in \mathfrak{A}$, we then get $\varphi(x, a^* y) = 0$, for every $y \in \mathfrak{A}_0$. Since $\varphi \in \mathcal{I}_{\mathfrak{A}_0}(\mathfrak{A})$, for each $b \in \mathfrak{A}$, there exists a sequence $y_n \in \mathfrak{A}_0$, such that $\|\lambda_\varphi(b) - \lambda_\varphi(y_n)\| \to 0$, as $n \to \infty$. This clearly implies that $\varphi(ax, b) = 0$, for every $b \in \mathfrak{A}$. Hence, $ax \in N_\varphi$. Thus, for every $a \in \mathfrak{A}$, the map $\pi_\varphi^\circ(a)$ is a well-defined linear operator from $\lambda_\varphi(\mathfrak{A}_0)$ into $\mathcal{H}_\varphi$. This fact together with the properties of φ listed in Definition 2.3.1 easily imply that π_φ° is a *-representation, whose restriction to $\mathfrak{A}_0$, maps $\lambda_\varphi(\mathfrak{A}_0)$ into itself. Since $(\mathfrak{A}, \mathfrak{A}_0)$ has a unit e, then (i) and (ii) follow from the very definitions. Denote by π_φ the closure of π_φ° and by $\mathcal{D}_{\pi_\varphi} \equiv \mathcal{D}_\varphi$ its domain. Then, it is easily seen that π_φ satisfies (2.4.7). It is also clear by the definition of π_φ that, since $(\mathfrak{A}, \mathfrak{A}_0)$ has a unit e, then $\xi_\varphi := \lambda_\varphi(e)$ is a cyclic vector for π_φ.

(ii) $\Rightarrow$ (i) From the equality (2.4.7) follows easily that

$$\varphi(a, a) \geq 0, \quad \forall\, a \in \mathfrak{A};$$

$$\varphi(ax, y) = \varphi(x, a^* y), \quad \forall\, a \in \mathfrak{A},\ x, y \in \mathfrak{A}_0.$$

Since $\pi_\varphi(\mathfrak{A}_0)\xi_\varphi$ is dense in $\mathcal{H}_\varphi$, for every $a \in \mathfrak{A}$, there exists a sequence $\{x_n\}$, $n \in \mathbb{N}$, in $\mathfrak{A}_0$, such that $\|\pi_\varphi(a)\xi_\varphi - \pi_\varphi(x_n)\xi_\varphi\| \to 0$, as $n \to \infty$. Therefore,

$$\|\lambda_\varphi(a) - \lambda_\varphi(x_n)\|^2 = \varphi(a - x_n, a - x_n) = \|\pi_\varphi(a)\xi_\varphi - \pi_\varphi(x_n)\xi_\varphi\|^2 \to 0,$$

as $n \to \infty$. This implies that $\lambda_\varphi(\mathfrak{A}_0)$ is dense in $\mathcal{H}_\varphi$; i.e., $\varphi \in \mathcal{I}_{\mathfrak{A}_0}(\mathfrak{A})$. □

Definition 2.4.2 The triple $(\pi_\varphi, \lambda_\varphi, \mathcal{H}_\varphi)$ constructed in Proposition 2.4.1 is called the *GNS construction* for φ and π_φ is called the *GNS representation* of $\mathfrak{A}$ corresponding to φ.

Proposition 2.4.3 *Let* $(\mathfrak{A}, \mathfrak{A}_0)$ *be a quasi *-algebra with unit e and* $\varphi \in \mathcal{I}_{\mathfrak{A}_0}(\mathfrak{A})$. *Then, the GNS construction* $(\pi_\varphi, \lambda_\varphi, \mathcal{H}_\varphi)$ *is unique up to unitary equivalence.*

Proof Let $(\pi'_\varphi, \lambda'_\varphi, \mathcal{H}'_\varphi)$ be another triple satisfying the same conditions as $(\pi_\varphi, \lambda_\varphi, \mathcal{H}_\varphi)$. Like above, we put $\xi_\varphi = \lambda_\varphi(e)$ and define

$$U\pi_\varphi(a)\xi_\varphi := \pi'_\varphi(a)\xi'_\varphi, \quad a \in \mathfrak{A}.$$

Then,

$$\langle U\pi_\varphi(a)\xi_\varphi | U\pi_\varphi(b)\xi_\varphi\rangle = \langle \pi'_\varphi(a)\xi'_\varphi | \pi'_\varphi(b)\xi'_\varphi\rangle = \varphi(a, b) = \langle \pi_\varphi(a)\xi_\varphi | \pi_\varphi(b)\xi_\varphi\rangle,$$

for all $a, b \in \mathfrak{A}$, which proves, at once, that $U : \mathcal{D}_\varphi \to \mathcal{D}'_\varphi$ is well-defined and preserves the inner product. Clearly U extends to a unitary operator from $\mathcal{H}_\varphi$ onto $\mathcal{H}'_\varphi$, which we denote by the same symbol. With $\mathcal{D}_{\pi_\varphi} \equiv \mathcal{D}_\varphi$ and $\mathcal{D}'_{\pi_\varphi} \equiv \mathcal{D}'_\varphi$, it is easily seen that $U\mathcal{D}_{\pi_\varphi} = \mathcal{D}'_{\pi_\varphi}$ and

$$U\pi_\varphi(a)\eta = \pi'_\varphi(a)U\eta, \quad \forall\, a \in \mathfrak{A},\ \eta \in \mathcal{D}_{\pi_\varphi}.$$

Hence, the two *-representations π_φ and π'_φ are unitarily equivalent. □

Definition 2.4.4 A sesquilinear form φ is called *admissible* if, for every $a \in \mathfrak{A}$, there exists $\gamma_a > 0$, such that

$$\varphi(ax, ax) \leq \gamma_a \varphi(x, x), \quad \forall\, x \in \mathfrak{A}_0.$$

Proposition 2.4.5 *The *-representation* π_φ *is bounded, if and only if,* φ *is admissible.*

Proof From the construction in Proposition 2.4.1, we obtain the equality

$$\|\pi_\varphi(a)\lambda_\varphi(x)\| = \|\lambda_\varphi(ax)\| = \varphi(ax, ax)^{1/2}, \quad \forall\, a \in \mathfrak{A},\ x \in \mathfrak{A}_0.$$

Thus, if π_φ is bounded, then for every $a \in \mathfrak{A}$, there exists $\gamma_a > 0$, such that

$$\varphi(ax, ax) = \|\pi(a)\lambda_\varphi(x)\|^2 \leq \gamma_a \|\lambda_\varphi(x)\| = \gamma_a \varphi(x, x), \quad \forall\, x \in \mathfrak{A}_0.$$

The converse is proved in a similar way. □

2.4.2 GNS-Like Construction with Linear Functionals

As announced at the beginning of this section, certain linear functionals over a quasi *-algebra $(\mathfrak{A}, \mathfrak{A}_0)$ allow a GNS-like construction.

Definition 2.4.6 Let $(\mathfrak{A}, \mathfrak{A}_0)$ be a quasi *-algebra and ω be a linear functional on $\mathfrak{A}$ satisfying the following conditions:

(L.1) $\omega(x^*x) \geq 0, \quad \forall\, x \in \mathfrak{A}_0$;
(L.2) $\omega(y^*a^*x) = \overline{\omega(x^*ay)}, \quad \forall\, a \in \mathfrak{A},\ x, y \in \mathfrak{A}_0$;
(L.3) $\forall\, a \in \mathfrak{A}$, there exists $\gamma_a > 0$, such that

$$|\omega(a^*x)| \leq \gamma_a \omega(x^*x)^{1/2}, \quad \forall\, x \in \mathfrak{A}_0.$$

Then, ω is called *representable linear functional on* $\mathfrak{A}$.

The family of representable linear functionals on $(\mathfrak{A}, \mathfrak{A}_0)$ is denoted by $\mathcal{R}(\mathfrak{A}, \mathfrak{A}_0)$.

Remark 2.4.7 The family $\mathcal{R}(\mathfrak{A}, \mathfrak{A}_0)$ enjoys the following properties:

(a) If $\omega_1, \omega_2 \in \mathcal{R}(\mathfrak{A}, \mathfrak{A}_0)$, then $\omega_1 + \omega_2 \in \mathcal{R}(\mathfrak{A}, \mathfrak{A}_0)$, as well as $\lambda\omega_1 \in \mathcal{R}(\mathfrak{A}, \mathfrak{A}_0)$, for every $\lambda \geq 0$. Indeed, the conditions (L.1) and (L.2) are obviously satisfied. As for (L.3), for every $a \in \mathfrak{A}$, there exist $\gamma_{a,1}, \gamma_{a,2} > 0$, such that

$$|\omega_1(a^*x)| \leq \gamma_{a,1}\omega_1(x^*x)^{1/2}, \quad |\omega_2(a^*x)| \leq \gamma_{a,2}\omega_2(x^*x)^{1/2}, \quad \forall\, x \in \mathfrak{A}_0,$$

hence,

$$\begin{aligned}
|(\omega_1 + \omega_2)(a^*x)| &\leq \gamma_{a,1}\omega_1(x^*x)^{1/2} + \gamma_{a,2}\omega_2(x^*x)^{1/2} \\
&\leq \max\{\gamma_{a,1}, \gamma_{a,2}\}\big(\omega_1(x^*x)^{1/2} + \omega_2(x^*x)^{1/2}\big) \\
&\leq \sqrt{2}\max\{\gamma_{a,1}, \gamma_{a,2}\}\big((\omega_1 + \omega_2)(x^*x)\big)^{1/2}.
\end{aligned}$$

(b) If $\omega \in \mathcal{R}(\mathfrak{A}, \mathfrak{A}_0)$ and $z \in \mathfrak{A}_0$, then the linear functional ω_z, defined by

$$\omega_z(a) := \omega(z^*az), \quad a \in \mathfrak{A},$$

is representable. We omit the easy proof.

Theorem 2.4.8 *Let* $(\mathfrak{A}, \mathfrak{A}_0)$ *be a quasi *-algebra with unit e and let* ω *be a representable linear functional on* $\mathfrak{A}$. *Then, there exists a closed cyclic *-representation* π_ω *of* $(\mathfrak{A}, \mathfrak{A}_0)$, *with a cyclic vector* η_ω, *such that*

$$\omega(a) = \langle \pi_\omega(a)\eta_\omega|\eta_\omega\rangle, \quad \forall\, a \in \mathfrak{A}.$$

This representation is unique up to unitary equivalence.

Proof We define $N_\omega = \{x \in \mathfrak{A}_0 : \omega(x^*x) = 0\}$. Then, N_ω is a left-ideal of $\mathfrak{A}_0$ and the quotient $\mathfrak{A}_0/N_\omega$ is a pre-Hilbert space with inner product

$$\langle \lambda_\omega(x)|\lambda_\omega(y)\rangle = \omega(y^*x), \quad x, y \in \mathfrak{A}_0,$$

where $\lambda_\omega(x)$, $x \in \mathfrak{A}_0$, denotes the coset in $\mathfrak{A}_0/N_\omega$ containing x. Let $\mathcal{H}_\omega$ be the completion of $\lambda_\omega(\mathfrak{A}_0)$.

If $a \in \mathfrak{A}$, we put $a^\omega(\lambda_\omega(x)) = \omega(a^*x)$. Then, by (L.3), it follows that a^ω is a well defined linear functional on $\lambda_\omega(\mathfrak{A}_0)$ and we have

$$|a^\omega(\lambda_\omega(x))| = |\omega(a^*x)| \le \gamma_a \omega(x^*x)^{1/2} = \gamma_a \|\lambda_\omega(x)\|, \quad \forall\, x \in \mathfrak{A}_0.$$

Thus, a^ω is bounded and by Riesz's lemma, there exists a unique $\xi_\omega^a \in \mathcal{H}_\omega$, such that

$$\omega(a^*x) = a^\omega(\lambda_\omega(x)) = \langle \lambda_\omega(x)|\xi_\omega^a\rangle, \quad \forall\, x \in \mathfrak{A}_0. \tag{2.4.9}$$

Furthermore, we put

$$\pi_\omega^\circ(a)\lambda_\omega(x) := \xi_\omega^{ax}, \quad x \in \mathfrak{A}_0. \tag{2.4.10}$$

Since,

$$\begin{aligned}\langle \lambda_\omega(y)|\pi_\omega^\circ(a)\lambda_\omega(x)\rangle &= \langle \lambda_\omega(y)|\xi_\omega^{ax}\rangle = (ax)^\omega(\lambda_\omega(y))\\ &= \omega(x^*a^*y) = \overline{\omega(y^*ax)}, \quad \forall\, y \in \mathfrak{A}_0,\end{aligned}$$

it follows from (L.3) that $\pi_\omega^\circ(a)$ is well-defined and maps $\lambda_\omega(\mathfrak{A}_0)$ into $\mathcal{H}_\omega$. In a similar way one can show the equality

$$\langle \pi_\omega^\circ(a^*)\lambda_\omega(y)|\lambda_\omega(x)\rangle = \overline{\omega(y^*ax)}, \quad \forall\, x, y \in \mathfrak{A}_0.$$

This implies that $\pi_\omega^\circ(a) \in \mathcal{L}^\dagger(\lambda_\omega(\mathfrak{A}_0), \mathcal{H}_\omega)$ and $\pi_\omega^\circ(a)^\dagger = \pi_\omega^\circ(a^*)$, for all $a \in \mathfrak{A}$.

By the uniqueness of $\xi_\omega^a \in \mathcal{H}_\omega$, we may extend the map λ_ω on $\mathfrak{A}$, keeping the same symbol; i.e.,

$$\lambda_\omega : \mathfrak{A} \to \mathcal{H}_\omega : a \mapsto \lambda_\omega(a) := \xi_\omega^a.$$

Then, from (2.4.10) we obtain

$$\pi_\omega^\circ(x)\lambda_\omega(y) = \lambda_\omega(xy), \quad \forall\, x, y \in \mathfrak{A}_0, \tag{2.4.11}$$

With analogous computations as before and taking into account the preceding equality, we conclude that

$$\langle \pi_\omega^\circ(ax)\lambda_\omega(y)|\lambda_\omega(z)\rangle = \langle \pi_\omega^\circ(x)\lambda_\omega(y)|\pi_\omega^\circ(a^*)\lambda_\omega(z)\rangle, \quad \forall\, y, z \in \mathfrak{A}_0.$$

This implies that $\pi_\omega^\circ(a)\Box\pi_\omega^\circ(x)$ (see beginning of Sect. 2.1.3) is well-defined and

$$\pi_\omega^\circ(ax) = \pi_\omega^\circ(a)\Box\pi_\omega^\circ(x), \quad \forall\, a \in \mathfrak{A},\ x \in \mathfrak{A}_0.$$

Thus, π_ω° is a *-representation. It is clear that $\pi_\omega^\circ(\mathfrak{A}_0)\eta_\omega$, $\eta_\omega := \lambda_\omega(e)$, is dense in $\mathcal{H}_\varphi$. Taking, finally, the closure π_ω of π_ω° we get the desired closed *-representation. The statement about the uniqueness comes again from a slight modification of the classical argument. □

Remark 2.4.9 Note that, for every $a \in \mathfrak{A}$, $\xi_\omega^a \in \mathcal{D}_{(\pi_\omega^\circ)^*} = \bigcap_{x\in\mathfrak{A}_0} D(\pi_\omega^\circ(x)^*)$ and $(\pi_\omega^\circ)^*(x)\xi_\omega^a = \xi_\omega^{xa}$, for every $x \in \mathfrak{A}_0$. Indeed, from (2.4.11) and (2.4.9), we obtain

$$\begin{aligned}\langle \lambda_\omega(y)|(\pi_\omega^\circ)^*(x)\xi_\omega^a\rangle &= \langle \pi_\omega^\circ(x^*)\lambda_\omega(y)|\xi_\omega^a\rangle \\ &= \langle \lambda_\omega(x^*y)|\xi_\omega^a\rangle = \omega\big((xa)^*y\big) = \langle \lambda_\omega(y)|\xi_\omega^{xa}\rangle, \quad \forall\, y \in \mathfrak{A}_0.\end{aligned}$$

So the proof of our claim is completed. The *-representation π_ω satisfies, therefore, the following properties:

$$\pi_\omega(a)\lambda_\omega(x) = \xi_\omega^{ax} \text{ and } \pi_\omega^*(x)\xi_\omega^a = \xi_\omega^{xa}, \quad \forall\, a \in \mathfrak{A},\ x \in \mathfrak{A}_0.$$

Remark 2.4.10 The *-representation π_ω is bounded, if and only if, for every $a \in \mathfrak{A}$ there exists $\gamma_a > 0$, such that

$$|\omega(y^*ax)| \le \gamma_a\omega(x^*x)^{1/2}\omega(y^*y)^{1/2}, \quad \forall\, x, y \in \mathfrak{A}_0.$$

Remark 2.4.11 It is not difficult to show that, for every $a \in \mathfrak{A}$, $\pi_\omega(a)$ maps $\lambda_\omega(\mathfrak{A}_0)$ into $\bigcap_{x\in\mathfrak{A}_0} D(\pi_\omega(x)^*)$.

The GNS construction given above also implies the following

Proposition 2.4.12 *Let $(\mathfrak{A}, \mathfrak{A}_0)$ be a quasi *-algebra with unit e and $\omega \in \mathcal{R}(\mathfrak{A}, \mathfrak{A}_0)$. Then, there exists a linear operator $T_\omega : \mathfrak{A} \to \mathcal{H}_\omega$, such that*

$$\omega(a) = \langle T_\omega a|\xi_\omega\rangle, \quad \forall\, a \in \mathfrak{A},$$

where ξ_ω is the cyclic vector of the GNS representation associated to ω.

Proof It is enough to put $T_\omega a := \xi_\omega^a$, as defined in the proof of Theorem 2.4.8. □

Example 2.4.13 Let $\mathcal{D}$ be a dense domain in a Hilbert space $\mathcal{H}$ and $\|\cdot\|_1$ a norm on $\mathcal{D}$, stronger than the Hilbert norm $\|\cdot\|$. Let $\mathsf{B}(\mathcal{D},\mathcal{D})$ denote the vector space of all *jointly* continuous sesquilinear forms on $\mathcal{D}\times\mathcal{D}$, with respect to $\|\cdot\|_1$. The map $\varphi \to \varphi^*$, with

$$\varphi^*(\xi,\eta) := \overline{\varphi(\eta,\xi)},$$

defines an involution in $\mathsf{B}(\mathcal{D},\mathcal{D})$.

As in (2.1.4), $\mathfrak{L}^\dagger(\mathcal{D})$ denotes the *-subalgebra of $\mathcal{L}^\dagger(\mathcal{D})$ consisting of all operators $A \in \mathcal{L}^\dagger(\mathcal{D})$ such that both A and $A^\dagger$ are continuous from $\mathcal{D}[\|\cdot\|_1]$ into itself.

Every $A \in \mathfrak{L}^\dagger(\mathcal{D})$ defines a sesquilinear form $\varphi_A \in \mathsf{B}(\mathcal{D},\mathcal{D})$ by

$$\varphi_A(\xi,\eta) := \langle A\xi|\eta\rangle, \quad \xi,\eta \in \mathcal{D}.$$

Indeed, for every $\xi,\eta\in\mathcal{D}$, there exists $\gamma > 0$, such that

$$|\varphi_A(\xi,\eta)| = |\langle A\xi|\eta\rangle| \le \|A\xi\|\|\eta\| \le \gamma\|A\xi\|_1\|\eta\|_1 \le \gamma'\|\xi\|_1\|\eta\|_1.$$

We put

$$\mathsf{B}^\dagger(\mathcal{D}) := \{\varphi_A : A \in \mathfrak{L}^\dagger(\mathcal{D})\}.$$

It is easily seen that $\varphi_A^* = \varphi_{A^\dagger}$, for every $A \in \mathfrak{L}^\dagger(\mathcal{D})$. Moreover, if $\varphi \in \mathsf{B}(\mathcal{D},\mathcal{D})$, then $\varphi_A \in \mathsf{B}^\dagger(\mathcal{D})$. We now define

$$(\varphi\circ\varphi_A)(\xi,\eta) := \varphi(A\xi,\eta), \quad \xi,\eta\in\mathcal{D},$$
$$(\varphi_A\circ\varphi)(\xi,\eta) := \varphi(\xi,A^\dagger\eta), \quad \xi,\eta\in\mathcal{D}.$$

Under the involution and the operations defined before, the pair $(\mathsf{B}(\mathcal{D},\mathcal{D}), \mathsf{B}^\dagger(\mathcal{D}))$ is a quasi *-algebra (see, also [2, Chap. 10], for a complete discussion).

For every $\xi\in\mathcal{D}$, we define

$$\omega_\xi(\varphi) := \varphi(\xi,\xi), \quad \varphi\in\mathsf{B}(\mathcal{D},\mathcal{D}).$$

Then, ω_ξ is a linear functional on $\mathsf{B}(\mathcal{D},\mathcal{D})$. Moreover,

$$\omega_\xi(\varphi_{A^\dagger}\circ\varphi_A) = (\varphi_A\circ\varphi_A)(\xi,\xi) = \langle A\xi|A\xi\rangle \ge 0.$$

$$\omega_\xi(\varphi_{B^\dagger}\circ\varphi\circ\varphi_A) = \varphi(A\xi,B\xi) = \overline{\omega_\xi(\varphi_{A^\dagger}\circ\varphi^*\circ\varphi_B^\dagger)}.$$

Hence, ω_ξ satisfies (L.1) and (L.2).

The functional ω_ξ satisfies (L.3), if and only if, for every $\varphi \in \mathsf{B}(\mathcal{D},\mathcal{D})$, there exists $\gamma_\varphi > 0$, such that

$$|\varphi(A\xi,\xi)| \le \gamma_\varphi \|A\xi\|, \quad \forall\, A \in \mathfrak{L}^\dagger(\mathcal{D}).$$

Indeed,

$$|\omega_\xi(\varphi^* \circ \varphi_A)| = |(\varphi^* \circ \varphi_A)(\xi,\xi)| = |\varphi(A\xi,\xi)| \le \gamma_\varphi \|A\xi\| = \gamma_\varphi \omega_\xi(\varphi_A^* \circ \varphi_A)^{1/2}.$$

The previous condition is clearly satisfied, if and only if, φ is bounded in the first variable on the subspace $\mathcal{M}_\xi = \{A\xi : A \in \mathfrak{L}^\dagger(\mathcal{D})\}$. If this is the case, then there exists $\zeta \in \mathcal{H}$, such that

$$\omega_\xi(\varphi \circ \varphi_A) = \langle A\xi|\zeta\rangle, \quad \forall\, A \in \mathfrak{L}^\dagger(\mathcal{D}).$$

Hence, *not every* ω_ξ *is representable.*

Let now $\varphi \in \mathcal{I}_{\mathfrak{A}_0}(\mathfrak{A})$; then the linear functional ω_φ, with $\omega_\varphi(a) := \varphi(a,e)$, $a \in \mathfrak{A}$, satisfies the conditions (L.1), (L.2) and (L.3) of Definition 2.4.6, therefore it is representable. Thus, Theorem 2.4.8 can be applied to get the *-representation $\widetilde{\pi}_{\omega_\varphi}$ constructed as shown above. On the other hand, we can also build up, as in Proposition 2.4.1, the closed *-representation π_φ, with cyclic vector ξ_φ. Since

$$\omega_\varphi(a) = \varphi(a,e) = \langle \pi_\varphi(a)\xi_\varphi|\xi_\varphi\rangle, \quad \forall\, a \in \mathfrak{A},$$

it turns out that $\widetilde{\pi}_{\omega_\varphi}$ and π_φ are unitarily equivalent.

Let $\varphi \in \mathcal{Q}_{\mathfrak{A}_0}(\mathfrak{A})$. Then, the linear functional ω_φ, with $\omega_\varphi(a) = \varphi(a,e)$, $a \in \mathfrak{A}$, is representable. Let π_{ω_φ} be the corresponding *-representation. If we define

$$\Omega_\varphi(a,b) := \langle \pi_{\omega_\varphi}(a)\xi_{\omega_\varphi}|\pi_{\omega_\varphi}(b)\xi_{\omega_\varphi}\rangle, \quad a,b \in \mathfrak{A},$$

it is easily seen that $\Omega_\varphi \in \mathcal{Q}_{\mathfrak{A}_0}(\mathfrak{A})$. But, in general, $\Omega_\varphi \neq \varphi$.

Proposition 2.4.14 *The following statements hold:*

(i) *for every* $\varphi \in \mathcal{Q}_{\mathfrak{A}_0}(\mathfrak{A})$, $\Omega_\varphi \in \mathcal{I}_{\mathfrak{A}_0}(\mathfrak{A})$;
(ii) *for every* $\varphi \in \mathcal{Q}_{\mathfrak{A}_0}(\mathfrak{A})$, *there exists* $\varphi_0 \in \mathcal{I}_{\mathfrak{A}_0}(\mathfrak{A})$, *which coincides with* φ *on all pairs* (a,x) *with* $a \in \mathfrak{A}$, $x \in \mathfrak{A}_0$ *and an* $\mathfrak{A}_0$*-singular form* s_φ (see Definition 2.3.4), *such that*

$$\varphi(a,b) = \varphi_0(a,b) + s_\varphi(a,b), \quad \forall\, a,b \in \mathfrak{A},$$

where $\varphi_0 = \Omega_\varphi$ *and* $s_\varphi = \varphi - \Omega_\varphi$;
(iii) $\Omega_\varphi = \varphi$, *if and only if,* $\lambda_\varphi(\mathfrak{A}_0)$ *is dense in* $\mathcal{H}_\varphi$, *i.e., if and only if,* $\varphi \in \mathcal{I}_{\mathfrak{A}_0}(\mathfrak{A})$.

Proof

(i) Since π_{ω_φ} is cyclic, then for every $a \in \mathfrak{A}$, there exists a sequence $\{x_n\} \subset \mathfrak{A}_0$, such that $\|\pi_{\omega_\varphi}(a - x_n)\xi_{\omega_\varphi}\| \to 0$, as $n \to \infty$. Then, we have

$$\|\lambda_{\Omega_\varphi}(a) - \lambda_{\Omega_\varphi}(x_n)\|^2 = \langle \pi_{\omega_\varphi}(a - x_n)\xi_{\omega_\varphi} | \pi_{\omega_\varphi}(a - x_n)\xi_{\omega_\varphi}\rangle \underset{n\to\infty}{\to} 0.$$

This implies that $\lambda_{\Omega_\varphi}(\mathfrak{A}_0)$ is dense in $\mathcal{H}_{\Omega_\varphi}$.

(ii) Put $\varphi_0 = \Omega_\varphi$ and $s_\varphi = \varphi - \Omega_\varphi$. Clearly, $s_\varphi(a, x) = 0$, for every $a \in \mathfrak{A}$, $x \in \mathfrak{A}_0$. Now we prove that $\Omega_\varphi(a, a) \le \varphi(a, a)$, for every $a \in \mathfrak{A}$. Indeed, we have

$$\Omega_\varphi(a, a) = \|\pi_{\omega_\varphi}(a)\lambda_{\omega_\varphi}(e)\|^2 = \|\lambda_{\omega_\varphi}(a)\|^2.$$

Notice that, by the construction in Theorem 2.4.8,

$$\begin{aligned}\|\lambda_{\omega_\varphi}(a)\| &= \sup\left\{|\omega_\varphi(a^*x)| : \omega_\varphi(x^*x) = 1\right\} \\ &= \sup\left\{|\varphi(a, x)| : \omega_\varphi(x^*x) = 1\right\} \le \varphi(a, a)^{1/2}.\end{aligned}$$

Hence, $s_\varphi \in \mathcal{Q}_{\mathfrak{A}_0}(\mathfrak{A})$. If $\lambda_\varphi(\mathfrak{A}_0)$ is not dense in $\mathcal{H}_\varphi$, there exists a sequence $\{a_n\}$ of elements of $\mathfrak{A}$ with the properties (see Proposition 2.3.2):

$$(a)\ \varphi(a_n, x) \underset{n\to\infty}{\to} 0, \quad \forall\, x \in \mathfrak{A}_0;$$

$$(b)\ \varphi(a_n - a_m, a_n - a_m) \underset{n\to\infty}{\to} 0;$$

$$(c)\ \lim_{n\to\infty} \varphi(a_n, a_n) = \alpha > 0.$$

Since $\lambda_{\Omega_\varphi}(\mathfrak{A})$ is dense in $\mathcal{H}_{\Omega_\varphi}$ and

$$\Omega_\varphi(a_n - a_m, a_n - a_m) \le \varphi(a_n - a_m, a_n - a_m) \to 0,$$

we have $\lim_{n\to\infty} \Omega_\varphi(a_n, a_n) = 0$. In conclusion, $s_\varphi(a_n, a_n) \underset{n\to\infty}{\to} \alpha > 0$. So that s_φ cannot be identically 0. This clearly implies that s_φ is $\mathfrak{A}_0$-singular.

(iii) is clear. □

If π is a *-representation of $\mathfrak{A}$ then, for every $\xi \in \mathcal{D}_\pi$, the *vector form* φ_ξ defined by

$$\varphi_\xi(a, b) := \langle \pi(a)\xi | \pi(b)\xi\rangle, \quad a, b \in \mathfrak{A}, \tag{2.4.12}$$

is an element of $\mathcal{Q}_{\mathfrak{A}_0}(\mathfrak{A})$, but it need not belong to $\mathcal{I}_{\mathfrak{A}_0}(\mathfrak{A})$. For this reason, we say that π is *regular* if, for every $\xi \in \mathcal{D}_\pi$, $\varphi_\xi \in \mathcal{I}_{\mathfrak{A}_0}(\mathfrak{A})$.

Proposition 2.4.15 *Let π be a *-representation of* $(\mathfrak{A}, \mathfrak{A}_0)$. *The following statements are equivalent:*

(i) *π is regular;*
(ii) *$\pi(a)\xi \in \overline{\pi(\mathfrak{A}_0)\xi}$, for every $a \in \mathfrak{A}$ and for every $\xi \in \mathcal{D}_\pi$;*
(iii) *for every $\xi \in \mathcal{D}_\pi$, $\pi_0 := \pi \lceil_{\mathcal{M}_\xi}$ is a *-representation of* $(\mathfrak{A}, \mathfrak{A}_0)$ *into* $\mathcal{L}^\dagger(\mathcal{M}_\xi, \overline{\mathcal{M}_\xi})$, *where* $\mathcal{M}_\xi = \pi(\mathfrak{A}_0)\xi$ (i.e., $\mathcal{M}_\xi$ is an invariant subspace for π).

Proof (i) $\Rightarrow$ (ii) Let π be regular and $\xi \in \mathcal{D}_\pi$. Then, $\varphi_\xi \in \mathcal{I}_{\mathfrak{A}_0}(\mathfrak{A})$, hence (Proposition 2.3.2) for every $a \in \mathfrak{A}$, there exists a sequence $\{x_n\} \subset \mathfrak{A}_0$, such that $\|\lambda_{\varphi_\xi}(a - x_n)\| \to 0$. Consequently,

$$\|(\pi(a) - \pi(x_n))\xi\|^2 = \|\lambda_{\varphi_\xi}(a - x_n)\|^2 \to 0.$$

This proves that $\pi(a)\xi \in \overline{\pi(\mathfrak{A}_0)\xi}$.

(ii) $\Rightarrow$ (iii) The assumption (ii) implies that, for every $a \in \mathfrak{A}$ and $\xi \in \mathcal{D}_\pi$, $\pi_0(a)$ maps $\pi(\mathfrak{A}_0)\xi$ into $\overline{\pi(\mathfrak{A}_0)\xi}$. Some simple calculations, which make use of the fact that π is a *-representation and that the module associativity holds, show that $\pi_0(a^*) = (\pi_0(a))^* \lceil_{\mathcal{M}_\xi}$ and that π_0 preserves the partial multiplication of $(\mathfrak{A}, \mathfrak{A}_0)$.

(iii) $\Rightarrow$ (i) The assumption (iii) yields that, for every $\xi \in \mathcal{D}_\pi$ and $a \in \mathfrak{A}$, $\pi(a)\xi \in \overline{\mathcal{M}_\xi}$. Therefore, for every $a \in \mathfrak{A}$, there exists a sequence $\{x_n\} \subset \mathfrak{A}_0$ such that $\|(\pi(a) - \pi(x_n))\xi\| \to 0$. Then, for φ_ξ, we have

$$\varphi_\xi(a - x_n, a - x_n) = \|\lambda_{\varphi_\xi}(a - x_n)\|^2 = \|(\pi(a) - \pi(x_n))\xi\|^2 \to 0.$$

Consequently, π is regular. □

If φ is a sesquilinear form on $\mathfrak{A} \times \mathfrak{A}$ and $x \in \mathfrak{A}_0$, we denote with φ_x the sesquilinear form on $\mathfrak{A} \times \mathfrak{A}$ defined by

$$\varphi_x(a, b) := \varphi(ax, bx), \quad a, b \in \mathfrak{A}, \tag{2.4.13}$$

and with ω_{φ_x} the corresponding linear functional on $\mathfrak{A}$ defined by

$$\omega_{\varphi_x}(a) := \varphi(ax, x), \quad a \in \mathfrak{A}. \tag{2.4.14}$$

It is readily checked that, if $\varphi \in \mathcal{Q}_{\mathfrak{A}_0}(\mathfrak{A})$, then $\varphi_x \in \mathcal{Q}_{\mathfrak{A}_0}(\mathfrak{A})$ too, for all $x \in \mathfrak{A}_0$.

Proposition 2.4.16 *Let* $(\mathfrak{A}, \mathfrak{A}_0)$ *be a quasi *-algebra with unit e,* $\varphi \in \mathcal{I}_{\mathfrak{A}_0}(\mathfrak{A})$ *and π_φ° the *-representation defined in* (2.4.8). *The following statements are equivalent:*

(i) *π_φ° is regular;*
(ii) $\varphi_x \in \mathcal{I}_{\mathfrak{A}_0}(\mathfrak{A}), \quad \forall\, x \in \mathfrak{A}_0$.

Proof If $\eta \in \mathcal{D}_\varphi = \lambda_\varphi(\mathfrak{A}_0)$, then $\eta = \lambda_\varphi(x)$, for some $x \in \mathfrak{A}_0$. Hence,

$$\varphi_\eta(a, b) = \langle \pi_\varphi^\circ(a)\lambda_\varphi(x) | \pi_\varphi^\circ(b)\lambda_\varphi(x)\rangle = \varphi(ax, bx) = \varphi_x(a, b), \quad \forall\, a, b \in \mathfrak{A}.$$

Thus, $\varphi_\eta = \varphi_x$. This equality clearly implies the equivalence of (i) and (ii). □

It is then useful to introduce the notation

$$\mathcal{I}^s_{\mathfrak{A}_0}(\mathfrak{A}) := \left\{\varphi \in \mathcal{I}_{\mathfrak{A}_0}(\mathfrak{A}) : \varphi_x \in \mathcal{I}_{\mathfrak{A}_0}(\mathfrak{A}), \quad \forall\, x \in \mathfrak{A}_0\right\}.$$

It is clear that for every $\varphi \in \mathcal{I}^s_{\mathfrak{A}_0}(\mathfrak{A})$, the corresponding *-representation π_φ° is regular.

Remark 2.4.17 For the GNS representation π_φ constructed by φ (i.e., for the closure of π_φ°) the implication (i) ⇒ (ii) still holds, in an obvious way; however, (ii) does not imply (i), in general.

Example 2.4.18 Let π be a regular *-representation of $(\mathfrak{A}, \mathfrak{A}_0)$ in $\mathcal{L}^\dagger(\mathcal{D}_\pi, \mathcal{H}_\pi)$. Let $\xi \in \mathcal{D}_\pi$ and let φ_ξ be the corresponding vector form (in the sense of (2.4.12)). Then, by definition, $\varphi_\xi \in \mathcal{I}_{\mathfrak{A}_0}(\mathfrak{A})$. Since, $\pi(x)\xi \in \mathcal{D}_\pi$, for every $x \in \mathfrak{A}_0$, one also has $(\varphi_\xi)_x \in \mathcal{I}_{\mathfrak{A}_0}(\mathfrak{A})$, for every $x \in \mathfrak{A}_0$. In this case, $\varphi_\xi \in \mathcal{I}^s_{\mathfrak{A}_0}(\mathfrak{A})$.

Notice that a bounded *-representation π of $\mathfrak{A}$ (i.e., $\overline{\pi(a)} \in \mathcal{B}(\mathcal{H})$, for every $a \in \mathfrak{A}$) need not be regular.

➡ In the sequel, we shall indicate by $\mathrm{Rep}(\mathfrak{A})$, $\mathrm{Rep}_r(\mathfrak{A})$ *the family of all *-representations* and *all regular *-representations of* $(\mathfrak{A}, \mathfrak{A}_0)$, *respectively*.

Chapter 3
Normed Quasi *-Algebras: Basic Theory and Examples

In this chapter we shall consider the case, where $\mathfrak{A}$ is endowed with a norm topology, making $(\mathfrak{A}, \mathfrak{A}_0)$ into a *normed quasi *-algebra* in the sense of Definition 3.1.1, below. This opens our discussion on locally convex quasi *-algebras, starting from the simplest situation. Nevertheless, as we shall see, *simple* does not mean *trivial* at all.

3.1 Basic Definitions and Facts

Definition 3.1.1 A quasi *-algebra $(\mathfrak{A}, \mathfrak{A}_0)$ is called a *normed quasi *-algebra* if $\mathfrak{A}$ is a normed space under a norm $\|\cdot\|$ satisfying the following properties:

(i) $\|a^*\| = \|a\|, \quad \forall\, a \in \mathfrak{A}$;
(ii) $\mathfrak{A}_0$ is dense in $\mathfrak{A}[\|\cdot\|]$;
(iii) for every $x \in \mathfrak{A}_0$, the map $R_x : a \in \mathfrak{A}[\|\cdot\|] \to ax \in \mathfrak{A}[\|\cdot\|]$ is continuous.

A normed quasi *-algebra will be denoted as $(\mathfrak{A}[\|\cdot\|], \mathfrak{A}_0)$. If $\mathfrak{A}[\|\cdot\|]$ is a Banach space, we say that $(\mathfrak{A}[\|\cdot\|], \mathfrak{A}_0)$ is a *Banach quasi *-algebra*.

The continuity of the involution implies that

(iv) for every $x \in \mathfrak{A}_0$, the map $L_x : a \in \mathfrak{A}[\|\cdot\|] \to xa \in \mathfrak{A}[\|\cdot\|]$ is continuous too.

If $(\mathfrak{A}, \mathfrak{A}_0)$ has no unit, it can always be embedded in a normed quasi *-algebra with unit e, using the standard procedure of *unitization*.

In what follows, we shall always assume that

(a) if $ax = 0$, for every $x \in \mathfrak{A}_0$, then $a = 0$;
(b) if $ax = 0$, for every $a \in \mathfrak{A}$, then $x = 0$.

Of course, both these conditions are automatically true if $(\mathfrak{A}, \mathfrak{A}_0)$ has a unit e.

M. Fragoulopoulou, C. Trapani, *Locally Convex Quasi *-Algebras and their Representations*, Lecture Notes in Mathematics 2257,
https://doi.org/10.1007/978-3-030-37705-2_3

If $(\mathfrak{A}[\|\cdot\|], \mathfrak{A}_0)$ is a normed quasi *-algebra, a norm topology can be defined on $\mathfrak{A}_0$ in the following way. For $x \in \mathfrak{A}_0$, the following functions

$$\|x\|_L := \sup_{\|a\|\leq 1} \|ax\| \quad \text{and} \quad \|x\|_R := \sup_{\|a\|\leq 1} \|xa\|, \quad x \in \mathfrak{A}_0,\ a \in \mathfrak{A}, \tag{3.1.1}$$

are well defined norms on $\mathfrak{A}_0$. It is easy to see that $\|x\|_L = \|x^*\|_R$ (and, of course, $\|x\|_R = \|x^*\|_L$), for every $x \in \mathfrak{A}_0$. Moreover, by (3.1.1) it follows that

$$\|ax\| \leq \|a\|\|x\|_L \quad \text{and} \quad \|xa\| \leq \|a\|\|x\|_R, \quad \forall\, a \in \mathfrak{A},\ x \in \mathfrak{A}_0. \tag{3.1.2}$$

Again by (3.1.1) and together with (3.1.2), we deduce that

$$\|xy\|_L \leq \|x\|_L\|y\|_L \quad \text{and} \quad \|xy\|_R \leq \|x\|_R\|y\|_R, \quad \forall\, x, y \in \mathfrak{A}_0. \tag{3.1.3}$$

Remark 3.1.2

1. It is clear from (3.1.3) that $\mathfrak{A}_0$ becomes a normed algebra under the preceding norms $\|\cdot\|_L$ and $\|\cdot\|_R$, respectively. Moreover, from (3.1.2) we conclude that the left module multiplication in $\mathfrak{A}$ is continuous, when $\mathfrak{A}_0$ is endowed with $\|\cdot\|_R$, while the right module multiplication in $\mathfrak{A}$ is continuous, when $\mathfrak{A}_0$ is endowed with $\|\cdot\|_L$. In other words, $\mathfrak{A}[\|\cdot\|]$ is a left normed module over the normed algebra $\mathfrak{A}_0[\|\cdot\|_R]$ and a right normed module over the normed algebra $\mathfrak{A}_0[\|\cdot\|_L]$.
2. Suppose that $(\mathfrak{A}[\|\cdot\|], \mathfrak{A}_0)$ is a normed quasi *-algebra with unit e. Then, conditions (iii) and (iv) of Definition 3.1.1 are equivalent with the following condition:

   (v) the normed space $\mathfrak{A}[\|\cdot\|]$ is a normed module over the normed *-algebra $\mathfrak{A}_0[\|\cdot\|_0]$, where

   $$\|x\|_0 := \max\left\{\|x\|_L, \|x\|_R\right\}, \quad \forall\, x \in \mathfrak{A}_0. \tag{3.1.4}$$

   Indeed: first note that by setting $a = e$ in (3.1.2), we see that the norms $\|\cdot\|$, $\|\cdot\|_0$ are comparable on $\mathfrak{A}_0[\|\cdot\|_0]$. Moreover, from (3.1.3) the vector norm $\|\cdot\|_0$ is submultiplicative, therefore $\mathfrak{A}_0[\|\cdot\|_0]$ is a normed *-algebra. Suppose now that (v) holds. Fix x in $\mathfrak{A}_0$. Then, the composition of the following continuous maps

   $$\mathfrak{A}[\|\cdot\|] \to \mathfrak{A}_0[\|\cdot\|_0] \times \mathfrak{A}[\|\cdot\|] \to \mathfrak{A}[\|\cdot\|] : a \mapsto (x, a) \mapsto xa,$$

   is exactly the map R_x, so that the condition (iii) of Definition 3.1.1 holds. In the same way, it is proved that (iv) is also true.

Conversely, suppose that (iii) and (iv) of Definition 3.1.1 are valid. From (3.1.2), (3.1.4) we readily obtain that the left and right module multiplications of the normed $\mathfrak{A}_0[\|\cdot\|_0]$-module $\mathfrak{A}[\|\cdot\|]$ are continuous; hence (v) follows.

Corollary 3.1.3 *If* $(\mathfrak{A}[\|\cdot\|], \mathfrak{A}_0)$ *is a normed quasi *-algebra, then* $\mathfrak{A}_0[\|\cdot\|_0]$ *is a normed *-algebra and we have*

$$\|xy\| \leq \|x\|\|y\|_0, \ \ \|yx\| \leq \|x\|\|y\|_0, \ \forall\, x, y \in \mathfrak{A}_0.$$

The previous inequalities follow from the submultiplicativity of $\|\cdot\|_0$ and (3.1.2), (3.1.4).

Remark 3.1.4 Note that having a unit e in $(\mathfrak{A}[\|\cdot\|], \mathfrak{A}_0)$, we may always suppose, without loss of generality, that $\|e\| = 1$, since otherwise we can simply define the equivalent norm $\|\cdot\|'$ on $\mathfrak{A}$, with $\|a\|' := \|a\|/\|e\|$, $a \in \mathfrak{A}$ and have $\|e\|' = 1$.

If $(\mathfrak{A}[\|\cdot\|], \mathfrak{A}_0)$ has no unit, then the norms $\|\cdot\|$, $\|\cdot\|_0$ cannot be compared, in general (see Example 3.1.6).

Definition 3.1.5 A Banach quasi *-algebra $(\mathfrak{A}[\|\cdot\|], \mathfrak{A}_0)$ is called a *BQ*-algebra* if $\mathfrak{A}_0[\|\cdot\|_0]$ is a Banach *-algebra and a *proper CQ*-algebra* if $\mathfrak{A}_0[\|\cdot\|_0]$ is a C*-algebra.

3.1.1 Examples

Example 3.1.6 (Banach Function Spaces) Many Banach function spaces provide examples of Banach quasi *-algebras since they often contain a dense *-algebra of functions. For instance, if $I = [0, 1]$ then $(L^p(I), C(I))$, where $C(I)$ denotes the C*-algebra of all continuous functions on I and $p \geq 1$, is a Banach quasi *-algebra (more precisely, as we shall see in Chap. 5, a proper CQ*-algebra [50], if $C(I)$ is endowed with the usual supremum norm $\|\cdot\|_\infty$; actually in this case, one has $\|\cdot\|_0 = \|\cdot\|_\infty$).

Similarly $(L^p(\mathbb{R}), C_c^0(\mathbb{R}))$ is a Banach quasi *-algebra without unit; $C_c^0(\mathbb{R})$ stands here for the *-algebra of continuous functions on $\mathbb{R}$ with compact support. As pointed out in Remark 3.1.4, it is clear that $\|\cdot\|_p$ and $\|\cdot\|_\infty$ do not compare on $C_c^0(\mathbb{R})$.

Other examples of Banach quasi *-algebras are easily found among Sobolev spaces, Besov spaces etc. (see e.g., [29])

Example 3.1.7 (Noncommutative L^p-Spaces) Let $\mathfrak{M}$ be a von Neumann algebra and ϱ a normal semifinite faithful trace [24, 25] on $\mathfrak{M}$. Then, the completion of the *-ideal

$$\mathcal{J}_p := \left\{X \in \mathfrak{M} : \varrho(|X|^p) < \infty\right\},$$

with respect to the norm

$$\|X\|_p := \varrho(|X|^p)^{1/p}, \quad X \in \mathfrak{M},$$

is usually denoted by $L^p(\varrho)$ [70, 76] and is a Banach space consisting of operators affiliated with $\mathfrak{M}$. Then, $(L^p(\varrho), \mathcal{J}_p)$ is a Banach quasi *-algebra (without unit). If ϱ is a finite trace then $(L^p(\varrho), \mathfrak{M})$ is a BQ*-algebra. We shall come back to this example in Sect. 5.6.

Example 3.1.8 (Hilbert Algebras) A Hilbert algebra [19, Section 11.7] is a *-algebra $\mathfrak{A}_0$, which is also a pre-Hilbert space with inner product $\langle\cdot|\cdot\rangle$, such that

(i) The map $y \mapsto xy$, $x, y \in \mathfrak{A}_0$, is continuous with respect to the norm defined by the inner product.
(ii) $\langle xy|z\rangle = \langle y|x^*z\rangle, \quad \forall\, x, y, z \in \mathfrak{A}_0$.
(iii) $\langle x|y\rangle = \langle y^*|x^*\rangle, \quad \forall\, x, y \in \mathfrak{A}_0$.
(iv) $\mathfrak{A}_0^2$ is total in $\mathfrak{A}_0$.

Let $\mathcal{H}$ denote the Hilbert space, which is the completion of $\mathfrak{A}_0$ with respect to the norm defined by the inner product. The involution of $\mathfrak{A}_0$ extends to the whole of $\mathcal{H}$, since (iii) implies that * is isometric. Then, $(\mathcal{H}, \mathfrak{A}_0)$ is a Banach quasi *-algebra, which we name *Hilbert quasi *-algebra.*

Remark 3.1.9 Let $(\mathfrak{A}[\|\cdot\|], \mathfrak{A}_0)$ be a Banach quasi *-algebra and suppose that the norm $\|\cdot\|$ of $\mathfrak{A}$ is a *Hilbertian norm*; that is, it satisfies the parallelogram law

$$\|a+b\|^2 + \|a-b\|^2 = 2\|a\|^2 + 2\|b\|^2, \quad \forall\, a, b \in \mathfrak{A}.$$

Then, as it is well known, an inner product and a norm can be introduced in $\mathfrak{A}$ by

$$\langle a|b\rangle = \frac{1}{4}\sum_{k=0}^{3} \|a + i^k b\|^2, \ \|a\| = \langle a|a\rangle^{1/2}, \quad \forall\, a, b \in \mathfrak{A}.$$

It is easy to see that the conditions (i) and (iii) of Example 3.1.8 are satisfied on $\mathfrak{A}_0$ and extend to $\mathfrak{A}$. Moreover, if $(\mathfrak{A}[\|\cdot\|], \mathfrak{A}_0)$ has a unit, then (iv) is also trivially fulfilled. As for condition (ii), the operator L_x of left multiplication by $x \in \mathfrak{A}_0$ is bounded, therefore it has an adjoint L_x^*, but we do not know if $L_x^* = L_{x^*}$. For this reason, we will call *Hilbertian quasi *-algebra*, a Banach quasi *-algebra $(\mathfrak{A}[\|\cdot\|], \mathfrak{A}_0)$ such that the norm $\|\cdot\|$ of $\mathfrak{A}$ is a Hilbertian norm, just to distinguish this case from that considered in Example 3.1.8. A deeper analysis of Hilbertian quasi *-algebra will be performed in Sect. 5.5.

Example 3.1.10 Let us consider again the situation described in Example 2.4.13. Let $\mathcal{D}^\times$ denote the Banach conjugate dual space of $\mathcal{D}[\|\cdot\|_1]$ endowed with the dual norm $\|\cdot\|_{-1}$, i.e.,

$$\|f\|_{-1} := \sup_{\|\xi\|_1 \leq 1} |f(\xi)|, \ f \in \mathcal{D}^\times.$$

The Hilbert space $\mathcal{H}$ is canonically identified with a subspace of $\mathcal{D}^\times$, by the map $\xi \to f_\xi$, where $f_\xi(\eta) = \langle\xi|\eta\rangle$, for every $\eta \in \mathcal{D}$. The form $b(\cdot,\cdot)$ that puts $\mathcal{D}$ and

$\mathcal{D}^\times$ in conjugate duality is an extension of the inner product of $\mathcal{D}$, so that we adopt the same symbol for both. The space $\mathfrak{L}(\mathcal{D}, \mathcal{D}^\times)$ of all continuous linear maps from $\mathcal{D}[\|\cdot\|_1]$ into $\mathcal{D}^\times[\|\cdot\|_{-1}]$ (see Sect. 2.1.6) carries a natural involution $X \to X^\dagger$, $X \in \mathfrak{L}(\mathcal{D}, \mathcal{D}^\times)$ and a norm $\|\cdot\|_{\mathfrak{L}}$, defined by

$$\langle \xi | X^\dagger \eta \rangle = \overline{\langle \eta | X\xi \rangle}, \quad \|X\|_{\mathfrak{L}} := \sup_{\|\xi\|_1 \leq 1} \|X\xi\|_{-1}, \quad \xi, \eta \in \mathcal{D}, \ X \in \mathfrak{L}(\mathcal{D}, \mathcal{D}^\times).$$

The involution $X \to X^\dagger$ is isometric and since $\mathcal{D}^\times[\|\cdot\|_{-1}]$ is complete, $\mathfrak{L}(\mathcal{D}, \mathcal{D}^\times)[\|\cdot\|_{\mathfrak{L}}]$ is a Banach space. The space $\mathsf{B}(\mathcal{D}, \mathcal{D})$ (see also Example 2.4.13) has a natural norm $\|\cdot\|_{\mathsf{B}}$ too defined by

$$\|\varphi\|_{\mathsf{B}} := \sup\left\{|\varphi(\xi, \eta)| : \|\xi\|_1 = \|\eta\|_1 = 1\right\}, \quad \varphi \in \mathsf{B}(\mathcal{D}, \mathcal{D}).$$

With this norm, $\mathsf{B}(\mathcal{D}, \mathcal{D})$ is a Banach space. Moreover, $\|\varphi^*\|_{\mathsf{B}} = \|\varphi\|_{\mathsf{B}}$, for every $\varphi \in \mathsf{B}(\mathcal{D}, \mathcal{D})$.

In particular, for each $\varphi \in \mathsf{B}(\mathcal{D}, \mathcal{D})$, there exists $X \in \mathfrak{L}(\mathcal{D}, \mathcal{D}^\times)$, such that $\varphi := \varphi_X$, where

$$\varphi_X(\xi, \eta) = \langle \xi | X\eta \rangle, \quad \xi, \eta \in \mathcal{D}.$$

It is easy to prove, in this case, that $\|\varphi\|_{\mathsf{B}} = \|X\|_{\mathfrak{L}}$.

If $\mathcal{D}[\|\cdot\|_1]$ is a Banach space, then the converse is also true, i.e., if $X \in \mathfrak{L}(\mathcal{D}, \mathcal{D}^\times)$, then $\varphi_X \in \mathsf{B}(\mathcal{D}, \mathcal{D})$ and the map $X \mapsto \varphi_X$ is an isometric *-isomorphism [2, Ch. 10].

Let $\mathfrak{M}$ be an *O*-algebra* on $\mathcal{D}$, (in the sense that $\mathfrak{M}$ is a *-subalgebra of $\mathcal{L}^\dagger(\mathcal{D})$), with the property that each $X \in \mathfrak{M}$ is continuous from $\mathcal{D}[\|\cdot\|_1]$ into itself (this is always true if $\mathcal{D}[\|\cdot\|_1]$ is a reflexive space). Then, $(\overline{\mathfrak{M}}, \mathfrak{M})$, where $\overline{\mathfrak{M}}$ denotes the closure of $\mathfrak{M}$ in $\mathfrak{L}(\mathcal{D}, \mathcal{D}^\times)[\|\cdot\|_{\mathfrak{L}}]$, is a Banach quasi *-algebra.

For instance, let us assume that $\mathcal{D} = D(S)$, where S is a positive selfadjoint operator with domain $D(S)$, dense in $\mathcal{H}$. If $S \geq \mathbb{I}$, then $\mathcal{D}$ is a Hilbert space with norm $\|\cdot\|_S$ defined by $\|\xi\|_S = \|S\xi\|$, $\xi \in \mathcal{D}$.

➽ From now on, *the symbol "$\simeq$" will denote a topological isomorphism.*

From the above we have, $\mathfrak{L}(\mathcal{D}, \mathcal{D}^\times) \simeq \mathsf{B}(\mathcal{D}, \mathcal{D})$ and $\mathfrak{L}^\dagger(\mathcal{D}) = \mathcal{L}^\dagger(\mathcal{D})$. If $A \in \mathfrak{L}^\dagger(\mathcal{D})$, then

$$\|\varphi_A\|_{\mathsf{B}} = \sup\left\{|\langle A\xi | \eta \rangle| : \|S\xi\| = \|S\eta\| = 1\right\} = \|S^{-1}AS^{-1}\|. \tag{3.1.5}$$

For every O*-algebra $\mathfrak{M}$ on $\mathcal{D}$, $(\overline{\mathfrak{M}}, \mathfrak{M})$ is a Banach quasi *-algebra.

Now we check that $\overline{\mathfrak{M}}$ is not a *-algebra, in general. From the above discussion it follows that the set $S^{-1}\mathfrak{M}S^{-1}$ is a *-invariant vector space of bounded operators on $\mathcal{H}$. We denote by $\mathcal{M}_S$ its norm closure in $\mathcal{B}(\mathcal{H})$. Let

$$\mathfrak{M}_S^\dagger := \left\{\varphi \in \mathsf{B}^\dagger(\mathcal{D}) : \varphi = \varphi_A, \ A \in \mathfrak{M}\right\}$$

and $\overline{\mathfrak{M}_S^\dagger}$ its closure in $\mathsf{B}(\mathcal{D}, D)[\|\cdot\|_{\mathsf{B}}]$. Then,

$$\overline{\mathfrak{M}_S^\dagger} \subseteq \left\{\varphi \in \mathsf{B}(\mathcal{D}, D) : \varphi(\xi, \eta) = \langle YS\xi|S\eta\rangle, \ \forall\, \xi, \eta \in \mathcal{D} \text{ and some } Y \in \mathcal{M}_S\right\}.$$

Indeed, if $\varphi \in \overline{\mathfrak{M}_S^\dagger}$, then there exists a sequence $\{\varphi_n\} \subset \mathfrak{M}_S^\dagger$ converging to φ. Clearly, $\varphi_n = \varphi_{A_n}$, $A_n \in \mathcal{L}^\dagger(\mathcal{D})$. Since the sequence $\{\varphi_n\}$ is Cauchy, (3.1.5) yields that $\|S^{-1}(A_n - A_m)S^{-1}\| \to 0$. Hence, $S^{-1}A_nS^{-1} \to Y$, for some $Y \in \mathcal{B}(\mathcal{H})$. Clearly, $Y \in \mathcal{M}_S$ and

$$\begin{aligned}\sup_{\|S\xi\|=\|S\eta\|=1} |\varphi(\xi, \eta) - \langle YS\xi|S\eta\rangle| \le & \sup_{\|S\xi\|=\|S\eta\|=1} |\varphi(\xi, \eta) - \langle A_n\xi|\eta\rangle| \\ & + \sup_{\|S\xi\|=\|S\eta\|=1} |\langle A_n\xi|\eta\rangle - \langle YS\xi|S\eta\rangle| \to 0,\end{aligned}$$

since

$$\begin{aligned}\sup_{\|S\xi\|=\|S\eta\|=1} |\langle A_n\xi|\eta\rangle - \langle YS\xi|S\eta\rangle| &= \sup_{\|\xi'\|=\|\eta'\|=1} |\langle A_nS^{-1}\xi'|S^{-1}\eta'\rangle - \langle Y\xi'|\eta'\rangle| \\ &= \|S^{-1}A_nS^{-1} - Y\|.\end{aligned}$$

On the other hand, it is easily seen that $\overline{\mathfrak{M}_S^\dagger} \simeq \overline{\mathfrak{M}}$. Hence, if $X \in \overline{\mathfrak{M}}$, then for some Y in $\mathcal{M}_S$, we have

$$\langle X\xi|\eta\rangle = \langle YS\xi|S\eta\rangle, \quad \forall\, \xi, \eta \in \mathcal{D}.$$

Thus, if $YS(\mathcal{D})$ is not a subset of $\mathcal{D}$, then X is neither an element of $\mathfrak{M}$ nor an operator on the Hilbert space $\mathcal{H}$, but a *true* element of $\mathfrak{L}(\mathcal{D}, \mathcal{D}^\times)$.

3.1.2 Auxiliary Seminorms

Now we come to the main topic of this section. We shall define some seminorms (one of them is, in fact, an unbounded C*-seminorm), closely related with families of sesquilinear forms [77, 78, 85] and examine their interplay with the family of *-representations of a given quasi *-algebra $(\mathfrak{A}, \mathfrak{A}_0)$. In the case, where $(\mathfrak{A}[\|\cdot\|], \mathfrak{A}_0)$ is a normed quasi *-algebra, some information on the structure of $(\mathfrak{A}[\|\cdot\|], \mathfrak{A}_0)$ can be obtained by means of them. To begin with, let us fix some terminology.

Let $(\mathfrak{A}, \mathfrak{A}_0)$ be a quasi *-algebra and $\mathfrak{s}$ a seminorm on $\mathfrak{A}$. A sesquilinear form φ on $\mathfrak{A} \times \mathfrak{A}$ is said to be $\mathfrak{s}$-*bounded* if there exists a positive constant γ, such that

$$|\varphi(a, b)| \le \gamma\mathfrak{s}(a)\mathfrak{s}(b), \quad \forall\, a, b \in \mathfrak{A}.$$

In this case, we put

$$\|\varphi\|_{\mathfrak{s}} := \sup_{\mathfrak{s}(a)=\mathfrak{s}(b)=1} |\varphi(a,b)| = \sup_{\mathfrak{s}(a)=1} \varphi(a,a).$$

If $\mathfrak{s}$ is exactly the norm of $\mathfrak{A}$, we say *bounded* instead of $\mathfrak{s}$-*bounded* and we write $\|\varphi\|$ instead of $\|\varphi\|_{\mathfrak{s}}$.

Let us now define

$$\mathfrak{q}_{\mathcal{I}}(a) := \sup\left\{\varphi(ax,ax)^{1/2} : \varphi \in \mathcal{I}_{\mathfrak{A}_0}(\mathfrak{A}),\ x \in \mathfrak{A}_0,\ \varphi(x,x) = 1\right\} \tag{3.1.6}$$

and

$$\mathcal{D}(\mathfrak{q}_{\mathcal{I}}) := \left\{a \in \mathfrak{A} : \mathfrak{q}_{\mathcal{I}}(a) < \infty\right\}.$$

Remark 3.1.11 If $(\mathfrak{A}, \mathfrak{A}_0)$ has a unit e, then one can easily check that

$$\mathfrak{q}_{\mathcal{I}}(a) = \sup\left\{\varphi(a,a)^{1/2} : \varphi \in \mathcal{I}_{\mathfrak{A}_0}(\mathfrak{A}),\ \varphi(e,e) = 1\right\}.$$

Proposition 3.1.12 *Let $(\mathfrak{A}, \mathfrak{A}_0)$ be a quasi *-algebra. For each $\varphi \in \mathcal{I}_{\mathfrak{A}_0}(\mathfrak{A})$, let π_φ denote the corresponding GNS representation. Then,*

$$\begin{aligned}\mathcal{D}(\mathfrak{q}_{\mathcal{I}}) &= \left\{a \in \mathfrak{A} : \overline{\pi_\varphi(a)} \in \mathcal{B}(\mathcal{H}_\varphi),\ \forall\, \varphi \in \mathcal{I}_{\mathfrak{A}_0}(\mathfrak{A}) \text{ and } \sup_{\varphi \in \mathcal{I}_{\mathfrak{A}_0}(\mathfrak{A})} \|\overline{\pi_\varphi(a)}\| < \infty\right\} \\ &= \left\{a \in \mathfrak{A} : \pi(a) \text{ isbounded},\ \forall\, \pi \in \mathrm{Rep}_r(\mathfrak{A}) \text{ and } \sup_{\pi \in \mathrm{Rep}_r(\mathfrak{A})} \|\overline{\pi(a)}\| < \infty\right\}\end{aligned}$$

and

$$\mathfrak{q}_{\mathcal{I}}(a) = \sup_{\varphi \in \mathcal{I}_{\mathfrak{A}_0}(\mathfrak{A})} \|\overline{\pi_\varphi(a)}\| = \sup_{\pi \in \mathrm{Rep}_r(\mathfrak{A})} \|\overline{\pi(a)}\|, \quad \forall\, a \in \mathcal{D}(\mathfrak{q}). \tag{3.1.7}$$

Proof Without loss of generality, we may assume that $(\mathfrak{A}, \mathfrak{A}_0)$ has a unit e. For shortness we put

$$\mathfrak{A}_1 = \left\{a \in \mathfrak{A} : \overline{\pi_\varphi(a)} \in \mathcal{B}(\mathcal{H}_\varphi),\ \forall\, \varphi \in \mathcal{I}_{\mathfrak{A}_0}(\mathfrak{A}) \text{ and } \sup_{\varphi \in \mathcal{I}_{\mathfrak{A}_0}(\mathfrak{A})} \|\overline{\pi_\varphi(a)}\| < \infty\right\}$$

and

$$\mathfrak{A}_2 = \left\{a \in \mathfrak{A} : \pi(a) \text{ is bounded},\ \forall\, \pi \in \mathrm{Rep}_r(\mathfrak{A}) \text{ and } \sup_{\pi \in \mathrm{Rep}_r(\mathfrak{A})} \|\overline{\pi(a)}\| < \infty\right\}.$$

Let $a \in \mathcal{D}(\mathfrak{q}_{\mathcal{I}})$. Then, if $\varphi \in \mathcal{I}_{\mathfrak{A}_0}(\mathfrak{A})$, we have

$$\varphi(ax, ax) \leq \mathfrak{q}_{\mathcal{I}}(a)^2 \varphi(x, x), \quad \forall\, x \in \mathfrak{A}_0.$$

Hence, $\pi_\varphi(a)$ is bounded and $\|\overline{\pi_\varphi(a)}\| \leq \mathfrak{q}_{\mathcal{I}}(a)$. Therefore, $a \in \mathfrak{A}_1$ and

$$\sup_{\varphi \in \mathcal{I}_{\mathfrak{A}_0}(\mathfrak{A})} \|\overline{\pi_\varphi(a)}\| \leq \mathfrak{q}_{\mathcal{I}}(a).$$

Let $a \in \mathfrak{A}_1$. Clearly, $\pi_\varphi(a)$ is bounded, if and only if, $\pi_\varphi^\circ(a)$ is bounded. Since π_φ° is regular (Proposition 2.4.16), we conclude that

$$\sup_{\varphi \in \mathcal{I}_{\mathfrak{A}_0}(\mathfrak{A})} \|\overline{\pi_\varphi(a)}\| \leq \sup_{\pi \in \mathrm{Rep}_r(\mathfrak{A})} \|\overline{\pi(a)}\|. \tag{3.1.8}$$

On the other hand, if $\pi \in \mathrm{Rep}_r(\mathfrak{A})$ then, for every $\xi \in \mathcal{D}_\pi$, we consider the corresponding vector form φ_ξ. The regularity of π implies that $\varphi_\xi \in \mathcal{I}_{\mathfrak{A}_0}(\mathfrak{A})$ (see discussion after (2.4.12)) and so the *-representation $\pi_{\varphi_\xi}^\circ$, with cyclic vector $\xi_\varphi = \lambda_{\varphi_\xi}(e)$, can be constructed. By assumption, the operator $\pi_{\varphi_\xi}^\circ(a)$ is bounded. Then, we have

$$\|\pi(a)\xi_\varphi\|^2 = \|\pi_{\varphi_\xi}^\circ(a)\lambda_{\varphi_\xi}(e)\|^2 \leq \|\pi_{\varphi_\xi}^\circ(a)\|^2 \|\xi_\varphi\|^2,$$

which implies that $\pi(a)$ is bounded and that the converse inequality in (3.1.8) holds; i.e., $\mathfrak{A}_1 \subseteq \mathfrak{A}_2$. Therefore, it is sufficient to prove the equalities

$$\mathcal{D}(\mathfrak{q}_{\mathcal{I}}) = \mathfrak{A}_2 \text{ and } \mathfrak{q}_{\mathcal{I}}(a) = \sup_{\pi \in \mathrm{Rep}_r(\mathfrak{A})} \|\overline{\pi(a)}\|, \quad \forall\, a \in \mathcal{D}(\mathfrak{q}_{\mathcal{I}}).$$

Now, let $\pi \in \mathrm{Rep}_r(\mathfrak{A})$ and, for $\xi \in \mathcal{D}_\pi$, define φ_ξ as in (2.4.12). From (3.1.6) it follows that

$$\varphi_\xi(ax, ax) \leq \mathfrak{q}_{\mathcal{I}}(a)^2 \varphi_\xi(x, x), \quad \forall\, a \in \mathcal{D}(\mathfrak{q}_{\mathcal{I}}),\ x \in \mathfrak{A}_0.$$

This implies that

$$\|\pi(a)\xi\|^2 = \varphi_\xi(a, a) \leq \mathfrak{q}_{\mathcal{I}}(a)^2 \varphi_\xi(e, e) = \mathfrak{q}_{\mathcal{I}}(a)^2 \|\xi\|^2, \quad \forall\, a \in \mathcal{D}(\mathfrak{q}_{\mathcal{I}}).$$

Thus, for every $a \in \mathcal{D}(\mathfrak{q}_{\mathcal{I}})$, $\pi(a)$ is a bounded operator and

$$\sup_{\pi \in \mathrm{Rep}_r(\mathfrak{A})} \|\overline{\pi(a)}\| \leq \mathfrak{q}_{\mathcal{I}}(a) < \infty.$$

Conversely, if $\pi(a)$ is bounded, for every $\pi \in \mathrm{Rep}_r(\mathfrak{A})$, and $\sup_{\pi \in \mathrm{Rep}_r(\mathfrak{A})} \|\overline{\pi(a)}\| < \infty$, this is in particular true for the representation π_φ° constructed by any $\varphi \in$

$\mathcal{I}_{\mathfrak{A}_0}(\mathfrak{A})$. Thus,

$$\varphi(ax, ax) = \|\pi_\varphi^\circ(a)\lambda_\varphi(x)\|^2 \leq \Big(\sup_{\pi \in \mathrm{Rep}_r(\mathfrak{A})} \|\overline{\pi(a)}\|^2\Big)\|\lambda_\varphi(x)\|^2$$

$$= \Big(\sup_{\pi \in \mathrm{Rep}_r(\mathfrak{A})} \|\overline{\pi(a)}\|^2\Big)\varphi(x, x), \quad \forall\, a \in \mathfrak{A},\ x \in \mathfrak{A}_0.$$

Therefore $a \in D(\mathsf{q}_{\mathcal{I}})$ and

$$\mathsf{q}_{\mathcal{I}}(a) \leq \sup_{\pi \in \mathrm{Rep}_r(\mathfrak{A})} \|\overline{\pi(a)}\|.$$

This completes the proof. □

Lemma 3.1.13 *Let* $(\mathfrak{A}[\|\cdot\|], \mathfrak{A}_0)$ *be a normed quasi *-algebra. Set* (see also Definition 2.3.1)

$$\mathcal{P}_{\mathfrak{A}_0}(\mathfrak{A}) := \{\varphi \in \mathcal{Q}_{\mathfrak{A}_0}(\mathfrak{A}) : \varphi \text{ is bounded}\}.$$

Then, $\mathcal{P}_{\mathfrak{A}_0}(\mathfrak{A}) \subseteq \mathcal{I}_{\mathfrak{A}_0}(\mathfrak{A})$.

Proof If $\varphi \in \mathcal{P}_{\mathfrak{A}_0}(\mathfrak{A})$, then the subspace $\lambda_\varphi(\mathfrak{A}_0)$ is dense in $\mathcal{H}_\varphi$. Indeed, if $a \in \mathfrak{A}$, there exists a sequence $\{x_n\}$, $x_n \in \mathfrak{A}_0$, such that $x_n \underset{n\to\infty}{\to} a$ in $\mathfrak{A}$. Then, we have

$$\|\lambda_\varphi(a) - \lambda_\varphi(x_n)\|^2 = \varphi(a - x_n, a - x_n) \leq \|\varphi\|^2\|a - x_n\|^2 \underset{n\to\infty}{\to} 0.$$ □

Definition 3.1.14 We define

$$\mathcal{S}_{\mathfrak{A}_0}(\mathfrak{A}) := \{\varphi \in \mathcal{P}_{\mathfrak{A}_0}(\mathfrak{A}) : \|\varphi\| \leq 1\}.$$

More explicitly, $\varphi \in \mathcal{S}_{\mathfrak{A}_0}(\mathfrak{A})$, if and only if, it satisfies the following conditions:

(i) $\varphi(a, a) \geq 0, \quad \forall\, a \in \mathfrak{A}$;
(ii) $\varphi(ax, y) = \varphi(x, a^*y), \quad \forall\, a \in \mathfrak{A},\ x, y \in \mathfrak{A}_0$;
(iii) $|\varphi(a, b)| \leq \|a\|\,\|b\|, \quad \forall\, a, b \in \mathfrak{A}$.

Remark 3.1.15 Of course, the possibility that $\mathcal{S}_{\mathfrak{A}_0}(\mathfrak{A}) = \{0\}$ is not excluded and examples, where this happens, can be given (see, e.g., Example 3.1.29 below).

If $(\mathfrak{A}[\|\cdot\|], \mathfrak{A}_0)$ is a Banach quasi *-algebra, we denote the Banach dual space of $\mathfrak{A}$ by $\mathfrak{A}^\star$. Then, $\mathfrak{A}^\star$ can be made into a Banach $\mathfrak{A}_0$-bimodule under the norm

$$\|f\|^\star := \sup_{\|x\|\leq 1} |f(x)|,\ f \in \mathfrak{A}^\star,$$

by defining, for $f \in \mathfrak{A}^\star$, $x \in \mathfrak{A}_0$, the module operations $(f, x) \mapsto f \circ x$ (resp. $(x, f) \mapsto x \circ f$), in the following way:

$$(f \circ x)(a) := f(ax), \quad (x \circ f)(a) := f(xa), \quad a \in \mathfrak{A}.$$

As usual, an involution $f \mapsto f^*$ can be defined on $\mathfrak{A}^\star$ by $f^*(a) := \overline{f(a^*)}$, $a \in \mathfrak{A}$. Under this notation we can easily prove the following (see, also [86]):

Proposition 3.1.16 *Let $(\mathfrak{A}[\|\cdot\|], \mathfrak{A}_0)$ be a Banach quasi *-algebra and φ a positive sesquilinear form on $\mathfrak{A} \times \mathfrak{A}$. The following statements are equivalent:*

(i) $\varphi \in \mathcal{S}_{\mathfrak{A}_0}(\mathfrak{A})$;
(ii) *there exists a bounded conjugate linear operator $T : \mathfrak{A} \to \mathfrak{A}^\star$ with the properties:*

  (ii.1) $(Ta)(a) \geq 0, \quad \forall\, a \in \mathfrak{A}$;
  (ii.2) $T(xa) = (Ta) \circ x^*, \quad \forall\, x \in \mathfrak{A}_0,\ a \in \mathfrak{A}$, *where $\circ$ as above;*
  (ii.3) $\|T\|_{\mathcal{B}(\mathfrak{A}, \mathfrak{A}^\star)} \leq 1$, *$\mathcal{B}(\mathfrak{A}, \mathfrak{A}^\star)$ standing for the algebra of all bounded conjugate linear operators from $\mathfrak{A}$ in $\mathfrak{A}^\star$;*
  (ii.4) $\varphi(a, b) = (Tb)(a), \quad \forall\, a, b \in \mathfrak{A}$.

3.1.3 Sufficient Family of Forms and *-Semisimplicity

Definition 3.1.17 Let $(\mathfrak{A}[\|\cdot\|], \mathfrak{A}_0)$ be a Banach quasi *-algebra. We say that $\mathcal{S}_{\mathfrak{A}_0}(\mathfrak{A})$ (see discussion before Remark 3.1.15) is *sufficient*, if $a \in \mathfrak{A}$ with $\varphi(a, a) = 0$, for each $\varphi \in \mathcal{S}_{\mathfrak{A}_0}(\mathfrak{A})$, implies that $a = 0$.

Lemma 3.1.19, just below, allows us to formulate in various ways the notion of sufficiency for $\mathcal{S}_{\mathfrak{A}_0}(\mathfrak{A})$. Before we state it, we give the definition of a seminorm $\mathfrak{p}$, needed in this lemma, but also in the sequel. Let $(\mathfrak{A}[\|\cdot\|], \mathfrak{A}_0)$ be a normed quasi *-algebra. We put

$$\mathfrak{p}(a) := \sup_{\varphi \in \mathcal{S}_{\mathfrak{A}_0}(\mathfrak{A})} \varphi(a, a)^{1/2}, \quad a \in \mathfrak{A}. \tag{3.1.9}$$

Then, $\mathfrak{p}$ is a seminorm on $\mathfrak{A}$, with $\mathfrak{p}(a) \leq \|a\|$, for every $a \in \mathfrak{A}$ (see Definition 3.1.14(iii)).

Proposition 3.1.18 *Let $(\mathfrak{A}[\|\cdot\|], \mathfrak{A}_0)$ be a Banach quasi *-algebra. Then, for every $a \in \mathfrak{A}$, there exists $\varphi \in \mathcal{S}_{\mathfrak{A}_0}(\mathfrak{A})$, such that $\varphi(a, a) = \mathfrak{p}(a)$.*

Proof Let $a \in \mathfrak{A}$ be fixed. By the definition of $\mathfrak{p}$ itself, it follows that if $\varepsilon > 0$ and $n \in \mathbb{N}$, there exists $\varphi_{\varepsilon,n} \in \mathcal{S}_{\mathfrak{A}_0}(\mathfrak{A})$, such that $\mathfrak{p}(a)^2 - \frac{\varepsilon}{2^n} < \varphi_{\varepsilon,n}(a, a)$. We define

$$\varphi(a', b') := \sum_{n=1}^{\infty} \frac{1}{2^n} \varphi_{\varepsilon,n}(a', b'), \quad \forall\, a', b' \in \mathfrak{A}.$$

It is easy to verify that $\varphi \in \mathcal{S}_{\mathfrak{A}_0}(\mathfrak{A})$; in particular

$$|\varphi(a', b')| = \left|\sum_{n=1}^{\infty} \frac{1}{2^n} \varphi_{\varepsilon,n}(a', b')\right| \leq \sum_{n=1}^{\infty} \frac{1}{2^n} |\varphi_{\varepsilon,n}(a', b')|$$
$$\leq \left(\sum_{n=1}^{\infty} \frac{1}{2^n}\right) \|a'\| \|b'\| = \|a'\| \|b'\|, \quad \forall\, a', b' \in \mathfrak{A}.$$

Moreover,

$$\varphi(a, a) = \sum_{n=1}^{\infty} \frac{1}{2^n} \varphi_{\varepsilon,n}(a, a) > \sum_{n=1}^{\infty} \frac{1}{2^n} (\mathfrak{p}(a)^2 - \frac{\varepsilon}{2^n}) = \mathfrak{p}(a)^2 - \frac{4}{3}\varepsilon.$$

Then, since ε is arbitrary, $\varphi(a, a) \geq \mathfrak{p}(a)^2$. But also $\varphi(b, b) \leq \mathfrak{p}(b)^2$, for every $b \in \mathfrak{A}$. Thus, we conclude that $\varphi(a, a) = \mathfrak{p}(a)^2$. □

Lemma 3.1.19 *Let $(\mathfrak{A}[\|\cdot\|], \mathfrak{A}_0)$ be a Banach quasi *-algebra with unit e. For an element $a \in \mathfrak{A}$, the following statements are equivalent:*

(i) $\mathfrak{p}(a) = 0$;
(ii) $\varphi(a, a) = 0, \quad \forall\, \varphi \in \mathcal{S}_{\mathfrak{A}_0}(\mathfrak{A})$;
(iii) $\varphi(a, b) = 0, \quad \forall\, \varphi \in \mathcal{S}_{\mathfrak{A}_0}(\mathfrak{A})$ *and* $b \in \mathfrak{A}$;
(iv) $\omega_\varphi(a) = 0, \quad \forall\, \varphi \in \mathcal{S}_{\mathfrak{A}_0}(\mathfrak{A})$;
(v) $\varphi(ax, x) = 0, \quad \forall\, \varphi \in \mathcal{S}_{\mathfrak{A}_0}(\mathfrak{A})$ *and* $x \in \mathfrak{A}_0$;
(vi) $\varphi(ax, y) = 0, \quad \forall\, \varphi \in \mathcal{S}_{\mathfrak{A}_0}(\mathfrak{A})$ *and* $x, y \in \mathfrak{A}_0$.

Definition 3.1.20 Let $(\mathfrak{A}[\|\cdot\|], \mathfrak{A}_0)$ be a normed quasi *-algebra with unit e and π a *-representation of $(\mathfrak{A}[\|\cdot\|], \mathfrak{A}_0)$ into $\mathcal{L}^\dagger(\mathcal{D}_\pi, \mathcal{H}_\pi)$ (see Definition 2.2.5 and discussion before Sect. 2.2). We say that π is

- *weakly continuous*, if π is continuous from $\mathfrak{A}[\|\cdot\|]$ into $\mathcal{L}^\dagger(\mathcal{D}_\pi, \mathcal{H}_\pi)[\mathsf{t}_w]$;
- *strongly continuous*, if π is continuous from $\mathfrak{A}[\|\cdot\|]$ into $\mathcal{L}^\dagger(\mathcal{D}_\pi, \mathcal{H}_\pi)[\mathsf{t}_s]$;
- *strongly* continuous*, if π is continuous from $\mathfrak{A}[\|\cdot\|]$ into $\mathcal{L}^\dagger(\mathcal{D}_\pi, \mathcal{H}_\pi)[\mathsf{t}_{s^*}]$.

It is easy to prove that a *-representation is strongly continuous, if and only if, it is strongly* continuous. We generalize now some familiar notions from Banach *-algebras theory, such as that of *-radical and of *-semisimplicity; for the Banach case, see Definition A.5.2.

Definition 3.1.21 If $(\mathfrak{A}[\|\cdot\|], \mathfrak{A}_0)$ is a normed quasi *-algebra, we define the **-radical* $\mathsf{R}_{\mathfrak{A}_0}(\mathfrak{A})$ of $(\mathfrak{A}[\|\cdot\|], \mathfrak{A}_0)$ as the intersection of the kernels of all of its strongly continuous qu*-representations (see Sect. 2.2).

Proposition 3.1.22 *Let $(\mathfrak{A}[\|\cdot\|], \mathfrak{A}_0)$ be a normed quasi *-algebra with unit e. For an element $a \in \mathfrak{A}$ the following statements are equivalent:*

(i) $a \in \mathsf{R}_{\mathfrak{A}_0}(\mathfrak{A})$;
(ii) $\varphi(a, a) = 0$, *for every* $\varphi \in \mathcal{S}_{\mathfrak{A}_0}(\mathfrak{A})$.

Proof Assume that (i) holds and let $\varphi \in \mathcal{S}_{\mathfrak{A}_0}(\mathfrak{A})$. Let $(\pi_\varphi, \lambda_\varphi, \mathcal{H}_\varphi)$ be the GNS construction of Proposition 2.4.1. Then, the restriction π_φ° of π_φ to $\lambda_\varphi(\mathfrak{A}_0)$ is a qu*-representation. Since,

$$\|\pi_\varphi^\circ(b)\lambda_\varphi(x)\|^2 = \varphi(bx, bx) \leq \|bx\|^2 \leq \|b\|^2\|x\|^2, \quad \forall\, b \in \mathfrak{A},\ x \in \mathfrak{A}_0,$$

π_φ° is strongly continuous. Hence, $\pi_\varphi^\circ(a) = 0$. Then,

$$\varphi(a, a) = \|\pi_\varphi^\circ(a)\lambda_\varphi(e)\|^2 = 0.$$

Conversely, assume that (ii) holds. Let π be any strongly continuous qu*-representation of $(\mathfrak{A}[\|\cdot\|], \mathfrak{A}_0)$. Then, there exists $\gamma_\pi > 0$, such that

$$\|\pi(a)\xi\| \leq \gamma_\pi \|a\|, \quad \forall\, a \in \mathfrak{A}.$$

Let $\xi \in \mathcal{D}_\pi$, with $\|\xi\| = 1$. Define

$$\varphi_\pi(a, b) := \langle \pi(a)\xi | \pi(b)\xi \rangle, \quad a, b \in \mathcal{D}_\pi.$$

Then,

$$\|\varphi_\pi(a, b)\| = |\langle \pi(a)\xi | \pi(b)\xi \rangle| \leq \|\pi(a)\xi\|\,\|\pi(b)\xi\| \leq \gamma_\pi^2 \|a\|\|b\|, \quad \forall\, a, b \in \mathcal{D}_\pi.$$

It is easy to prove that $\varphi_\pi/\gamma_\pi^2 \in \mathcal{S}_{\mathfrak{A}_0}(\mathfrak{A})$. Thus, $\varphi_\pi(a, a) = 0$ and this implies that $\|\pi(a)\xi\| = 0$. Since, ξ is arbitrary, we conclude that $\pi(a) = 0$. □

Definition 3.1.23 A normed quasi *-algebra $(\mathfrak{A}[\|\cdot\|], \mathfrak{A}_0)$ is called **-semisimple* if

$$\mathsf{R}_{\mathfrak{A}_0}(\mathfrak{A}) = \{0\}.$$

An immediate consequence of the previous Proposition 3.1.22 is the following

Corollary 3.1.24 *A normed quasi *-algebra* $(\mathfrak{A}[\|\cdot\|], \mathfrak{A}_0)$ *is *-semisimple, if and only if,* $\mathsf{S}_{\mathfrak{A}_0}(\mathfrak{A})$ *is sufficient.*

Let $(\mathfrak{A}[\|\cdot\|], \mathfrak{A}_0)$ be a Banach quasi *-algebra. We denote by $\mathfrak{A}_0^+$ the set of *positive* elements of $\mathfrak{A}_0$; i.e.,

$$\mathfrak{A}_0^+ := \left\{ \sum_{k=1}^{n} x_k^* x_k : x_k \in \mathfrak{A}_0,\ k = 1, \ldots, n,\ n \in \mathbb{N} \right\}.$$

We put $\mathfrak{A}^+ := \overline{\mathfrak{A}_0^+}^{\|\cdot\|}$, the closure of $\mathfrak{A}_0^+$ in the norm topology of $\mathfrak{A}$. Elements of $\mathfrak{A}^+$ are called *positive* too and we often write $a \geq 0$ instead of $a \in \mathfrak{A}^+$. If $a \in \mathfrak{A}^+$, then $a = a^*$, as it follows immediately by the definition. Moreover, $\mathfrak{A}^+$ is a convex cone.

A linear functional ω on $\mathfrak{A}$ is called *positive* if $\omega(a) \geq 0$, for every $a \in \mathfrak{A}^+$.

We have already denoted by $\mathfrak{A}^\star$ the Banach dual space of $\mathfrak{A}[\|\cdot\|]$. Then, the set of positive elements of $\mathfrak{A}^\star$ (that is, the set of *bounded* positive functionals) is denoted by $\mathfrak{A}^\star_+$.

In this regard, we have the following

Proposition 3.1.25 *Let* $(\mathfrak{A}[\|\cdot\|], \mathfrak{A}_0)$ *be a Banach quasi *-algebra with unit e. If the linear span of the set*

$$\mathfrak{A}^\star_{\mathcal{P}} := \{\omega_\varphi : \varphi \in \mathcal{P}_{\mathfrak{A}_0}(\mathfrak{A})\}$$

is weakly-dense in* $\mathfrak{A}^\star$, *then* $\mathcal{S}_{\mathfrak{A}_0}(\mathfrak{A})$ *is sufficient or, equivalently,* $(\mathfrak{A}[\|\cdot\|], \mathfrak{A}_0)$ *is *-semisimple. Conversely, if* $\mathfrak{A}[\|\cdot\|]$ *is a reflexive Banach space and* $\mathcal{S}_{\mathfrak{A}_0}(\mathfrak{A})$ *is sufficient, then the linear span of* $\mathfrak{A}^\star_{\mathcal{P}}$ *is weakly*-dense in* $\mathfrak{A}^\star$.

Proof Assume that $\mathcal{S}_{\mathfrak{A}_0}(\mathfrak{A})$ is not sufficient. Then, there exists $a \in \mathfrak{A}$, $a \neq 0$, such that for every $\varphi \in \mathcal{S}_{\mathfrak{A}_0}(\mathfrak{A})$, $\varphi(a,a) = 0$. This implies that $\omega_\varphi(a) = 0$, for each $\varphi \in \mathcal{S}_{\mathfrak{A}_0}(\mathfrak{A})$. Thus, the non-zero continuous linear functional f_a on $\mathfrak{A}^\star$, defined by $f_a(\omega) := \omega(a)$, is zero all over the set $\mathfrak{A}^\star_{\mathcal{P}}$ and therefore, on its linear span. Hence, this set is not weakly*-dense in $\mathfrak{A}^\star$.

Conversely, assume that the linear span of $\mathfrak{A}^\star_{\mathcal{P}}$ is not weakly*-dense in $\mathfrak{A}^\star$. Then, by the reflexivity of $\mathfrak{A}[\|\cdot\|]$, there would exist an element $a \in \mathfrak{A}$, $a \neq 0$, such that $\omega_\varphi(a) = \varphi(a,e) = 0$, for each $\varphi \in \mathcal{P}_{\mathfrak{A}_0}(\mathfrak{A})$ [28, p. 186, Corollary 1]. Thus, by Lemma 3.1.19, we obtain $\varphi(a,a) = 0$, for each $\varphi \in \mathfrak{A}^\star_{\mathcal{P}}$. This implies $a = 0$, a contradiction. □

➨ Here and in Appendix A, *for simplicity's sake*, we shall *denote the unit ball of* $\mathfrak{A}^\star$ by $\mathcal{U}_{\mathfrak{A}^\star}$, *instead of* $\mathcal{U}(\mathfrak{A}^\star)$, which is our standard symbol for the unit ball of a normed space (see discussion before Proposition 4.1.28).

Let $\mathcal{U}_{\mathfrak{A}^\star} := \{\omega \in \mathfrak{A}^\star : \|\omega\|^\star \leq 1\}$ be the unit ball of $\mathfrak{A}^\star$ and $\mathfrak{A}^\star_{\mathcal{S}} = \{\omega_\varphi : \varphi \in \mathcal{S}_{\mathfrak{A}_0}(\mathfrak{A})\}$. Of course, it is truly possible that $\mathcal{S}_{\mathfrak{A}_0}(\mathfrak{A}) = \{0\}$ (or, equivalently, $\mathfrak{A}^\star_{\mathcal{S}} = \{0\}$). It is, clearly, much more interesting to consider Banach quasi *-algebras, for which the set $\mathcal{S}_{\mathfrak{A}_0}(\mathfrak{A})$ is sufficiently rich (Sect. 3.1.3).

Proposition 3.1.26 *Assume that* $(\mathfrak{A}[\|\cdot\|], \mathfrak{A}_0)$ *has a unit and that* $\mathcal{S}_{\mathfrak{A}_0}(\mathfrak{A}) \neq \{0\}$. *Then, the following statements hold:*

(i) $\mathfrak{A}^\star_{\mathcal{S}}$ *is a convex, weakly*-compact subset of* $\mathcal{U}_{\mathfrak{A}^\star}$;
(ii) $\mathfrak{A}^\star_{\mathcal{S}}$ *has extreme points. If* ω_φ *is extreme, then* $\|\varphi\| = 1$;
(iii) ω_φ *is extreme in* $\mathfrak{A}^\star_{\mathcal{S}}$, *if and only if,* φ *is extreme in* $\mathcal{S}_{\mathfrak{A}_0}(\mathfrak{A})$.

The proof is very simple and we omit it.

Proposition 3.1.27 *Let* $(\mathfrak{A}[\|\cdot\|], \mathfrak{A}_0)$ *be a *-semisimple Banach quasi *-algebra with unit. Let* $a \in \mathfrak{A}$. *Then, the following hold:*

(i) $a = a^*$, *if and only if,* $\omega_\varphi(a) \in \mathbb{R}$, *for each* $\varphi \in \mathcal{S}_{\mathfrak{A}_0}(\mathfrak{A})$;
(ii) *if* $a \geq 0$, *then* $\omega_\varphi(a) \geq 0$, *for each* $\varphi \in \mathcal{S}_{\mathfrak{A}_0}(\mathfrak{A})$;
(iii) $a \in \mathfrak{A}^+ \cap \{-\mathfrak{A}^+\}$, *if and only if,* $a = 0$.

Proof

(i) Assume that, for each $\varphi \in \mathcal{S}_{\mathfrak{A}_0}(\mathfrak{A})$, $\omega_\varphi(a) \in \mathbb{R}$. Then, we have

$$\omega_\varphi(a - a^*) = \omega_\varphi(a) - \omega_\varphi(a^*) = \omega_\varphi(a) - \overline{\omega_\varphi(a)} = 0,$$

for every $\varphi \in \mathcal{S}_{\mathfrak{A}_0}(\mathfrak{A})$. By Lemma 3.1.19 one has $\varphi(a - a^*, a - a^*) = 0$, for every $\varphi \in \mathcal{S}_{\mathfrak{A}_0}(\mathfrak{A})$. Hence, $a = a^*$. The converse implication is obvious. Lemma 3.1.19 is also used in the proof of (ii) and (iii).

(ii) If $a \geq 0$, then by definition a is the limit of a sequence of elements of $\mathfrak{A}_0^+$ and every ω_φ is positive on $\mathfrak{A}_0$ and continuous on $\mathfrak{A}[\| \cdot \|]$ (for the latter, see Definition 3.1.14).

(iii) Assume that $a \in \mathfrak{A}^+ \cap \{-\mathfrak{A}^+\}$; then, by (ii) it follows that $\omega_\varphi(a) = 0$, for every $\varphi \in \mathcal{S}_{\mathfrak{A}_0}(\mathfrak{A})$. From this we conclude that $a = 0$. □

Proposition 3.1.28 *Let* $(\mathfrak{A}[\| \cdot \|], \mathfrak{A}_0)$ *be a *-semisimple BQ*-algebra. Then,* $\mathfrak{A}_0$ *is a *-semisimple Banach *-algebra.*

Proof It suffices to show that if $x \in \mathfrak{A}_0$ and $\omega(x^*x) = 0$, for each positive linear functional ω on $\mathfrak{A}_0$, then $x = 0$. If this assumption is satisfied then, in particular, we will have $\omega_\varphi(x^*x) = 0$, for each $\varphi \in \mathcal{S}_{\mathfrak{A}_0}(\mathfrak{A})$. This implies that $\varphi(x, x) = 0$, for every $\varphi \in \mathcal{S}_{\mathfrak{A}_0}(\mathfrak{A})$ and thus $x = 0$. □

The case of a *-semisimple normed quasi *-algebra (i.e., with trivial *-radical) is particularly interesting. The main reason is that, for a *-semisimple normed quasi *-algebra, it is possible to define a *refinement* of the partial multiplication. In this way, the lattices of multipliers become nontrivial. We will now sketch the construction that makes of any *-semisimple Banach quasi *-algebra a nontrivial partial *-algebra. But before going forth let us examine an example.

Example 3.1.29 In this example we shall discuss the *-semisimplicity of the Banach quasi *-algebra $(L^p(I), C(I))$, where $I = [0, 1]$ and $p \geq 1$ (Example 3.1.6). We consider here the spaces L^p with respect to the Lebesgue measure dt. However, it is not difficult to realize that the same argument holds if in the place of I is a compact Hausdorff space X with a Borel measure μ. The following two well-known facts will be needed:

(lp.i) Let y be a measurable function on I, and assume that $xy \in L^r(I)$, for all $x \in L^p(I)$ with $1 \leq r \leq p$. Then, $y \in L^q(I)$, with $p^{-1} + q^{-1} = r^{-1}$.

(lp.ii) Let $p, q, r \geq 1$, such that $p^{-1} + q^{-1} = r^{-1}$. Let $w \in L^q(I)$. Then, the linear operator $T_w : x \in L^p(I) \mapsto xw \in L^r(I)$ is bounded and $\|T_w\|_{p,r} = \|w\|_q$, where $\|T_w\|_{p,r}$ denotes the norm of T_w as bounded operator from $L^p(I)$ into $L^r(I)$.

We put

$$\mathcal{B}_+^p = \left\{ v \in L^{p/(p-2)}(I),\ \ v \geq 0 \text{ and } \|v\|_{p/(p-2)} \leq 1 \right\}.$$

If $p = 2$, we set $\frac{p}{p-2} = \infty$.

Statement 1 Let $p \geq 2$. Then, $\varphi \in \mathcal{S}_{C(I)}(L^p(I))$, if and only if, there exists $v \in \mathcal{B}_+^p$, such that

$$\varphi(x, y) = \int_I x(t)\overline{y(t)}v(t)\,dt, \quad \forall\, x, y \in L^p(I). \tag{3.1.10}$$

The sufficiency is straightforward. As for the necessity, we first notice that any bounded sesquilinear form φ on $L^p(I) \times L^p(I)$ can be represented as

$$\varphi(x, y) = \int_I x(t)\overline{(Ty)(t)}\,dt, \quad x, y \in L^p(I), \tag{3.1.11}$$

where T is a bounded linear operator from $L^p(I)$ into its dual space $L^{p'}(I)$, with $p^{-1} + p'^{-1} = 1$ [17, Vol. II, §40]. From (lp.ii) and Eq. (3.1.11) it follows easily that

$$Ty = yTu, \quad \forall\, y \in L^p(I),$$

where $u(t) = 1$, for each $t \in I$. Set $v := Tu$; from (lp.i) we obtain $v \geq 0$.

By the positivity of φ, we get $v \in L^{p/(p-2)}(I)$. Making use of (lp.ii), it is also easy to check that $\|v\|_{p/(p-2)} \leq 1$.

Statement 2 If $1 \leq p < 2$, then $\mathcal{S}_{C(I)}(L^p(I)) = \{0\}$. Indeed, let $v \neq 0$. Then, we can choose $\alpha > 0$, in such a way, that the set $I_\alpha = \{t \in I : v(t) > \alpha\}$ has positive measure. Let $x \in L^p(I_\alpha) \setminus L^2(I_\alpha)$ (such a function always exists because of the assumption on p). Now define

$$\widetilde{x}(t) = \begin{cases} x(t), & \text{if } t \in I_\alpha \\ 0, & \text{if } t \in I \setminus I_\alpha \end{cases}$$

Clearly, $\widetilde{x} \in L^p(I)$. Moreover,

$$\varphi(\widetilde{x}, \widetilde{x}) = \int_I |\widetilde{x}(t)|^2 v(t)dt = \int_{I_\alpha} |x(t)|^2 v(t)dt \geq \alpha \int_{I_\alpha} |x(t)|^2 dt = \infty$$

and this is a contradiction.

Statement 3 If $p \geq 2$, then $(L^p(I), C(I))$ is *-semisimple.

We show first that for each $x \in L^p(I)$, there exists $\widetilde{\varphi} \in \mathcal{S}_{C(I)}(L^p(I))$, such that $\widetilde{\varphi}(x, x) = \|x\|_p^2$. This is achieved by setting

$$\widetilde{\varphi}(y, z) = \|x\|_p^{2-p} \int_I y\overline{z}|x|^{p-2}\,dt, \quad y, z \in L^p(I),$$

where $\widetilde{\varphi} \in \mathcal{S}_{C(I)}(L^p(I))$, since the function $v = |x|^{p-2}\|x\|_p^{2-p}$ belongs to the set $\mathcal{B}_+^p$. A direct calculation shows that $\widetilde{\varphi}(x, x) = \|x\|_p^2$.

Let us now suppose that $\varphi(x, x) = 0$, for all $\varphi \in \mathcal{S}_{C(I)}(L^p(I))$. Then, in particular, $\widetilde{\varphi}(x, x) = \|x\|_p^2 = 0$. Therefore, $x = 0$ and so $(L^p(I), C(I))$ is *-semisimple.

In a *-semisimple Banach quasi *-algebra the multiplication can be refined (in the sense that we can define an extension of the partial multiplication) as follows (see also [39]).

Definition 3.1.30 Let $(\mathfrak{A}[\|\cdot\|], \mathfrak{A}_0)$ be a *-semisimple normed quasi *-algebra. We say that the *weak* multiplication, $a \Box b$, $a, b \in \mathfrak{A}$, is well-defined if there exists $c \in \mathfrak{A}$, such that

$$\varphi(bx, a^* y) = \varphi(cx, y), \quad \forall\, x, y \in \mathfrak{A}_0 \text{ and } \varphi \in \mathcal{S}_{\mathfrak{A}_0}(\mathfrak{A}).$$

In this case, we put $a \Box b := c$.

The following result is immediate.

Proposition 3.1.31 *Let $(\mathfrak{A}[\|\cdot\|], \mathfrak{A}_0)$ be a *-semisimple normed quasi *-algebra. Then, $\mathfrak{A}$ is also a partial *-algebra with respect to the weak multiplication.*

We shall denote by $R_w(\mathfrak{A})$ the space of universal right weak multipliers of $\mathfrak{A}$; i.e., the space of all $b \in \mathfrak{A}$, such that $a \Box b$ is well-defined, for every $a \in \mathfrak{A}$. In the same way $L_w(\mathfrak{A})$ is defined. Clearly, $\mathfrak{A}_0 \subseteq R_w(\mathfrak{A})$ (resp. $\mathfrak{A}_0 \subseteq L_w(\mathfrak{A})$). Given a fixed $a \in \mathfrak{A}$, denote by $R_w(a)$ the set of all $b \in \mathfrak{A}$, such that $a \Box b$ is well-defined. Similarly, we define $L_w(a)$.

Remark 3.1.32 The sesquilinear forms of $\mathcal{S}_{\mathfrak{A}_0}(\mathfrak{A})$ define on $\mathfrak{A}$, the topologies $\tau_w, \tau_s, \tau_{s^*}$ generated, respectively, by the following families of seminorms:

$$\begin{aligned}
\tau_w&: \quad a \mapsto |\varphi(ax, y)|, \quad a \in \mathfrak{A},\ \varphi \in \mathcal{S}_{\mathfrak{A}_0}(\mathfrak{A}),\ x, y \in \mathfrak{A}_0;\\
\tau_s&: \quad a \mapsto \varphi(a, a)^{1/2}, \quad a \in \mathfrak{A},\ \varphi \in \mathcal{S}_{\mathfrak{A}_0}(\mathfrak{A});\\
\tau_{s^*}&: \quad a \mapsto \max\left\{\varphi(a, a)^{1/2}, \varphi(a^*, a^*)^{1/2}\right\}, \quad a \in \mathfrak{A},\ \varphi \in \mathcal{S}_{\mathfrak{A}_0}(\mathfrak{A}).
\end{aligned}$$

From continuity of $\varphi \in \mathcal{S}_{\mathfrak{A}_0}(\mathfrak{A})$, it follows that the topologies $\tau_w, \tau_s, \tau_{s^*}$ are coarser than the initial norm topology of $\mathfrak{A}$.

Now we have the following

Proposition 3.1.33 *The following statements are equivalent:*

(i) *the weak product $a \Box b$ is well-defined;*
(ii) *there exists a sequence $\{y_n\}$ in $\mathfrak{A}_0$ and $c \in \mathfrak{A}$, such that $\|y_n - b\| \to 0$ and $a y_n \xrightarrow{\tau_w} c$;*
(iii) *there exists a sequence $\{x_n\}$ in $\mathfrak{A}_0$ and $c \in \mathfrak{A}$, such that $\|x_n - a\| \to 0$ and $x_n b \xrightarrow{\tau_w} c$.*

Proof We prove only that (i) $\Leftrightarrow$ (ii). The proof of (i) $\Leftrightarrow$ (iii) is very similar. Assume that $a \Box b$ is defined. By the $\|\cdot\|$-density of $\mathfrak{A}_0$, there exists a sequence $\{y_n\}$ in $\mathfrak{A}_0$

approximating b. Then, for every $z, z' \in \mathfrak{A}_0$

$$\varphi\big((ay_n)z, z'\big) = \varphi(y_n z, a^* z') \to \varphi(bz, a^* z') = \varphi\big((a\Box b)z, z'\big),$$

i.e., $ay_n \xrightarrow{\tau_w} a\Box b$. Conversely, assume the existence of a sequence $\{y_n\}$ in $\mathfrak{A}_0$ approximating b such that $ay_n \xrightarrow{\tau_w} c \in \mathfrak{A}$. Then, for every $z, z' \in \mathfrak{A}_0$, we have

$$\varphi(bz, a^* z') = \lim_{n\to\infty} \varphi(y_n z, a^* z') = \lim_{n\to\infty} \varphi\big((ay_n)z, z'\big) = \varphi(cz, z'),$$

i.e., $a\Box b$ is well-defined. □

Let $(\mathfrak{A}[\|\cdot\|], \mathfrak{A}_0)$ be a Banach quasi *-algebra. To every $a \in \mathfrak{A}$ we may correspond the linear maps L_a and R_a defined as follows

$$x \in \mathfrak{A}_0 \to L_a x := ax \in \mathfrak{A}, \quad x \in \mathfrak{A}_0 \to R_a x := xa \in \mathfrak{A}. \tag{3.1.12}$$

If $(\mathfrak{A}[\|\cdot\|], \mathfrak{A}_0)$ is a *-semisimple Banach quasi *-algebra, then the weak multiplication □ allows us to extend L_a, (resp., R_a) to the set $R_w(a)$ (resp., $L_w(a)$). Let us denote by $\widehat{L}_a$, (resp., $\widehat{R}_a$) these extensions. Then, $\widehat{L}_a b = a\Box b$, for every $b \in R_w(a)$ and $\widehat{R}_a c = c\Box a$, for every $c \in L_w(a)$.

Proposition 3.1.34 *Let $(\mathfrak{A}[\|\cdot\|], \mathfrak{A}_0)$ be a *-semisimple Banach quasi *-algebra. Then, for every $a \in \mathfrak{A}$, $\widehat{L}_a$, $\widehat{R}_a$ are closed linear maps in $\mathfrak{A}[\|\cdot\|]$.*

Proof We prove the statement only for L_a. Let $\|b_n - b\| \to 0$, with $b_n \in R_w(a)$ and $\|a\Box b_n - c\| \to 0$. Then, for every $\varphi \in \mathcal{S}_{\mathfrak{A}_0}(\mathfrak{A})$ and for all $z, z' \in \mathfrak{A}_0$,

$$\begin{aligned}\varphi\big((a\Box b_n - c)z, z'\big) &= \varphi\big((a\Box b_n)z, z'\big) - \varphi(cz, z') \\ &= \varphi(b_n z, a^* z') - \varphi(cz, z') \\ &\to \varphi(bz, a^* z') - \varphi(cz, z') = 0.\end{aligned}$$

By the *-semisimplicity, these relations show that $b \in R_w(a)$ and $c = a\Box b$; i.e., $\widehat{L}_a$ is closed. □

Example 3.1.35 We give here an example, where $\mathcal{I}_{\mathfrak{A}_0}(\mathfrak{A}) \supsetneq \mathcal{P}_{\mathfrak{A}_0}(\mathfrak{A})$. We consider the Banach quasi *-algebra $(L^1(I), L^\infty(I))$, $I = [0, 1]$. For every $x \in L^1(I)$ we denote with x_0 its restriction to $I_a := [0, a]$, with $0 < a < 1$. Define

$$\mathfrak{A} := \{x \in L^1(I) : x_0 \in L^2(I_a)\}.$$

Clearly $(\mathfrak{A}, L^\infty(I))$ is a normed quasi *-algebra, when $\mathfrak{A}$ is endowed with the norm induced by $L^1(I)$. It is easily shown that the positive sesquilinear form φ defined by

$$\varphi(x, y) = \int_0^a x_0(t)\overline{y_0(t)}dt, \quad x, y \in \mathfrak{A},$$

is an element of $\mathcal{I}_{\mathfrak{A}_0}(\mathfrak{A})$. In fact, in this case,

$$\mathfrak{A}/N_\varphi \simeq L^2(I \setminus I_a) \text{ and } \lambda_\varphi(L^\infty(I)) \simeq L^\infty(I \setminus I_a), \text{ which is dense in } L^2(I \setminus I_a).$$

As shown in Example 3.1.29, $\mathcal{S}_{C(I)}(L^p(I)) = \{0\}$; then, $\mathcal{P}_{C(I)}(L^1(I)) = \{0\}$ too. Therefore, $\varphi \notin \mathcal{P}_{C(I)}(L^1(I)) = \{0\}$.

Example 3.1.36 Let us consider again a Banach quasi *-algebra of the type $(\overline{\mathfrak{M}}, \mathfrak{M})$ constructed in Example 3.1.10. We put

$$\mathcal{D}_0(\mathfrak{M}) = \{\xi \in \mathcal{D} : X\xi \in \mathcal{H}, \forall X \in \overline{\mathfrak{M}}\}.$$

For $\xi \in \mathcal{D}_0(\mathfrak{M})$, we define

$$\varphi_\xi(X, Y) = \langle X\xi | Y\eta \rangle, \quad X, Y \in \overline{\mathfrak{M}}.$$

Then, it is easy to see that $\varphi_\xi \in \mathcal{Q}_{\mathfrak{M}}(\overline{\mathfrak{M}})$.

Following the definitions, it is easily seen that

- $\varphi_\xi \in \mathcal{I}_{\mathfrak{M}}(\overline{\mathfrak{M}}) \Leftrightarrow \xi \in \mathcal{D}_0(\mathfrak{M})$ and $X\xi \in \overline{\mathfrak{M}\xi}$, $\forall X \in \overline{\mathfrak{M}}$, where $\overline{\mathfrak{M}\xi}$ denotes the closure of $\mathfrak{M}\xi$ in $\mathcal{H}$;
- $\varphi_\xi \in \mathcal{P}_{\mathfrak{M}}(\overline{\mathfrak{M}}) \Leftrightarrow \xi \in \mathcal{D}_0(\mathfrak{M})$ and $\sup_{\|X\|_{\mathfrak{L}} \leq 1} \|X\xi\| < \infty$.

It is worth mentioning the fact that one can construct examples, where $\mathcal{D}_0(\mathfrak{M}) = \{0\}$. For instance, let $\mathcal{H} = L^2(I)$, with $I = [0, 1]$ and $\mathcal{D} = L^p(I)$, with $p > 2$. If η is a measurable function, denote by M_η the operator of multiplication by η. Consider as $\mathfrak{M}$ the O*-algebra of multiplication operators by a function $\phi \in L^\infty(I)$, i.e.,

$$\mathfrak{M} = \{M_\phi : \phi \in L^\infty(I)\}.$$

Then, it is easily seen that

$$\overline{\mathfrak{M}} = \{M_\phi : \phi \in L^{p/(p-2)}(I)\} \text{ and } \|M_\phi\|_{\mathfrak{L}} = \|\phi\|_{p/(p-2)}.$$

Moreover, the following hold:

- If $2 < p < 4$, then $\mathcal{D}_0(\mathfrak{M}) = L^{2p/(4-p)}(I)$ and every φ_ξ, $\xi \in \mathcal{D}_0(\mathfrak{M})$, is bounded.
- If $p = 4$, then $\mathcal{D}_0(\mathfrak{M}) = L^\infty(I)$ and again every φ_ξ, $\xi \in \mathcal{D}_0(\mathfrak{M})$, is bounded.
- If $p > 4$, then $\mathcal{D}_0(\mathfrak{M}) = \{0\}$.

Lemma 3.1.37 below, will be often used in what follows. We remind that an m^*-seminorm $\mathfrak{s}$ on a *-algebra $\mathfrak{A}_0$ is a seminorm satisfying the properties:

(i) $\mathfrak{s}(x^*) = \mathfrak{s}(x), \quad \forall x \in \mathfrak{A}_0$;
(ii) $\mathfrak{s}(xy) \leq \mathfrak{s}(x)\mathfrak{s}(y), \quad \forall x, y \in \mathfrak{A}_0$.

Lemma 3.1.37 *Let $\mathfrak{A}_0$ be a *-algebra and ω a positive linear functional on $\mathfrak{A}_0$. Assume that there exists an m*-seminorm $\mathfrak{s}$ on $\mathfrak{A}_0$, such that*

$$\forall\, y \in \mathfrak{A}_0,\ \exists\, \gamma_y > 0:\ |\omega(y^*xy)| \leq \gamma_y \mathfrak{s}(x), \quad \forall\, x \in \mathfrak{A}_0.$$

Then,

$$|\omega(y^*xy)| \leq \mathfrak{s}(x)\omega(y^*y), \quad \forall\, x \in \mathfrak{A}_0.$$

Proof The argument for proof is based on Kaplansy's inequality, using the assumption and applying repeatedly the Cauchy–Schwarz inequality. □

Let us now define a new seminorm $\mathfrak{q}$ as follows

$$\mathfrak{q}(a) := \sup\left\{\varphi(ax, ax)^{1/2} : \varphi \in \mathcal{P}_{\mathfrak{A}_0}(\mathfrak{A}),\ x \in \mathfrak{A}_0,\ \varphi(x, x) = 1\right\}, \qquad (3.1.13)$$

for all $a \in \mathfrak{A}$ and

$$\mathcal{D}(\mathfrak{q}) := \left\{a \in \mathfrak{A} : \mathfrak{q}(a) < \infty\right\}.$$

Clearly, $\mathcal{D}(\mathfrak{q}_{\mathcal{I}}) \subseteq \mathcal{D}(\mathfrak{q})$ and $\mathfrak{q}(a) \leq \mathfrak{q}_{\mathcal{I}}(a)$, for every $a \in \mathcal{D}(\mathfrak{q}_{\mathcal{I}})$ (see (3.1.6)). As in the case of $\mathfrak{q}_{\mathcal{I}}$ one can easily check that if $(\mathfrak{A}, \mathfrak{A}_0)$ has a unit e, then

$$\mathfrak{q}(a) = \sup\left\{\varphi(a, a)^{1/2} : \varphi \in \mathcal{P}_{\mathfrak{A}_0}(\mathfrak{A}),\ \varphi(e, e) = 1\right\},\ a \in \mathcal{D}(\mathfrak{q}).$$

In order to obtain a description of $\mathcal{D}(\mathfrak{q})$ similar to that of $\mathcal{D}(\mathfrak{q}_{\mathcal{I}})$, we give the following

Definition 3.1.38 Let $(\mathfrak{A}[\|\cdot\|], \mathfrak{A}_0)$ be a normed quasi *-algebra and π a *-representation of $(\mathfrak{A}[\|\cdot\|], \mathfrak{A}_0)$ with domain $\mathcal{D}_\pi$. We say that π is *completely regular* if, for every $\xi \in \mathcal{D}_\pi$, the positive sesquilinear form φ_ξ is bounded. The set of all completely regular *-representations of $(\mathfrak{A}[\|\cdot\|], \mathfrak{A}_0)$ is denoted by $\mathrm{Rep}_{cr}(\mathfrak{A})$.

Clearly, if π is completely regular, then it is regular.

Proposition 3.1.39 *Let $(\mathfrak{A}[\|\cdot\|], \mathfrak{A}_0)$ be a normed quasi *-algebra. For each $\varphi \in \mathcal{P}_{\mathfrak{A}_0}(\mathfrak{A})$, let π_φ denote the corresponding GNS representation. Then,*

$$\begin{aligned}\mathcal{D}(\mathfrak{q}) &= \left\{a \in \mathfrak{A} : \overline{\pi_\varphi(a)} \in \mathcal{B}(\mathcal{H}_\varphi),\ \forall\, \varphi \in \mathcal{P}_{\mathfrak{A}_0}(\mathfrak{A}) \text{ and } \sup_{\varphi \in \mathcal{P}_{\mathfrak{A}_0}(\mathfrak{A})} \|\overline{\pi_\varphi(a)}\| < \infty\right\} \\ &= \left\{a \in \mathfrak{A} : \overline{\pi(a)} \text{ is bounded},\ \forall\, \pi \in \mathrm{Rep}_{cr}(\mathfrak{A}) \text{ and } \sup_{\pi \in \mathrm{Rep}_{cr}(\mathfrak{A})} \|\overline{\pi(a)}\| < \infty\right\}\end{aligned}$$

and

$$\mathfrak{q}(a) = \sup_{\varphi \in \mathcal{P}_{\mathfrak{A}_0}(\mathfrak{A})} \|\overline{\pi_\varphi(a)}\| = \sup_{\pi \in \mathrm{Rep}_{cr}(\mathfrak{A})} \|\overline{\pi(a)}\|, \quad \forall\, a \in \mathcal{D}(\mathfrak{q}). \qquad (3.1.14)$$

Proof The proof is very similar to that of Proposition 3.1.12, so we do not repeat all the details. The only point to be taken into account is that if $\varphi \in \mathcal{P}_{\mathfrak{A}_0}(\mathfrak{A})$, then the corresponding representation π_φ° is completely regular. Indeed, if $\xi = \lambda_\varphi(x)$, then (see (3.1.2), (3.1.4) and Definition 3.1.14(iii)) there exists $\gamma > 0$, such that

$$\varphi_\xi(a, a) = \langle \pi_\varphi^\circ(a)\lambda_\varphi(x)|\pi_\varphi^\circ(a)\lambda_\varphi(x)\rangle = \varphi(ax, ax) \leq \gamma \|x\|_0^2 \|a\|^2, \quad a \in \mathfrak{A},\ x \in \mathfrak{A}_0.$$

Hence, φ_ξ is bounded. □

In what follows we will show that $\mathfrak{q}$ plays, together with $\mathfrak{p}$, a crucial role for the structure of a normed or Banach quasi *-algebra.

The following preliminary proposition has been given in [80]. The statement (i) below follows from Lemma 3.1.37, while (ii) can be easily deduced from Proposition 3.1.39. The statements (iii) and (iv) follow by the very definitions.

Proposition 3.1.40 *The following statements hold:*

(i) $\mathfrak{A}_0 \subseteq \mathcal{D}(\mathfrak{q})$ *and* $\mathfrak{q}(x) \leq \|x\|_0, \quad \forall\, x \in \mathfrak{A}_0$;
(ii) $\mathfrak{q}$ *is an extended C*-seminorm on* $(\mathfrak{A}, \mathfrak{A}_0)$ (i.e., $\mathfrak{q}(a^*) = \mathfrak{q}(a),\ \forall\, a \in \mathcal{D}(\mathfrak{q})$ and $\mathfrak{q}(x^*x) = \mathfrak{q}(x)^2,\ \forall\, x \in \mathfrak{A}_0$; see [77]);
(iii) $\mathfrak{p}(ax) \leq \mathfrak{q}(a)\mathfrak{p}(x), \quad \forall\, a \in \mathcal{D}(\mathfrak{q}),\ x \in \mathfrak{A}_0$;
(iv) $\mathfrak{p}(xa) \leq \|x\|_0 \mathfrak{p}(a), \quad \forall\, a \in \mathfrak{A},\ x \in \mathfrak{A}_0$.

Remark 3.1.41 If $(\mathfrak{A}, \mathfrak{A}_0)$ has a unit e, then from (iii) it follows that $\mathfrak{p}(a) \leq \mathfrak{q}(a)$, for every $a \in \mathcal{D}(\mathfrak{q})$.

Now, we put

$$N(\mathfrak{p}) := \{a \in \mathfrak{A} : \mathfrak{p}(a) = 0\}.$$

Then, by (iv) of Proposition 3.1.40 it follows that $N(\mathfrak{p})$ is a left qu-ideal of $(\mathfrak{A}, \mathfrak{A}_0)$.

We consider $N_0(\mathfrak{p}) = N(\mathfrak{p}) \cap \mathfrak{A}_0$. Then, the quotient $\mathfrak{A}_0^{\mathfrak{p}} := \mathfrak{A}_0/N_0(\mathfrak{p})$ is a normed space with norm $\|x + N_0(\mathfrak{p})\|_{\mathfrak{p}} := \mathfrak{p}(x)$, $x \in \mathfrak{A}_0$. Let us denote by $\mathfrak{A}_{\mathfrak{p}}$ the completion of $\mathfrak{A}_0^{\mathfrak{p}}[\|\cdot\|_{\mathfrak{p}}]$. Then, we have the following

Proposition 3.1.42 *The quotient* $\mathfrak{A}/N(\mathfrak{p})$ *can be identified with a dense subspace of* $\mathfrak{A}_{\mathfrak{p}}$. *Moreover,*

(i) *if* $\mathfrak{p}(a) = \mathfrak{p}(a^*)$, *for every* $a \in \mathfrak{A}$, *then* $\mathfrak{A}_{\mathfrak{p}}$ *is a Banach space with isometric involution extending the natural involution of* $\mathfrak{A}_0^{\mathfrak{p}}$; $\mathfrak{A}_0^{\mathfrak{p}}$ *is a *-algebra and* $(\mathfrak{A}_{\mathfrak{p}}, \mathfrak{A}_0^{\mathfrak{p}})$ *can be made into a Banach quasi *-algebra.*
(ii) *If* $\mathfrak{p}(xa) \leq \mathfrak{p}(x)\mathfrak{p}(a)$, *for every* $x \in \mathfrak{A}_0$ *and* $a \in \mathfrak{A}$, *then* $\mathfrak{A}_{\mathfrak{p}}$ *is a Banach algebra.*
(iii) *If* $\mathfrak{p}$ *is an m*-seminorm on* $\mathfrak{A}_0$, *then* $\mathfrak{A}_{\mathfrak{p}}$ *is a Banach *-algebra.*
(iv) *If* $\mathfrak{p}$ *is a C*-seminorm on* $\mathfrak{A}_0$, *then* $\mathfrak{A}_{\mathfrak{p}}$ *is a C*-algebra.*

Proof Let $a \in \mathfrak{A}$. Then, there exists a sequence $\{x_n\}$ of elements of $\mathfrak{A}_0$, such that $\|a - x_n\| \to 0$, as $n \to \infty$. From the properties of $\mathfrak{p}$ (see, after Definition 3.1.17) this implies that $\mathfrak{p}(a - x_n) \to 0$, as $n \to \infty$. We define $\widehat{a} := \|\cdot\|_{\mathfrak{p}} - \lim_{n\to\infty}(x_n +$

$N_0(\mathfrak{p})$). By the construction of the completion, $\widehat{a}$ does not depend on the choice of the sequence $\{x_n\}$. Moreover, the map

$$j : a + N(\mathfrak{p}) \in \mathfrak{A}/N(\mathfrak{p}) \to \widehat{a} \in \mathfrak{A}_{\mathfrak{p}}$$

is well-defined. Indeed, if $a, a' \in \mathfrak{A}$, such that $a - a' \in N(\mathfrak{p})$ and $y_n \to a - a'$ with respect to the norm of $\mathfrak{A}$, then $\mathfrak{p}(y_n) \to 0$ and so $j(a - a') = 0$, where j is injective. Indeed, assume that $\widehat{a} = 0$ and let $\{x_n\}$ be a sequence of elements of $\mathfrak{A}_0$, such that $\|a - x_n\| \to 0$, as $n \to \infty$. Then, $\|\cdot\|_{\mathfrak{p}} - \lim_{n\to\infty}(x_n + N_0(\mathfrak{p})) = 0$. Hence, $\mathfrak{p}(x_n) \to 0$, as $n \to \infty$. This, in turn, implies that $\mathfrak{p}(a) = 0$ and so $a \in N(\mathfrak{p})$.

The proofs of (ii), (iii) and (iv) are easily checked. As for (i), from the definition of $\mathfrak{p}$ and the assumptions it follows that $N_0(\mathfrak{p})$ is a *-ideal of $\mathfrak{A}_0$, therefore $\mathfrak{A}_0/N_0(\mathfrak{p})$ is a *-algebra. For defining the multiplication that makes of $(\mathfrak{A}_{\mathfrak{p}}, \mathfrak{A}_0^{\mathfrak{p}})$ a quasi *-algebra, we proceed as follows: if $z \in \mathfrak{A}_{\mathfrak{p}}$, then $z = \|\cdot\|_{\mathfrak{p}} - \lim_{n\to\infty}(x_n + N_0(\mathfrak{p}))$, with $x_n \in \mathfrak{A}_0$; so taking also $x \in \mathfrak{A}_0$ and using (iv) of Proposition 3.1.40, we obtain that the sequence $(xx_n + N_0(\mathfrak{p}))$ is $\|\cdot\|_{\mathfrak{p}}$-Cauchy. Thus, we can define $xz = \|\cdot\|_{\mathfrak{p}} - \lim_{n\to\infty}(xx_n + N_0(\mathfrak{p}))$. □

Remark 3.1.43 We shall show below that (iii) and (iv) are indeed equivalent.

Example 3.1.44 Let us consider the Banach quasi *-algebra $(L^p(I), C(I))$ with $I = [0, 1]$. In this case, one has (see Example 3.1.29) that

$$\mathcal{P}_{C(I)}(L^p(I)) = \begin{cases} \{\varphi_w : w \in L^{p/(p-2)}(I),\ w \geq 0\}, & \text{if } p \geq 2 \\ \{0\}, & \text{if } 1 \leq p < 2, \end{cases}$$

where

$$\varphi_w(x, y) = \int_I x(t)\overline{y(t)}w(t)dt, \quad x, y \in L^p(I).$$

If $1 \leq p < 2$ both $\mathfrak{p}$ and $\mathfrak{q}$ are identically zero. If $p \geq 2$, then one can prove that $\mathfrak{p}(x) = \|x\|_p$ and $\mathfrak{q}(x) = \sup\{\varphi_w(x, x)^{1/2} : w \in L^{p/(p-2)}(I),\ \|w\|_1 \leq 1\}$, which is finite, if and only if, $x \in L^\infty(I)$. In fact, we obtain that

$$\mathfrak{q}(x) = \|x\|_\infty, \quad \forall\, x \in L^\infty(I).$$

Example 3.1.45 Let $\mathfrak{A}_0$ be a Hilbert algebra (Example 3.1.8) and $\mathcal{H}$ its Hilbert space completion. As we have seen $(\mathcal{H}, \mathfrak{A}_0)$ is a Banach quasi *-algebra.

Since the inner product of $\mathcal{H}$ is an element of $\mathcal{P}_{\mathfrak{A}_0}(\mathcal{H})$, one has $\mathfrak{p}(a) = \|a\|$, for every $a \in \mathcal{H}$. As for $\mathfrak{q}$ it is easily seen that

$$\mathcal{D}(\mathfrak{q}) = \{a \in \mathcal{H} : L_a \text{ is bounded}\} = \{a \in \mathcal{H} : R_a \text{ is bounded}\}$$

and

$$\mathfrak{q}(a) = \|L_a\| = \|R_a\|, \quad \forall\, a \in \mathcal{D}(\mathfrak{q}),$$

where $L_a : x \in \mathfrak{A}_0 \to ax \in \mathcal{H}$ and $R_a : x \in \mathfrak{A}_0 \to xa \in \mathcal{H}$. Thus, $\mathcal{D}(\mathfrak{q})$ coincides, as already shown in [80], with the set of *bounded* elements of $\mathcal{H}$.

We conclude this section by discussing, in the case that $(\mathfrak{A}[\|\cdot\|], \mathfrak{A}_0)$ is a Banach quasi *-algebra, some results on the automatic continuity for the classes of positive linear functionals and positive sesquilinear forms introduced so far. For this we need some lemmas (Lemmas 3.1.47 and 3.1.48), that are already known in slight different situations. For the sake of completeness we give the proofs adapted to the cases under consideration.

We first remind the reader the definition of closed and closable sesquilinear form.

Definition 3.1.46 Let $\mathfrak{B}[\|\cdot\|]$ be a Banach space and φ a positive sesquilinear form defined on $D(\varphi) \times D(\varphi)$, where $D(\varphi)$ is a dense subspace of $\mathfrak{B}[\|\cdot\|]$. Then, φ is said to be *closed* in $D(\varphi)$ if $D(\varphi)$ is complete under the norm $\|\cdot\|_\varphi$ defined by

$$\|a\|_\varphi := \left(\|a\|^2 + \varphi(a,a)\right)^{1/2}, \quad a \in D(\varphi).$$

A form φ, with domain $D(\varphi)$ is said to be *closable* if it has a closed extension $\overline{\varphi}$ to a domain $D(\overline{\varphi}) \subseteq \mathfrak{B}$.

In other words, φ is closed if, whenever a sequence $\{a_n\}$ of elements of $D(\varphi)$ converges to $a \in \mathfrak{B}$ and $\varphi(a_n - a_m, a_n - a_m) \to 0$, as $n, m \to \infty$, one has

$$a \in D(\varphi) \ \text{ and } \ \varphi(a_n - a, a_n - a) \to 0, \ \text{ as } n \to \infty.$$

The following lemma characterizes closed and everywhere defined forms.

Lemma 3.1.47 *Let $\mathfrak{B}[\|\cdot\|]$ be a Banach space and φ a positive sesquilinear form on $\mathfrak{B} \times \mathfrak{B}$. The following statements are equivalent:*

(i) *φ is lower semicontinuous; i.e., if $\{a_n\} \subset \mathfrak{B}$ is a sequence converging to $a \in \mathfrak{B}$, with respect to $\|\cdot\|$, one has $\varphi(a,a) \leq \liminf_{n\to\infty} \varphi(a_n, a_n)$;*
(ii) *φ is closed;*
(iii) *φ is bounded.*

Proof (i) ⇒ (ii) Let $\{a_n\}$ be a Cauchy sequence in $\mathfrak{B}$, with respect to $\|\cdot\|_\varphi$. Then, for every $\varepsilon > 0$, there exists $n_\varepsilon \in \mathbb{N}$, such that

$$\|a_n - a_m\|^2 + \varphi(a_n - a_m, a_n - a_m) < \varepsilon^2, \quad \forall\, n, m > n_\varepsilon.$$

The completeness of $\mathfrak{B}[\|\cdot\|]$ implies the existence of an element $a \in \mathfrak{A}$, such that $\lim_{n\to\infty} \|a - a_n\| = 0$. Now, fix $m > n_\varepsilon$ and let $n \to \infty$. We obtain

$$\|a - a_m\|^2 + \varphi(a - a_m, a - a_m) \leq \|a - a_m\|^2 + \liminf_{n\to\infty} \varphi(a_n - a_m, a_n - a_m) \leq \varepsilon^2.$$

Hence, $\|a - a_m\|_\varphi \to 0$. This proves that $\mathfrak{B}$ is complete with respect to $\|\cdot\|_\varphi$.

(ii) $\Rightarrow$ (iii) Since $\mathfrak{B}[\|\cdot\|]$ is a Banach space and $\|a\| \leq \|a\|_\varphi$, for every $a \in \mathfrak{B}$, it follows that $\|\cdot\|$ and $\|\cdot\|_\varphi$ are equivalent (by the inverse mapping theorem) and therefore φ is bounded.

(iii) $\Rightarrow$ (i) is straightforward. □

For the terminology and notation applied just below, see discussion before Proposition 3.1.25.

Lemma 3.1.48 *Let $(\mathfrak{A}[\|\cdot\|], \mathfrak{A}_0)$ be a Banach quasi *-algebra. Then, every positive linear functional ω on $\mathfrak{A}$ is bounded on positive elements; i.e., there exists $\gamma > 0$, such that*

$$\omega(a) \leq \gamma \|a\|, \quad \forall\, a \in \mathfrak{A}^+.$$

Proof Suppose that ω is not bounded on $\mathfrak{A}^+$. Then, there would exist a sequence $\{a_n\}$ of positive elements of $\mathfrak{A}$, such that $\|a_n\| \leq 2^{-n}$ and $\omega(a_n) \to \infty$. Let $b = \sum_{k=1}^{\infty} a_k$. Then,

$$\omega(b) = \omega\left(\sum_{k=1}^{\infty} a_k\right) \geq \omega\left(\sum_{k=1}^{n} a_k\right) = \sum_{k=1}^{n} \omega(a_k) \to \infty,$$

a contradiction. □

Theorem 3.1.49 *Let $(\mathfrak{A}[\|\cdot\|], \mathfrak{A}_0)$ be a Banach quasi *-algebra satisfying the following condition:*

(D) *every $a = a^* \in \mathfrak{A}$ can be uniquely decomposed as $a = a_+ - a_-$, with $a_+, a_- \in \mathfrak{A}^+$ and $\|a\| = \|a_+\| + \|a_-\|$.*

Then, every $\varphi \in \mathcal{I}_{\mathfrak{A}_0}(\mathfrak{A})$, such that ω_{φ_x} is positive, for every $x \in \mathfrak{A}_0$, is bounded.

Proof Let $x \in \mathfrak{A}_0$. Since ω_{φ_x}, defined as in (2.4.14), is positive, Lemma 3.1.48 implies that ω_{φ_x} is bounded on positive elements; i.e., there exists $\gamma > 0$, such that

$$\omega_{\varphi_x}(a) \leq \gamma \|a\|, \quad \forall\, a \in \mathfrak{A}^+.$$

Condition (D) then yields that, for every $a = a^* \in \mathfrak{A}$,

$$|\omega_{\varphi_x}(a)| = |\omega_{\varphi_x}(a_+ - a_-)| \leq \omega_{\varphi_x}(a_+) + \omega_{\varphi_x}(a_-) \leq \gamma(\|a_+\| + \|a_-\|) = \gamma \|a\|.$$

The general statement is easily obtained by decomposing every $z \in \mathfrak{A}$ as $z = a+ib$, with $a = a^*$, $b = b^*$. Using the polarization identity, one proves easily that, for every $x, y \in \mathfrak{A}_0$, the linear functional $L_{x,y}(a) := \varphi(ax, y)$, $a \in \mathfrak{A}$, is bounded.

Let now $\{a_n\}$ be a sequence in $\mathfrak{A}$ and $a \in \mathfrak{A}$ with $\lim_{n\to\infty} \|a_n - a\| = 0$. For every $y \in \mathfrak{A}_0$, by the Cauchy–Schwarz inequality, we have

$$|\varphi(a_n, y)| \leq \varphi(a_n, a_n)^{1/2}\varphi(y, y)^{1/2}.$$

Taking the lim inf in both sides, we obtain

$$|\varphi(a, y)| \leq \liminf_{n\to\infty} \varphi(a_n, a_n)^{1/2}\varphi(y, y)^{1/2}.$$

Now, since $\varphi \in \mathcal{I}_{\mathfrak{A}_0}(\mathfrak{A})$ (see Proposition 2.3.2 and discussion after it), there exists a sequence $\{x_k\}$ of elements of $\mathfrak{A}_0$ such that $\varphi(a - x_k, a - x_k) \to 0$. This implies that

$$\lim_{k\to\infty} \varphi(a, x_k) = \varphi(a, a) \quad \text{and} \quad \lim_{k\to\infty} \varphi(x_k, x_k) = \varphi(a, a).$$

Then, from

$$|\varphi(a, x_k)| \leq \liminf_{n\to\infty} \varphi(a_n, a_n)^{1/2}\varphi(x_k, x_k)^{1/2},$$

for $k \to \infty$, we conclude that

$$\varphi(a, a) \leq \left(\liminf_{n\to\infty} \varphi(a_n, a_n)\right)^{1/2} \varphi(a, a)^{1/2}.$$

Hence,

$$\varphi(a, a) \leq \liminf_{n\to\infty} \varphi(a_n, a_n);$$

i.e., φ is lower semicontinuous. The statement then follows from Lemma 3.1.47. □

3.2 Continuity of Representable Linear Functionals

As we have seen in Chap. 2 another way to construct a GNS-like representation of a quasi *-algebra $(\mathfrak{A}, \mathfrak{A}_0)$ is provided by representable linear functionals (Definition 2.4.6). Of course, for a normed or Banach quasi *-algebra, additional properties of the family $\mathcal{R}(\mathfrak{A}, \mathfrak{A}_0)$ can be given and it is, moreover, of true interest considering the subclass $\mathcal{R}_c(\mathfrak{A}, \mathfrak{A}_0)$ of *continuous* (or, equivalently, *bounded*) *representable linear functionals* on $(\mathfrak{A}[\|\cdot\|], \mathfrak{A}_0)$. In particular, the question as to whether every representable linear functional is continuous deserves an answer. Investigations on this subject have been done in [31], to which we refer for more details.

Theorem 3.2.1 *Let* $(\mathfrak{A}[\|\cdot\|], \mathfrak{A}_0)$ *be a normed quasi *-algebra with unit e and* ω *a linear functional on* $\mathfrak{A}$ *satisfying the conditions* (L.1) *and* (L.2) of Definition 2.4.6. *Then, the following statements are equivalent:*

(i) ω *is representable;*
(ii) *there exists a *-representation* π *defined on a dense domain* $\mathcal{D}_\pi$ *of a Hilbert space* $\mathcal{H}_\pi$ *and a vector* $\zeta \in \mathcal{D}_\pi$, *such that* $\omega(a) = \langle \pi(a)\zeta | \zeta \rangle$, *for all* $a \in \mathfrak{A}$;
(iii) *there exists a sesquilinear form* $\Omega \in \mathcal{Q}_{\mathfrak{A}_0}(\mathfrak{A})$, *such that* $\omega(a) = \Omega(a, e)$, *for all* $a \in \mathfrak{A}$;
(iv) *there exists a Hilbert seminorm p on* $\mathfrak{A}$, *that is a seminorm satisfying the property*

$$p(a+b)^2 + p(a-b)^2 = 2p(a)^2 + 2p(b)^2, \quad \forall\, a, b \in \mathfrak{A},$$

such that $|\omega(a^*x)| \leq p(a)\omega(x^*x)^{1/2}$, *for all* $x \in \mathfrak{A}_0$.

Proof (i) implies (ii) by Theorem 2.4.8. Suppose now that (ii) holds and define

$$\Omega(a, b) := \langle \pi(a)\zeta | \pi(b)\zeta \rangle, \quad a, b \in \mathfrak{A}.$$

It is then easy to check that Ω has the desired properties. This proves (iii).

Now suppose that (iii) holds and define $p(a) := \Omega(a, a)^{1/2}$. Then, (iv) follows immediately from the Cauchy–Schwarz inequality.

Finally suppose that (iv) holds. Then,

$$|\omega(a^*x)| \leq p(a)\omega(x^*x)^{\frac{1}{2}} \leq \gamma_a \omega(x^*x)^{1/2}$$

where, for instance, $\gamma_a = \left(1 + p(a)^2\right)^{1/2}$. Hence, ω is representable. □

To every $\omega \in \mathcal{R}(\mathfrak{A}, \mathfrak{A}_0)$, we can associate two sesquilinear forms, which are useful for our discussion. The first one Ω (already introduced in (iii) of Theorem 3.2.1), can be directly and equivalently computed by using the GNS representation π_ω, with cyclic vector ξ_ω. Then, we can write the everywhere defined sesquilinear form Ω^ω as follows

$$\Omega^\omega(a, b) := \langle \pi_\omega(a)\xi_\omega | \pi_\omega(b)\xi_\omega \rangle, \quad a, b \in \mathfrak{A}. \tag{3.2.15}$$

Thus, $\Omega^\omega \in \mathcal{Q}_{\mathfrak{A}_0}(\mathfrak{A})$ and $\omega(a) = \Omega^\omega(a, e)$, for every $a \in \mathfrak{A}$.

The second sesquilinear form, which we denote by φ_ω is defined only on $\mathfrak{A}_0 \times \mathfrak{A}_0$. To every $\omega \in \mathcal{R}(\mathfrak{A}, \mathfrak{A}_0)$ a sesquilinear form φ_ω is associated, defined by

$$\varphi_\omega(x, y) := \omega(y^*x), \quad x, y \in \mathfrak{A}_0. \tag{3.2.16}$$

It is clear that Ω^ω extends φ_ω. It is easy to see that

(i) $\varphi_\omega(x, x) \geq 0$, for every $x \in \mathfrak{A}_0$;
(ii) $\varphi_\omega(xy, z) = \varphi_\omega(y, x^*z)$, for every $x, y, z \in \mathfrak{A}_0$.

Proposition 3.2.2 *Let $(\mathfrak{A}[\|\cdot\|], \mathfrak{A}_0)$ be a Banach quasi *-algebra, with unit e. For every $\omega \in \mathcal{R}_c(\mathfrak{A}, \mathfrak{A}_0)$, the positive sesquilinear form φ_ω has a bounded extension $\overline{\varphi}_\omega$ to $\mathfrak{A} \times \mathfrak{A}$ and $\overline{\varphi}_\omega = \Omega^\omega$.*

Proof Since ω is representable, there exists a Hilbert space $\mathcal{H}_\omega$, a linear map $\lambda_\omega : \mathfrak{A}_0 \to \mathcal{H}_\omega$ and a *-representation π_ω with values in $\mathcal{L}^\dagger(\lambda_\omega(\mathfrak{A}_0), \mathcal{H}_\omega)$ such that

$$\omega(y^* a x) = \langle \pi_\omega(a)\lambda_\omega(x)|\lambda_\omega(y)\rangle, \quad \forall\, a \in \mathfrak{A},\ x, y \in \mathfrak{A}_0.$$

By the continuity of ω, (3.1.4) and the discussion before (3.1.4), we obtain that for every $a \in \mathfrak{A}$ and $x, y \in \mathfrak{A}_0$,

$$|\langle \pi_\omega(a)\lambda_\omega(x)|\lambda_\omega(y)\rangle| = |\omega(y^* a x)| \le \gamma \|a\| \|x\|_0 \|y\|_0,\ \gamma > 0. \tag{3.2.17}$$

Now, consider the sesquilinear form Ω^ω defined in (3.2.15). As already noticed, Ω^ω extends φ_ω. It remains to show that Ω^ω is closable.

Suppose that $\{a_n\}$ is a sequence in $\mathfrak{A}$, such that

$$\|a_n\| \to 0 \text{ and } \Omega^\omega(a_n - a_m, a_n - a_m) = \|\pi_\omega(a_n - a_m)\xi_\omega\|^2 \to 0.$$

Then, the sequence $\pi_\omega(a_n)\xi_\omega$ converges to a vector $\zeta \in \mathcal{H}_\omega$. Thus,

$$\langle \pi_\omega(a_n)\xi_\omega|\lambda_\omega(y)\rangle \to \langle \zeta|\lambda_\omega(y)\rangle, \quad \forall\, y \in \mathfrak{A}_0.$$

Using (3.2.17) and the density of $\lambda_\omega(\mathfrak{A}_0)$ in $\mathcal{H}_\omega$, we obtain $\zeta = 0$. Hence, $\Omega^\omega(a_n, a_n) \to 0$; i.e., Ω^ω is closable (Definition 3.1.46). Thus, Ω^ω is closed and everywhere defined, hence bounded (Lemma 3.1.47). It is clear that $\Omega^\omega = \overline{\varphi}_\omega$. □

It appears natural for a normed quasi *-algebra to give a stronger notion of a representable linear functional by requiring a better control on the constant γ_a, which appears in the condition (L.3) of Definition 2.4.6. For this reason we will define *uniformly representable linear functionals*. In the Banach case, as we shall see this new notion is equivalent to the continuity of the representable linear functional under consideration.

Definition 3.2.3 Let $(\mathfrak{A}[\|\cdot\|], \mathfrak{A}_0)$ be a normed quasi *-algebra. We say that $\omega \in \mathcal{R}(\mathfrak{A}, \mathfrak{A}_0)$ is *uniformly representable* if there exists $\gamma > 0$, such that

$$|\omega(a^* x)| \le \gamma \|a\| \omega(x^* x)^{1/2}, \quad \forall\, a \in \mathfrak{A},\ x \in \mathfrak{A}_0. \tag{3.2.18}$$

The set of uniformly representable linear functionals is denoted by $\mathcal{R}_c^u(\mathfrak{A}, \mathfrak{A}_0)$.

Proposition 3.2.4 *Let $(\mathfrak{A}[\|\cdot\|], \mathfrak{A}_0)$ be a Banach quasi *-algebra with unit e and $\omega \in \mathcal{R}(\mathfrak{A}, \mathfrak{A}_0)$. The following statements are equivalent:*

(i) *$\omega \in \mathcal{R}_c(\mathfrak{A}, \mathfrak{A}_0)$; i.e., ω is bounded;*
(ii) *$\omega \in \mathcal{R}_c^u(\mathfrak{A}, \mathfrak{A}_0)$; i.e., ω is uniformly representable.*

Proof (i) ⇒ (ii) Consider the operator T_ω of Proposition 2.4.12, which is defined by $T_\omega(a) := \xi_\omega^a$, for all $a \in \mathfrak{A}$. Since T_ω is everywhere defined, it is enough to prove that T_ω is closable and use the closed graph theorem. Let $\{a_n\}$ be a sequence in $\mathfrak{A}$, such that $\|a_n\| \to 0$ and $T_\omega a_n \to \xi \in \mathcal{H}_\omega$. Then, from (2.4.9), we have

$$\langle \lambda_\omega(x)|\xi\rangle = \lim_{n\to\infty} \langle \lambda_\omega(x)|T_\omega a_n\rangle = \lim_{n\to\infty} \langle \lambda_\omega(x)|\xi_\omega^{a_n}\rangle = \lim_{n\to\infty} \omega(a_n^* x), \quad \forall\, x \in \mathfrak{A}_0.$$

But, for every $x \in \mathfrak{A}_0$, the continuity of ω and (3.1.4), give

$$|\omega(a_n^* x)| \leq c\|a_n^* x\| \leq c\|a_n\|\|x\|_0 \to 0, \; c > 0.$$

Hence,

$$\langle \lambda_\omega(x)|\xi\rangle = 0, \quad \forall\, x \in \mathfrak{A}_0.$$

Then, the density of $\lambda_\omega(\mathfrak{A}_0)$ in $\mathcal{H}_\omega$ implies that $\xi = 0$. Hence, T_ω is closable. The closed graph theorem then implies that T_ω is bounded. Furthermore, using again (2.4.9), we have

$$|\omega(a^* x)| = |\langle \lambda_\omega(x)|T_\omega a\rangle| \leq \|\lambda_\omega(x)\|\|T_\omega a\| \leq \|T_\omega\|\|a\|\omega(x^* x)^{1/2}, \quad \forall\, a \in \mathfrak{A},\; x \in \mathfrak{A}_0.$$

That is $\omega \in \mathcal{R}_c^u(\mathfrak{A}, \mathfrak{A}_0)$.

(ii) ⇒ (i) is obvious. □

We define a *partial order* in $\mathcal{R}(\mathfrak{A}, \mathfrak{A}_0)$ as follows. If $\omega, \theta \in \mathcal{R}(\mathfrak{A}, \mathfrak{A}_0)$ we say that $\omega \leq \theta$ if $\Omega^\omega(a, a) \leq \Omega^\theta(a, a)$, for every $a \in \mathfrak{A}$.

By a slight modification of standard arguments (see, e.g., [2, Proposition 9.2.3]), one can prove the following

Lemma 3.2.5 *Let $\omega, \rho \in \mathcal{R}(\mathfrak{A}, \mathfrak{A}_0)$ and let π_ω denote the GNS representation associated to ω. If $\rho \leq \omega$, there exists an operator $T \in (\pi_\omega(\mathfrak{A}), \mathcal{D}_\omega)'_{\mathrm{w}}$, with $0 \leq T \leq I$, such that*

$$\Omega^\rho(a, b) = \langle \pi_\omega(a)\xi_\omega|T\pi_\omega(b)\xi_\omega\rangle, \quad \forall\, a, b \in \mathfrak{A}.$$

Proof Using the Cauchy–Schwarz inequality, we get

$$\begin{aligned}|\Omega^\rho(a, b)|^2 &\leq \Omega^\rho(a, a)\Omega^\rho(b, b)\\ &\leq \Omega^\omega(a, a)\Omega^\omega(b, b)\\ &= \|\pi_\omega(a)\xi_\omega\|^2\|\pi_\omega(b)\xi_\omega\|^2, \quad \forall\, a, b \in \mathfrak{A}.\end{aligned}$$

Thus, the equality $\Omega^\rho(a, b) := \langle \pi_\omega(a)\xi_\omega|\pi_\omega(b)\xi_\omega\rangle$, $a, b \in \mathfrak{A}$, gives a densely defined, bounded sesquilinear form on $\mathfrak{A} \times \mathfrak{A}$ and there exists a unique bounded operator T in $\mathcal{H}_\omega$ [20, p. 44, Corollary], such that

$$\Omega^\rho(a,b) = \langle \pi_\omega(a)\xi_\omega | T\pi_\omega(b)\xi_\omega\rangle, \quad \forall\, a,b \in \mathfrak{A}.$$

The fact that $\rho \le \omega$ easily implies that $0 \le T \le I$. Moreover, for every $a \in \mathfrak{A}$, $x, y \in \mathfrak{A}_0$,

$$\begin{aligned}\langle \pi_\omega(a)\pi_\omega(x)\xi_\omega | T\pi_\omega(y)\xi_\omega\rangle &= \Omega^\rho(ax,y) = \Omega^\rho(x,a^*y)\\ &= \langle \pi_\omega(x)\xi_\omega | T\pi_\omega(a^*)\pi_\omega(y)\xi_\omega\rangle\\ &= \langle \pi_\omega(x)\xi_\omega | T\pi_\omega(a)^\dagger\pi_\omega(y)\xi_\omega\rangle.\end{aligned}$$

Therefore,

$$\langle T\pi_\omega(a)\pi_\omega(x)\xi_\omega | \pi_\omega(y)\xi_\omega\rangle = \langle T\pi_\omega(x)\xi_\omega | \pi_\omega(a)^\dagger\pi_\omega(y)\xi_\omega\rangle, \quad \forall\, a \in \mathfrak{A},\ x,y \in \mathfrak{A}_0.$$

Hence, $T \in (\pi_\omega(\mathfrak{A}), \mathcal{D}_\omega)'_{\mathrm{w}}$ (see discussion before Proposition 2.2.6). □

Lemma 3.2.6 *Let $\omega, \theta \in \mathcal{R}(\mathfrak{A}, \mathfrak{A}_0)$ with $\omega \le \theta$. Then, $\theta - \omega \in \mathcal{R}(\mathfrak{A}, \mathfrak{A}_0)$.*

Proof The conditions (L.1) and (L.2) of Definition 2.4.6 are obviously satisfied. We check (L.3). For every $a \in \mathfrak{A}$ and $x \in \mathfrak{A}_0$, using the Cauchy–Schwarz inequality for the positive sesquilinear form $\Omega^\theta - \Omega^\omega$, we obtain

$$\begin{aligned}|(\theta-\omega)(a^*x)| &= |(\Omega^\theta - \Omega^\omega)(a^*x, e)| = |(\Omega^\theta - \Omega^\omega)(x,a)|\\ &\le (\Omega^\theta - \Omega^\omega)(x,x)^{1/2}(\Omega^\theta - \Omega^\omega)(a,a)^{1/2}\\ &= (\Omega^\theta - \Omega^\omega)(a,a)^{1/2}(\theta-\omega)(x^*x)^{1/2}.\end{aligned}$$ □

As mentioned at the beginning of this section, $\mathcal{R}_c(\mathfrak{A}, \mathfrak{A}_0)$ denotes the set of all bounded representable linear functionals on $(\mathfrak{A}[\|\cdot\|], \mathfrak{A}_0)$; i.e., $\omega \in \mathcal{R}_c(\mathfrak{A}, \mathfrak{A}_0)$, if and only if, there exists $\gamma > 0$, such that

$$|\omega(a)| \le \gamma\|a\|, \quad \forall\, a \in \mathfrak{A}.$$

It is easily seen that if $\omega \in \mathcal{R}_c(\mathfrak{A}, \mathfrak{A}_0)$, then for every $x \in \mathfrak{A}_0$, the linear functional ω_x defined by

$$\omega_x(a) := \omega(x^*ax), \quad a \in \mathfrak{A}, \tag{3.2.19}$$

is bounded (see, for instance, (3.2.17)); thus $\omega_x \in \mathcal{R}_c(\mathfrak{A}, \mathfrak{A}_0)$, for every $x \in \mathfrak{A}_0$.

Theorem 3.2.7 *Let $(\mathfrak{A}[\|\cdot\|], \mathfrak{A}_0)$ be a Banach quasi *-algebra with unit e. The following statements are equivalent:*

(i) *every $\omega \in \mathcal{R}(\mathfrak{A}, \mathfrak{A}_0)$ is bounded; i.e., $\mathcal{R}(\mathfrak{A}, \mathfrak{A}_0) = \mathcal{R}_c(\mathfrak{A}, \mathfrak{A}_0)$;*
(ii) *every *-representation π of $\mathfrak{A}[\|\cdot\|]$ into $\mathcal{L}^\dagger(\mathcal{D}_\pi, \mathcal{H}_\pi)[\mathsf{t}_w]$ is continuous;*
(iii) *for every $\omega \in \mathcal{R}(\mathfrak{A}, \mathfrak{A}_0)$, $\omega \neq 0$, there exists a nonzero $\theta \in \mathcal{R}_c(\mathfrak{A}, \mathfrak{A}_0)$, such that $\theta \leq \omega$.*

Proof (i) $\Rightarrow$ (ii) For the definition of t_w, see end of Sect. 2.1.3. Let π be a *-representation of $(\mathfrak{A}[\|\cdot\|], \mathfrak{A}_0)$. Then, for every $\xi \in \mathcal{D}_\pi$ the linear functional $\omega(a) := \langle \pi(a)\xi|\xi\rangle$, $a \in \mathfrak{A}$ is representable, therefore bounded. This easily implies, using the polarization identity, that π is weakly continuous.

(ii) $\Rightarrow$ (iii) Let $\omega \in \mathcal{R}(\mathfrak{A}, \mathfrak{A}_0)$ and π_ω the corresponding GNS representation, which is weakly continuous by assumption and ξ_ω the corresponding cyclic vector. Then, for every $\xi, \eta \in \mathcal{D}_\omega$, there exists $\gamma_{\xi,\eta} > 0$, such that

$$|\langle \pi_\omega(a)\xi|\eta\rangle| \leq \gamma_{\xi,\eta}\|a\|, \quad \forall\, a \in \mathfrak{A}.$$

In particular, for the cyclic vector ξ_ω, we have

$$|\omega(a)| = |\langle \pi_\omega(a)\xi_\omega|\xi_\omega\rangle| \leq \gamma_{\xi_\omega,\xi_\omega}\|a\|, \quad \forall\, a \in \mathfrak{A}.$$

Then, (iii) holds with the obvious choice of $\theta = \omega$.

(iii) $\Rightarrow$ (i) By assumption, the set $\mathcal{K}_\omega = \{\theta \in \mathcal{R}_c(\mathfrak{A}, \mathfrak{A}_0) : \theta \leq \omega\}$ is a nonempty partially ordered (by $\leq$) set. Let $\mathcal{W}$ be a totally ordered subset of $\mathcal{K}_\omega$. Then, $\lim_{\theta\in\mathcal{W}} \theta(a)$ exists, for every $a \in \mathfrak{A}$. Indeed, the set of numbers $\{\Omega^\theta(a,a) : \theta \in \mathcal{W}\}$ is increasing and bounded from above by $\Omega^\omega(a,a)$. We set,

$$\Lambda(a,a) = \lim_{\theta\in\mathcal{W}} \Omega^\theta(a,a), \quad \forall\, a \in \mathfrak{A}.$$

Then, Λ satisfies the equality

$$\Lambda(a+b, a+b) + \Lambda(a-b, a-b) = 2\Lambda(a,a) + 2\Lambda(b,b), \quad \forall\, a, b \in \mathfrak{A}.$$

We can then define Λ on $\mathfrak{A} \times \mathfrak{A}$ using the polarization identity.

Now we put $\omega^\circ(a) := \Lambda(a,e)$, $a \in \mathfrak{A}$. It is easy to see that $\omega^\circ(a) = \lim_{\theta\in\mathcal{W}} \theta(a)$, $a \in \mathfrak{A}$. We prove that $\omega^\circ \in \mathcal{R}_c(\mathfrak{A}, \mathfrak{A}_0)$. It is clear that ω° is a linear functional on $\mathfrak{A}$ and $\omega^\circ \leq \omega$. The conditions (L.1) and (L.2) of Definition 2.4.6 are obviously satisfied. We prove (L.3). Let $a \in \mathfrak{A}$ and $x \in \mathfrak{A}_0$. Then,

$$\begin{aligned}|\omega^\circ(a^*x)| &= \lim_{\theta\in\mathcal{W}} |\theta(a^*x)| \leq \lim_{\theta\in\mathcal{W}} (1+\Omega^\theta(a,a))^{1/2} \lim_{\theta\in\mathcal{W}} \theta(x^*x)^{1/2} \\ &\leq (1+\Lambda(a,a))^{1/2}\omega^\circ(x^*x)^{1/2}.\end{aligned}$$

We show now that ω° is bounded. For every $a \in \mathfrak{A}$ the set $\{|\theta(a)| : \theta \in \mathcal{W}\}$ is bounded; indeed, for every $\theta \in \mathcal{W}$, we obtain

$$|\theta(a)| = |\Omega^\theta(a, e)| \le \Omega^\theta(a, a)^{1/2}\Omega^\theta(e, e)^{1/2} \le \Omega^\omega(a, a)^{1/2}\Omega^\omega(e, e)^{1/2}, \quad a \in \mathfrak{A}.$$

By the uniform boundedness principle, we conclude that there exists $\gamma > 0$, such that $|\theta(a)| \le \gamma \|a\|$, for every $\theta \in \mathcal{W}$ and for every $a \in \mathfrak{A}$. Hence,

$$|\omega^\circ(a)| = \lim_{\theta \in \mathcal{W}} |\theta(a)| \le \gamma \|a\|, \quad \forall a \in \mathfrak{A}.$$

Thus $\mathcal{W}$ has an upper bound. Then, by Zorn's lemma, $\mathcal{K}_\omega$ has a maximal element. Let us call it $\omega^\bullet$. It remains to prove that $\omega = \omega^\bullet$. Assume, on the contrary that $\omega > \omega^\bullet$. Let us consider the functional $\omega - \omega^\bullet$, which is representable by Lemma 3.2.6 and nonzero. Then, there exists $\sigma \in \mathcal{R}_c(\mathfrak{A}, \mathfrak{A}_0)$, $\sigma \ne 0$, such that $\omega - \omega^\bullet \ge \sigma$. Consequently, $\omega \ge \omega^\bullet + \sigma$, contradicting the maximality of $\omega^\bullet$. Thus, $\omega = \omega^\bullet$ and therefore ω is continuous. □

Remark 3.2.8 The equivalence of (i) and (ii) of the previous theorem holds also in the case when $(\mathfrak{A}[\|\cdot\|], \mathfrak{A}_0)$ is only a normed quasi *-algebra. The proof of (iii) ⇒ (i) is similar to a known result in the theory of Banach *-algebras (see [7, Lemma 5.5.5]).

3.3 Continuity of *-Representations

The seminorms $\mathfrak{p}$ and $\mathfrak{q}$, introduced in Sect. 3.1.2 (see, (3.1.9) and (3.1.13), respectively), play an interesting role also in the study of the continuity of a *-representation. As we shall see at the end of this section, the most favourable situation occurs when $\mathfrak{p}$ is an m*-seminorm. In fact, in this case, $\mathfrak{A}$ may be viewed (up to a quotient) as a subspace of the C*-algebra $\mathfrak{A}_\mathfrak{p}$ and (as expected) any regular *-representation will be bounded and norm-continuous. But this is, in a sense, a rather extreme situation rarely realized in practice and Banach quasi *-algebras having unbounded *-representations do really exist. For this reason, we begin with looking for conditions that guarantee the strong continuity of any regular *-representation.

Proposition 3.3.1 *Let $\mathfrak{A}$ be a normed quasi *-algebra. The following statements hold:*

(i) *every strongly continuous *-representation is regular;*
(ii) *every completely regular *-representation is strongly continuous.*

Proof

(i) Let π be strongly continuous. Then, for every $\xi \in \mathcal{D}_\pi$, there exists $\gamma_\xi > 0$, such that

$$\|\pi(a)\xi\| \leq \gamma_\xi \|a\|, \quad \forall a \in \mathfrak{A}. \tag{3.3.20}$$

From the previous inequality and from the density of $\mathfrak{A}_0$ in $\mathfrak{A}$ it follows that, for every $a \in \mathfrak{A}$, $\pi(a)\xi \in \overline{\pi(\mathfrak{A}_0)\xi}$. The statement then follows from Proposition 2.4.15.

(ii) Let π be a completely regular *-representation of $(\mathfrak{A}[\|\cdot\|], \mathfrak{A}_0)$ (see Definition 3.1.38). Then, for every $\xi \in \mathcal{D}_\pi$, the vector form φ_ξ is bounded. Therefore, for some $\gamma_\xi > 0$,

$$\|\pi(a)\xi\|^2 = \varphi_\xi(a, a) \leq \gamma_\xi \|a\|^2, \quad \forall a \in \mathfrak{A}.$$

Hence π is strongly continuous. □

The statements (i) and (ii) of Proposition 3.3.1 are not, in general, equivalent unless (see discussion after Proposition 2.4.16) $\mathcal{I}^s_{\mathfrak{A}_0}(\mathfrak{A}) = \mathcal{P}_{\mathfrak{A}_0}(\mathfrak{A})$, as the next theorem shows.

Theorem 3.3.2 *Let $(\mathfrak{A}[\|\cdot\|], \mathfrak{A}_0)$ be a normed quasi *-algebra with unit e. The following statements are equivalent:*

(i) *every $\varphi \in \mathcal{I}^s_{\mathfrak{A}_0}(\mathfrak{A})$ is bounded; i.e., $\mathcal{I}^s_{\mathfrak{A}_0}(\mathfrak{A}) = \mathcal{P}_{\mathfrak{A}_0}(\mathfrak{A})$;*
(ii) *every regular *-representation π is completely regular;*
(iii) *every regular *-representation π of $(\mathfrak{A}[\|\cdot\|], \mathfrak{A}_0)$ is strongly continuous.*

*If $(\mathfrak{A}[\|\cdot\|], \mathfrak{A}_0)$ is a Banach quasi *-algebra with unit, then* (i), (ii) *and* (iii) *are also equivalent to the following statement:*

(iv) *every $\varphi \in \mathcal{I}^s_{\mathfrak{A}_0}(\mathfrak{A})$ is lower semicontinuous.*

Proof (i) ⇒ (ii) Let π be a regular *-representation of $(\mathfrak{A}[\|\cdot\|], \mathfrak{A}_0)$. Then, for every $\xi \in \mathcal{D}_\pi$, $\varphi_\xi \in \mathcal{I}^s_{\mathfrak{A}_0}(\mathfrak{A})$. Thus, by (i), φ_ξ is bounded.

(ii) ⇒ (iii) This follows immediately from (ii) of Proposition 3.3.1.

(iii) ⇒ (i) Let $\varphi \in \mathcal{I}^s_{\mathfrak{A}_0}(\mathfrak{A})$. Then π°_φ is a regular *-representation (see Proposition 2.4.16). Hence, it is strongly continuous from our assumption. Thus, for some $\gamma_\varphi > 0$, we have

$$\varphi(a, a) = \langle \pi^\circ_\varphi(a)\lambda_\varphi(e) | \pi^\circ_\varphi(a)\lambda_\varphi(e) \rangle = \|\pi^\circ_\varphi(a)\lambda_\varphi(e)\|^2 \leq \gamma_\varphi \|a\|^2, \quad \forall a \in \mathfrak{A}.$$

Finally, if $\mathfrak{A}[\|\cdot\|]$ is complete, then (iv) ⇒ (i) follows from Lemma 3.1.47. The implication (i) ⇒ (iv) is obvious. □

If π is strongly continuous, then by (3.3.20), we can define a new norm on $\mathcal{D}_\pi$ by

$$\|\xi\|_\pi := \sup_{\|a\|\leq 1} \|\pi(a)\xi\|, \quad \xi \in \mathcal{D}_\pi.$$

Since $(\mathfrak{A}[\|\cdot\|], \mathfrak{A}_0)$ has a unit, we have that $\|\xi\| \leq \|\xi\|_\pi$, for every $\xi \in \mathcal{D}_\pi$. With this definition one has, of course,

$$\|\pi(a)\xi\| \leq \|a\| \|\xi\|_\pi, \quad \forall\, a \in \mathfrak{A}, \xi \in \mathcal{D}_\pi.$$

We put

$$|||\pi(a)||| := \sup_{\|\xi\|_\pi \leq 1} \|\pi(a)\xi\|, \quad \forall\, a \in \mathfrak{A}.$$

By the definition itself it follows that $|||\pi(a)||| \leq \|a\|$, for every $a \in \mathfrak{A}$.

We denote with $\mathrm{Rep}_{sc}(\mathfrak{A})$ the set of all strongly continuous *-representations of $(\mathfrak{A}[\|\cdot\|], \mathfrak{A}_0)$.

The next theorem shows that the seminorm $\mathfrak{p}$ (see (3.1.9)) on $\mathfrak{A}$ is determined by $\mathrm{Rep}_{sc}(\mathfrak{A})$.

Theorem 3.3.3 *Let* $(\mathfrak{A}[\|\cdot\|], \mathfrak{A}_0)$ *be a normed quasi *-algebra with unit e. Then,*

$$\mathfrak{p}(a) = \sup_{\pi \in \mathrm{Rep}_{sc}(\mathfrak{A})} |||\pi(a)|||, \quad \forall\, a \in \mathfrak{A}.$$

Proof Let $\pi \in \mathrm{Rep}_{sc}(\mathfrak{A})$. For any $\xi \in \mathcal{D}_\pi$ we define, as before, φ_ξ by

$$\varphi_\xi(a, b) = \langle \pi(a)\xi | \pi(b)\xi \rangle, \quad \forall\, a, b \in \mathfrak{A}.$$

Then, $\varphi_\xi \in \mathcal{P}_{\mathfrak{A}_0}(\mathfrak{A})$, since from the discussion before present theorem, we conclude

$$|\varphi_\xi(a, b)| \leq \|a\| \|b\| \|\xi\|_\pi^2, \quad \forall\, a, b \in \mathfrak{A}.$$

Clearly, $\varphi_\xi \in \mathcal{S}_{\mathfrak{A}_0}(\mathfrak{A})$, if $\|\xi\|_\pi \leq 1$. Therefore,

$$|||\pi(a)|||^2 = \sup_{\|\xi\|_\pi \leq 1} \|\pi(a)\xi\|^2 = \sup_{\|\xi\|_\pi \leq 1} \varphi_\xi(a, a) \leq \mathfrak{p}(a)^2, \quad \forall\, a \in \mathfrak{A}. \tag{3.3.21}$$

On the other hand, if $\varphi \in \mathcal{S}_{\mathfrak{A}_0}(\mathfrak{A})$ and π_φ is the corresponding GNS representation, then (see (2.4.8) and (2.3.5))

$$\|\pi_\varphi^\circ(a)\lambda_\varphi(x)\|^2 = \varphi(ax, ax) = \varphi_x(a, a) \leq \|\varphi_x\| \|a\|^2, \quad \forall\, a \in \mathfrak{A} \text{ and } x \in \mathfrak{A}_0.$$

Therefore, π_φ° is strongly continuous. Hence,

$$\|\lambda_\varphi(x)\|_{\pi_\varphi^\circ} = \sup_{\|a\|\le 1} \|\pi_\varphi^\circ(a)\lambda_\varphi(x)\| = \|\varphi_x\|^{1/2}, \quad \forall\, a \in \mathfrak{A} \text{ and } x \in \mathfrak{A}_0,$$

so

$$|||\pi_\varphi^\circ(a)||| = \sup\left\{\|\pi_\varphi^\circ(a)\lambda_\varphi(x)\| : x \in \mathfrak{A}_0 \text{ with } \|\varphi_x\| \le 1\right\}, \quad a \in \mathfrak{A}.$$

This equality implies that

$$\begin{aligned}\sup_{\varphi\in\mathcal{P}_{\mathfrak{A}_0}(\mathfrak{A})} |||\pi_\varphi^\circ(a)||| &= \sup\left\{\varphi_x(a,a)^{1/2} : x \in \mathfrak{A}_0 \text{ with } \|\varphi_x\| \le 1\right\} \\ &= \sup_{\varphi\in\mathcal{S}_{\mathfrak{A}_0}(\mathfrak{A})} \varphi(a,a)^{1/2} = \mathfrak{p}(a), \quad \forall\, a \in \mathfrak{A}.\end{aligned}$$

Consequently (see also (3.3.21)),

$$\mathfrak{p}(a) = \sup_{\varphi\in\mathcal{P}_{\mathfrak{A}_0}(\mathfrak{A})} |||\pi_\varphi^\circ(a)||| \le \sup_{\pi\in \mathrm{Rep}_{sc}(\mathfrak{A})} |||\pi(a)||| \le \mathfrak{p}(a), \quad \forall\, a \in \mathfrak{A}.$$

This completes the proof. □

Example 3.3.4 A simple example of a Banach quasi *-algebra having a strongly continuous unbounded *-representation is provided by $(L^p(I), C(I))$, $I = [0, 1]$, with $p \ge 2$. If we put

$$(\pi(x)\xi)(t) = x(t)\xi(t), \quad x \in L^p(I),\ \xi \in C(I),\ t \in I,$$

then π is a *-representation of $(L^p(I), C(I))$ in the Hilbert space $L^2(I)$. It is easily seen that $\pi(x)$ is bounded, if and only if, $x \in L^\infty(I)$. This *-representation is strongly continuous. Indeed,

$$\|\pi(x)\xi\|_2 \le \|\xi\|_\infty \|x\|_2 \le \|\xi\|_\infty \|x\|_p, \quad x \in L^p(I),\ \xi \in C(I).$$

A criterion for $(\mathfrak{A}[\|\cdot\|], \mathfrak{A}_0)$ to have only bounded strongly continuous *-representations is given by the following

Proposition 3.3.5 *Let $(\mathfrak{A}[\|\cdot\|], \mathfrak{A}_0)$ be a Banach quasi *-algebra with unit e. The following statements are equivalent:*

(i) *$\mathcal{D}(\mathfrak{q}) = \mathfrak{A}$;*
(ii) *every strongly continuous *-representation π of $(\mathfrak{A}[\|\cdot\|], \mathfrak{A}_0)$ is bounded;*
(iii) *every $\varphi \in \mathcal{P}_{\mathfrak{A}_0}(\mathfrak{A})$ is admissible,* in the sense of Definition 2.4.4.

Proof (i) ⇒ (ii) Assume that there exists a strongly continuous unbounded *-representation π of $(\mathfrak{A}[\|\cdot\|], \mathfrak{A}_0)$. Then, for some $a \in \mathfrak{A}$, $\pi(a)$ is an unbounded operator. This implies that there exists a sequence $\{\xi_n\}$ of vectors in $\mathcal{D}_\pi$ with the properties

$$\|\xi_n\| = 1, \quad \|\pi(a)\xi_n\| \underset{n\to\infty}{\to} \infty.$$

As before, put $\varphi_{\xi_n}(b, c) = \langle \pi(b)\xi_n | \pi(c)\xi_n \rangle$, $b, c \in \mathfrak{A}$, $n \in \mathbb{N}$. The strong continuity of π implies the existence of $\gamma > 0$, such that

$$\varphi_{\xi_n}(b, b) = \|\pi(b)\xi_n\|^2 \leq \gamma \|b\|^2 \, \|\xi_n\|^2 = \gamma \|b\|^2, \quad \forall\, b \in \mathfrak{A}.$$

Thus, for every $n \in \mathbb{N}$, φ_{ξ_n} is bounded. As it is easily seen, $\varphi_{\xi_n}(e, e) = 1$. Hence (see also (3.1.13)),

$$\mathfrak{q}(a)^2 \geq \sup_{n\in\mathbb{N}} \varphi_{\xi_n}(a, a) = \|\pi(a)\xi_n\|^2 \underset{n\to\infty}{\to} \infty, \quad a \in \mathfrak{A}.$$

Consequently, $a \notin \mathcal{D}(\mathfrak{q})$, a contradiction.

(ii) $\Rightarrow$ (iii) Let $\varphi \in \mathcal{P}_{\mathfrak{A}_0}(\mathfrak{A})$. Then, for some $\gamma > 0$, $\varphi(a, a) \leq \gamma \|a\|^2$, for every $a \in \mathfrak{A}$. If π_φ is the corresponding GNS representation, we obtain

$$\|\pi_\varphi(a)\lambda_\varphi(x)\|^2 = \varphi(ax, ax) \leq \gamma \|a\|^2 \|x\|_0^2, \quad \forall\, a \in \mathfrak{A},\ x \in \mathfrak{A}_0.$$

Hence, π_φ is strongly continuous and thus bounded. The conclusion now results from Proposition 2.4.5.

(iii) $\Rightarrow$ (i) Suppose that there exists $a \in \mathfrak{A}$, such that $a \notin \mathcal{D}(\mathfrak{q})$. Then,

$$\|\pi_\varphi(a)\lambda_\varphi(e)\| = \varphi(a, a)^{1/2} \geq \mathfrak{q}(a) \geq \infty,$$

which is a contradiction according to Proposition 2.4.5. This completes the proof. □

Remark 3.3.6 If $\mathfrak{q}(a) = 0$ implies that $a = 0$, then by the previous proposition it follows that $\mathfrak{A}$ is contained in the C*-algebra $\mathfrak{A}_\mathfrak{q}$, obtained by completing $\mathfrak{A}_0$ with respect to the norm $\mathfrak{q}$. We don't know if the identity map is necessarily continuous from $\mathfrak{A}[\|\cdot\|]$ into $\mathfrak{A}_\mathfrak{q}[\mathfrak{q}]$. If this is the case, then the next Proposition shows that $\mathfrak{p}(a) = \mathfrak{q}(a)$, for every $a \in \mathfrak{A}$. Namely, we have the following

Proposition 3.3.7 *Let $(\mathfrak{A}[\|\cdot\|], \mathfrak{A}_0)$ be a normed quasi *-algebra with unit e. The following statements are equivalent:*

(i) *$\mathfrak{p}$ is an m*-seminorm on $\mathfrak{A}_0$;*
(ii) *for each $\varphi \in \mathcal{P}_{\mathfrak{A}_0}(\mathfrak{A})$, $\|\varphi\| = \varphi(e, e)$;*
(iii) *$\mathcal{D}(\mathfrak{q}) = \mathfrak{A}$ and $\mathfrak{p}(a) = \mathfrak{q}(a)$, for every $a \in \mathfrak{A}$;*
(iv) *$\mathfrak{p}$ is a C*-seminorm on $\mathfrak{A}_0$;*
(v) *$\mathcal{D}(\mathfrak{q}) = \mathfrak{A}$ and $\mathfrak{q}(a) \leq \|a\|$, for every $a \in \mathfrak{A}$.*

Proof (i) $\Rightarrow$ (ii) Let $\varphi \in \mathcal{P}_{\mathfrak{A}_0}(\mathfrak{A})$. We define a linear functional $\widehat{\omega}_\varphi$ on $\mathfrak{A}_0/N_0(\mathfrak{p})$ by

$$\widehat{\omega}_\varphi(x + N_0(\mathfrak{p})) = \varphi(x, e), \quad x \in \mathfrak{A}_0.$$

Then, $\widehat{\omega}_\varphi$ is $\|\cdot\|_\mathfrak{p}$-bounded and positive, since $\varphi \in \mathcal{P}_{\mathfrak{A}_0}(\mathfrak{A})$, respectively,

$$\widehat{\omega}_\varphi\big((x+N_0(\mathfrak{p}))^*(x+N_0(\mathfrak{p}))\big) = \widehat{\omega}_\varphi(x^*x+N_0(\mathfrak{p})) = \varphi(x^*x, e) = \varphi(x, x) \geq 0, \quad x \in \mathfrak{A}_0.$$

We denote with $\widetilde{\omega}_\varphi$ the unique $\|\cdot\|_\mathfrak{p}$-bounded extension of $\widehat{\omega}_\varphi$ to $\mathfrak{A}_\mathfrak{p}$. If $y \in \mathfrak{A}_\mathfrak{p}$, then there exists a sequence $\{x_n\} \subset \mathfrak{A}_0$, such that $y = \|\cdot\|_\mathfrak{p} - \lim(x_n + N_0(\mathfrak{p}))$. Hence,

$$\widetilde{\omega}_\varphi(y^*y) = \lim_{n\to\infty} \widehat{\omega}_\varphi\big((x_n + N_0(\mathfrak{p}))^*(x_n + N_0(\mathfrak{p}))\big) \geq 0.$$

Consequently, $\widetilde{\omega}_\varphi$ is positive on $\mathfrak{A}_\mathfrak{p}$. Then, for the norm $\|\widetilde{\omega}_\varphi\|_\mathfrak{p}^\star$ of the linear functional $\widetilde{\omega}_\varphi$, one has

$$\|\widetilde{\omega}_\varphi\|_\mathfrak{p}^\star = \widetilde{\omega}_\varphi(e) = \varphi(e, e).$$

Therefore, for all $x, y \in \mathfrak{A}_0$,

$$\begin{aligned}|\varphi(x, y)| = |\widehat{\omega}_\varphi(y^*x + N_0(\mathfrak{p}))| &\leq \varphi(e, e)\mathfrak{p}(y^*x)\\ &\leq \varphi(e, e)\mathfrak{p}(x)\mathfrak{p}(y) \leq \varphi(e, e)\|x\|\|y\|.\end{aligned}$$

This inequality implies that $\|\varphi\| \leq \varphi(e, e)$ and since $\varphi(e, e) \leq \|\varphi\|$, we obtain the equality.

(ii) $\Rightarrow$ (iii) follows immediately from the definition of $\mathfrak{q}$ and from (ii).

(iii) $\Rightarrow$ (iv) follows from the equality $\mathfrak{p} = \mathfrak{q}$ (see also (3.1.14)).

(iii) $\Rightarrow$ (v) immediately.

(v) $\Rightarrow$ (i) Since

$$\varphi(ax, ax) \leq \mathfrak{q}(a)^2\varphi(x, x), \quad \forall\, a \in \mathcal{D}(\mathfrak{q}),\ x \in \mathfrak{A}_0,$$

then

$$\varphi(ax, ax) \leq \|a\|^2\varphi(x, x), \quad \forall\, a \in \mathfrak{A},\ x \in \mathfrak{A}_0.$$

For $x = e$ this gives

$$\varphi(a, a) \leq \|a\|^2\varphi(e, e), \quad \forall\, a \in \mathfrak{A}.$$

Therefore, $\varphi(e, e) \geq \|\varphi\|$. Since one always has $\varphi(e, e) \leq \|\varphi\|$, we conclude that $\|\varphi\| = \varphi(e, e)$. Thus, if $\varphi(e, e) = 1$, then $\varphi \in \mathcal{S}_{\mathfrak{A}_0}(\mathfrak{A})$. This implies that $\mathfrak{q}(a) \leq$

$\mathfrak{p}(a)$, for every $a \in \mathfrak{A}$, by the very definitions. Hence, applying now Lemma 3.1.40, (iii), we conclude that $\mathfrak{p}$ is an m^*-seminorm on $\mathfrak{A}_0$. □

Corollary 3.3.8 *Let* $(\mathfrak{A}[\|\cdot\|], \mathfrak{A}_0)$ *be a normed quasi *-algebra with unit e. The following statements are equivalent:*

(i) $\mathfrak{p}$ *is an* m^**-seminorm on* $\mathfrak{A}_0$*;*
(ii) *every regular *-representation* π *of* $(\mathfrak{A}[\|\cdot\|], \mathfrak{A}_0)$ *in a Hilbert space* $\mathcal{H}$ *is bounded and continuous from* $\mathfrak{A}[\|\cdot\|]$ *into* $\mathcal{B}(\mathcal{H})$*, such that* $\|\pi(a)\| \leq \|a\|$*, for every* $a \in \mathfrak{A}$*.*

Proof It follows immediately from Propositions 3.1.39 and 3.3.7. □

As seen in previous examples, there exist Banach quasi *-algebras $(\mathfrak{A}[\|\cdot\|], \mathfrak{A}_0)$ (with unit) for which $\mathcal{P}_{\mathfrak{A}_0}(\mathfrak{A}) = \{0\}$. If this is the case, there is no strongly continuous *-representation of $(\mathfrak{A}[\|\cdot\|], \mathfrak{A}_0)$, apart from the trivial one. This unpleasant feature is prevented if we require that $\mathcal{P}_{\mathfrak{A}_0}(\mathfrak{A})$ is *sufficient* by which we mean that if $a \in \mathfrak{A}$ and $\varphi(a, a) = 0$, for every $\varphi \in \mathcal{P}_{\mathfrak{A}_0}(\mathfrak{A})$, then $a = 0$. Clearly, if $\mathcal{P}_{\mathfrak{A}_0}(\mathfrak{A})$ is sufficient, then $(\mathfrak{A}[\|\cdot\|], \mathfrak{A}_0)$ has a sufficient family of strongly continuous *-representations, where *sufficient* means in this case, that for every $a \in \mathfrak{A}$, $a \neq 0$, there exists a strongly continuous *-representation π, such that $\pi(a) \neq 0$. In particular, if $\mathfrak{p}(a) = \|a\|$, for every $a \in \mathfrak{A}$, then $\mathcal{P}_{\mathfrak{A}_0}(\mathfrak{A})$ is clearly sufficient and, as shown in [80, Theorem 3.26], $\mathcal{D}(\mathfrak{q})$ is a C*-algebra, consisting of all *bounded* elements of $\mathfrak{A}$ (see Example 3.1.45).

In this case, it follows from Proposition 3.3.5, that a *genuine* Banach quasi *-algebra $(\mathfrak{A}[\|\cdot\|], \mathfrak{A}_0)$, in the sense, that $\mathfrak{A} \supsetneq \mathfrak{A}_0$, necessarily has strongly continuous *unbounded* *-representations.

Banach quasi *-algebras, with a sufficient $\mathcal{P}_{\mathfrak{A}_0}(\mathfrak{A})$, have been studied in more details in [49] and in [80]. The question, of characterizing in terms of the original norm of $\mathfrak{A}$, the existence of sufficiently many positive invariant sesquilinear forms, still remains open.

Chapter 4
Normed Quasi *-Algebras: Bounded Elements and Spectrum

Bounded elements of a Banach quasi *-algebra are intended to be those, whose images under every *-representation are bounded operators in a Hilbert space. This rough idea can be developed in several ways, as we shall see in the present chapter. These notions lead us to discuss a convenient concept of spectrum of an element in this context.

4.1 The *-Algebra of Bounded Elements

4.1.1 Bounded Elements

Definition 4.1.1 Let $(\mathfrak{A}[\|\cdot\|], \mathfrak{A}_0)$ be a Banach quasi *-algebra and $a \in \mathfrak{A}$. We say that a is *left bounded* if there exists $\gamma_a > 0$, such that

$$\|ax\| \leq \gamma_a \|x\|, \quad \forall\, x \in \mathfrak{A}_0.$$

The set of all left bounded elements of $\mathfrak{A}$ is denoted by $\mathfrak{A}_{\rm lb}$. Analogously, we say that a is *right bounded* if there exists $\gamma'_a > 0$, such that

$$\|xa\| \leq \gamma'_a \|x\|, \quad \forall\, x \in \mathfrak{A}_0.$$

The set of all right bounded elements of $\mathfrak{A}$ is denoted by $\mathfrak{A}_{\rm rb}$.

The motivation of this terminology is made clearer by considering, for each $a \in \mathfrak{A}$, the linear maps (3.1.12) from $\mathfrak{A}_0$ into $\mathfrak{A}$:

$$x \in \mathfrak{A}_0 \to L_a x := ax \in \mathfrak{A}, \quad x \in \mathfrak{A}_0 \to R_a x := xa \in \mathfrak{A}. \tag{4.1.1}$$

M. Fragoulopoulou, C. Trapani, *Locally Convex Quasi *-Algebras and their Representations*, Lecture Notes in Mathematics 2257,
https://doi.org/10.1007/978-3-030-37705-2_4

For instance, if a is left bounded, then L_a is *bounded*, as a map from $\mathfrak{A}_0[\|\cdot\|]$ into $\mathfrak{A}[\|\cdot\|]$, and so it has a bounded extension $\overline{L}_a$ to $\mathfrak{A}[\|\cdot\|]$. We put

$$\|a\|_{\mathrm{lb}} := \|\overline{L}_a\|, \quad a \in \mathfrak{A}.$$

Analogously, we define a norm on $\mathfrak{A}_{\mathrm{rb}}$ by

$$\|a\|_{\mathrm{rb}} := \|\overline{R}_a\|, \quad a \in \mathfrak{A}.$$

Remark 4.1.2 Observe that an element $a \in \mathfrak{A}_{\mathrm{lb}}$ is not necessarily right bounded.

We put $\mathfrak{A}_{\mathrm{b}} = \mathfrak{A}_{\mathrm{lb}} \cap \mathfrak{A}_{\mathrm{rb}}$. On $\mathfrak{A}_{\mathrm{b}}$ we define the norm

$$\|a\|_{\mathrm{b}} := \max\left\{\|\overline{L}_a\|, \|\overline{R}_a\|\right\}, \quad a \in \mathfrak{A}_{\mathrm{b}}. \tag{4.1.2}$$

It is easily seen that, if $a \in \mathfrak{A}_{\mathrm{b}}$, then $a^* \in \mathfrak{A}_{\mathrm{b}}$. Moreover, by the very definitions, (3.1.4) and the discussion around it, we see that $\mathfrak{A}_0 \subseteq \mathfrak{A}_{\mathrm{b}}$, $\|x\|_{\mathrm{b}} = \|x\|_0$, for all $x \in \mathfrak{A}_0$ and also an analogous result to that of Corollary 3.1.3 holds for $\|\cdot\|_{\mathrm{b}}$.

Remark 4.1.3 If $(\mathfrak{A}[\|\cdot\|], \mathfrak{A}_0)$ has a unit e, then obviously $e \in \mathfrak{A}_{\mathrm{b}}$ and since we may suppose that $\|e\| = 1$ (Remark 3.1.4), we obtain

$$\|a\|_{\mathrm{lb}} = \|\overline{L}_a\| \geq \|a\| \quad \forall\, a \in \mathfrak{A}_{\mathrm{lb}}.$$

Of course, an analogous inequality holds for $\|\cdot\|_{\mathrm{rb}}$. Thus, it turns out that, in the unital case, $\|\cdot\| \leq \|\cdot\|_{\mathrm{b}}$, on $\mathfrak{A}_{\mathrm{b}}$.

As usual, we denote with $\mathcal{B}(\mathfrak{A})$ the Banach algebra of bounded operators on the Banach space $\mathfrak{A}[\|\cdot\|]$. From the definition itself it follows that $\mathfrak{A}_{\mathrm{lb}}$, as well as $\mathfrak{A}_{\mathrm{rb}}$, can be identified with a subspace of $\mathcal{B}(\mathfrak{A})$.

Let $a \in \mathfrak{A}_{\mathrm{lb}}$ and $b \in \mathfrak{A}$. Then, we put

$$a \blacktriangleright b := \overline{L}_a b. \tag{4.1.3}$$

More explicitly, $a \blacktriangleright b$ is defined by picking a sequence $\{y_n\} \subset \mathfrak{A}_0$ converging to b and setting

$$a \blacktriangleright b := \overline{L}_a b = \lim_{n\to\infty} L_a y_n = \lim_{n\to\infty} a y_n.$$

Similarly, if $b \in \mathfrak{A}_{\mathrm{rb}}$ and $a \in \mathfrak{A}$, we put

$$a \blacktriangleleft b := \overline{R}_b a. \tag{4.1.4}$$

In this case, for an explicit definition of $a \blacktriangleleft b$, we have to select a sequence $\{x_n\} \subset \mathfrak{A}_0$ converging to a and put

$$a \blacktriangleleft b := \overline{R}_b a = \lim_{n\to\infty} R_b x_n = \lim_{n\to\infty} x_n b.$$

Then, it is clear that, if $a, b \in \mathfrak{A}_b$, we have that both $a \blacktriangleright b$ and $a \blacktriangleleft b$ are well defined, but, in general, $a \blacktriangleright b \neq a \blacktriangleleft b$. We shall discuss this point, in more details, below.

Remark 4.1.4 Let $a, b \in \mathfrak{A}_{\text{lb}}$. Then, $\overline{L}_a, \overline{L}_b \in \mathcal{B}(\mathfrak{A})$. Hence, $\overline{L}_a \overline{L}_b \in \mathcal{B}(\mathfrak{A})$ and $\overline{L}_a \overline{L}_b = \overline{L}_{a \blacktriangleright b}$.

Indeed, let $\{y_n\}$ be a sequence in $\mathfrak{A}_0$, $\|\cdot\|$-converging to b. Then, for each $x \in \mathfrak{A}_0$,

$$(\overline{L}_a \overline{L}_b)x = \overline{L}_a(\overline{L}_b x) = \lim_{n\to\infty} a(y_n x) = \overline{L}_a(bx).$$

On the other hand,

$$\overline{L}_{a \blacktriangleright b} x = (\overline{L}_a b)x = \lim_{n\to\infty} (a y_n)x = \lim_{n\to\infty} a(y_n x) = \overline{L}_a(bx), \quad x \in \mathfrak{A}_0. \qquad (4.1.5)$$

It is easy to show that if $a, b \in \mathfrak{A}_{\text{lb}}$ and $\mu \in \mathbb{C}$, then both $a + b$ and μa belong to $\mathfrak{A}_{\text{lb}}$.

Proposition 4.1.5 *If* $(\mathfrak{A}[\|\cdot\|], \mathfrak{A}_0)$ *is a Banach quasi *-algebra with unit, then the set* $\mathfrak{A}_{\text{lb}}$ *of all left bounded elements is a Banach algebra, with respect to the multiplication* $\blacktriangleright$ *and the norm* $\|\cdot\|_{\text{lb}}$.

Proof We prove that if $a, b \in \mathfrak{A}_{\text{lb}}$ then $a \blacktriangleright b \in \mathfrak{A}_{\text{lb}}$ and

$$\|a \blacktriangleright b\|_{\text{lb}} \leq \|a\|_{\text{lb}} \, \|b\|_{\text{lb}}.$$

Indeed, using the associativity properties of the multiplication in $\mathfrak{A}$, one has that

$$(\overline{L}_a b)x = \lim_{m\to\infty} (a y_m)x = \lim_{m\to\infty} a(y_m x) = \overline{L}_a(bx) = \overline{L}_a(L_b x), \quad \forall\, x \in \mathfrak{A}_0,$$

where $\{y_m\}$ is a sequence in $\mathfrak{A}_0$, $\|\cdot\|$-converging to b. Therefore, we have (see Remark 4.1.4)

$$\|\overline{L}_{a \blacktriangleright b} x\| = \|(\overline{L}_a b)x\| \leq \|\overline{L}_a\| \, \|\overline{L}_b\| \|x\|, \quad \forall\, x \in \mathfrak{A}_0.$$

Hence, $a \blacktriangleright b \in \mathfrak{A}_{\text{lb}}$ and moreover (see also discussion after (4.1.1)), we obtain

$$\|\overline{L}_{a \blacktriangleright b}\| \leq \|\overline{L}_a\| \, \|\overline{L}_b\| \leq \|a\|_{\text{lb}} \|b\|_{\text{lb}}.$$

From the latter inequality, together with (4.1.3) and (4.1.5), it follows that

$$\|a \blacktriangleright b\|_{\text{lb}} = \|\overline{L}_{a \blacktriangleright b}\| \leq \|a\|_{\text{lb}} \, \|b\|_{\text{lb}}, \quad \forall\, a, b \in \mathfrak{A}_{\text{lb}}.$$

Thus, $\mathfrak{A}_{\text{lb}}$, endowed with $\|\cdot\|_{\text{lb}}$, is a normed algebra. We shall now show that $\mathfrak{A}_{\text{lb}}[\|\cdot\|_{\text{lb}}]$ is complete. Let $\{a_n\}$ be a Cauchy sequence in $\mathfrak{A}_{\text{lb}}[\|\cdot\|_{\text{lb}}]$. Then, $\{\overline{L}_{a_n}\}$ is a Cauchy sequence in $\mathcal{B}(\mathfrak{A})$. Thus, there exists $L \in \mathcal{B}(\mathfrak{A})$, such that $\overline{L}_{a_n} \to L$, with respect to the natural norm of $\mathcal{B}(\mathfrak{A})$. By Remark 4.1.3, $\|a_n - a_m\| \to 0$, so there exists $a \in \mathfrak{A}$, such that $\|a_n - a\| \underset{n\to\infty}{\to} 0$. Since, the right multiplication by x is continuous in $\mathfrak{A}$, it follows that $a_n x \underset{n\to\infty}{\to} ax = \overline{L}_a x$, in the norm of $\mathfrak{A}$. This implies that $\overline{L}_a = L$. From these facts, it follows easily that a is left bounded and $a_n \underset{n\to\infty}{\to} a$, with respect to $\|\cdot\|_{\text{lb}}$. □

A similar result can be proved for $\mathfrak{A}_{\text{rb}}$, taking into account the following facts concerning the involution * of $\mathfrak{A}$:

(1*) $a \in \mathfrak{A}_{\text{lb}} \Leftrightarrow a^* \in \mathfrak{A}_{\text{rb}}$;
(2*) $\|a^*\|_{\text{rb}} = \|a\|_{\text{lb}}$, for every $a \in \mathfrak{A}_{\text{lb}}$;
(3*) $(a{\blacktriangleright}b)^* = b^*{\blacktriangleleft}a^*$, for every $a, b \in \mathfrak{A}_{\text{lb}}$.

These facts, as well as the next lemma, follow easily by the very definitions.

Lemma 4.1.6 *If $(\mathfrak{A}[\|\cdot\|], \mathfrak{A}_0)$ is a Banach quasi *-algebra, the following statements hold:*

(i) *if $a \in \mathfrak{A}_{\text{lb}}$ and $b \in \mathfrak{A}$, then $\|a{\blacktriangleright}b\| \le \|a\|_{\text{lb}}\,\|b\|$;*
(ii) *if $b \in \mathfrak{A}_{\text{rb}}$ and $a \in \mathfrak{A}$, then $\|a{\blacktriangleleft}b\| \le \|a\|\|b\|_{\text{rb}}$.*

4.1.2 Normal Banach Quasi *-Algebras

As mentioned before, if $a, b \in \mathfrak{A}_{\text{b}}$ then, both $a{\blacktriangleright}b$ and $a{\blacktriangleleft}b$ are well defined; but, in general, $a{\blacktriangleright}b \ne a{\blacktriangleleft}b$. We want to analyze, more carefully, this situation. First of all, if $a, b \in \mathfrak{A}_{\text{b}}$, then $\overline{L}_a, \overline{L}_b \in \mathcal{B}(\mathfrak{A})$. As shown in Remark 4.1.4, $\overline{L}_a\overline{L}_b = \overline{L}_{a{\blacktriangleright}b}$. Similarly, if $a, b \in \mathfrak{A}_{\text{b}}$, then $\overline{R}_b\overline{R}_a = \overline{R}_{a{\blacktriangleleft}b}$.

Recall that we denote with $\mathfrak{A}^\star$ the Banach dual space of $\mathfrak{A}[\|\cdot\|]$ and by $\|f\|^\star$ the norm of $f \in \mathfrak{A}^\star$.

Proposition 4.1.7 *If $(\mathfrak{A}[\|\cdot\|], \mathfrak{A}_0)$ is a Banach quasi *-algebra, the following statements are equivalent:*

(i) *$a{\blacktriangleright}b = a{\blacktriangleleft}b$, for every $a, b \in \mathfrak{A}_{\text{b}}$;*
(ii) *$a{\blacktriangleright}b$ is right bounded and $\|a{\blacktriangleright}b\| \le \|a\|\|b\|_{\text{rb}}$, for every $a, b \in \mathfrak{A}_{\text{b}}$;*
(iii) *$a{\blacktriangleleft}b$ is left bounded and $\|a{\blacktriangleleft}b\| \le \|a\|_{\text{lb}}\|b\|$, for every $a, b \in \mathfrak{A}_{\text{b}}$;*
(iv) *for any pair $\{x_n\}$, $\{y_n\}$ of sequences of elements of $\mathfrak{A}_0$, $\|\cdot\|$-converging to elements of $\mathfrak{A}_{\text{b}}$, one has*

$$\lim_{n\to\infty}\lim_{m\to\infty} x_n y_m = \lim_{m\to\infty}\lim_{n\to\infty} x_n y_m;$$

(v) *there exists a weakly*-dense subspace* $\mathcal{M}$ *of* $\mathfrak{A}^\star$, *such that for any pair* $\{x_n\}$, $\{y_n\}$ *of sequences of elements of* $\mathfrak{A}_0$, $\|\cdot\|$*-converging to elements of* $\mathfrak{A}_b$, *one has*

$$\lim_{n\to\infty}\lim_{m\to\infty} f(x_n y_m) = \lim_{m\to\infty}\lim_{n\to\infty} f(x_n y_m), \quad \forall\, f \in \mathcal{M}.$$

Proof (i) $\Rightarrow$ (ii) It follows directly from the assumption $a \blacktriangleright b = a \blacktriangleleft b$ in (i) and Lemma 4.1.6(ii).

(ii) $\Rightarrow$ (iii) Use the fact that $\|a^*\| = \|a\|$, for all $a \in \mathfrak{A}$ and apply (ii) and (1*), (2*), (3*), before Lemma 4.1.6.

(iii) $\Rightarrow$ (i) Assume that, for every $a, b \in \mathfrak{A}_b$, $a \blacktriangleleft b$ is left bounded and $\|a \blacktriangleleft b\| \leq \|a\|_{lb}\|b\|$. Let $\{y_n\} \subset \mathfrak{A}_0$ be a sequence, such that $\|b - y_n\| \to 0$, as $n \to \infty$. Then, since $\mathfrak{A}_0 \subseteq \mathfrak{A}_b$ and $\mathfrak{A}_b$ is a vector space, we get

$$\|a \blacktriangleleft b - a y_n\| = \|a \blacktriangleleft b - a \blacktriangleleft y_n\| = \|a \blacktriangleleft (b - y_n)\| \leq \|a\|_{lb}\|b - y_n\| \to 0.$$

Hence,

$$a \blacktriangleleft b = \lim_{n\to\infty} a y_n = \overline{L_a} b = a \blacktriangleright b.$$

(i) $\Rightarrow$ (iv) Let $\{x_n\}$, $\{y_n\}$ be sequences of elements of $\mathfrak{A}_0$, with $\|a - x_n\| \to 0$, $\|b - y_n\| \to 0$, as $n \to \infty$, where $a, b \in \mathfrak{A}_b$. Then, we have

$$a \blacktriangleright b = \overline{L_a} b = \lim_{m\to\infty} a y_m = \lim_{m\to\infty}\lim_{n\to\infty} x_n y_m.$$

On the other hand,

$$a \blacktriangleleft b = \overline{R_b} a = \lim_{n\to\infty} x_n b = \lim_{n\to\infty}\lim_{m\to\infty} x_n y_m.$$

The equality $a \blacktriangleright b = a \blacktriangleleft b$ then implies that the two iterated limits coincide.

(iv) $\Rightarrow$ (v) It is clear.

(v) $\Rightarrow$ (i) Assume that (i) fails. Then, there exists $f \in \mathfrak{A}^\star$, such that $f(a \blacktriangleright b) \neq f(a \blacktriangleleft b)$. Since $\mathcal{M}$ is weakly*-dense in $\mathfrak{A}^\star$, we may suppose that $f \in \mathcal{M}$. Then, if $\{x_n\}$, $\{y_n\}$ are sequences in $\mathfrak{A}_0$, $\|\cdot\|$-converging, respectively, to a and b, we have

$$\lim_{m\to\infty}\lim_{n\to\infty} f(x_n y_m) = f(a \blacktriangleright b) \neq f(a \blacktriangleleft b) = \lim_{n\to\infty}\lim_{m\to\infty} f(x_n y_m),$$

a contradiction. This completes the proof. □

If anyone of the equivalent conditions of Proposition 4.1.7 holds, we put

$$a \bullet b := a \blacktriangleright b = a \blacktriangleleft b, \quad a, b \in \mathfrak{A}_b,$$

and $a \bullet b$ is called the *product of a and b*.

Definition 4.1.8 A Banach quasi *-algebra $(\mathfrak{A}[\|\cdot\|], \mathfrak{A}_0)$, such that $a{\blacktriangleright}b = a{\blacktriangleleft}b$, for all a, b in $\mathfrak{A}_b$, is called *normal*.

Corollary 4.1.9 *If* $(\mathfrak{A}[\|\cdot\|], \mathfrak{A}_0)$ *is a Banach quasi *-algebra with unit, the following statements hold:*

(i) $(\mathfrak{A}[\|\cdot\|], \mathfrak{A}_0)$ *is normal, if and only if,* $\mathfrak{A}_b$ *is a *-algebra, with respect to* $\blacktriangleright$ (or, equivalently, with respect to $\blacktriangleleft$)*;*
(ii) *if* $(\mathfrak{A}[\|\cdot\|], \mathfrak{A}_0)$ *is a normal Banach quasi *-algebra, then* $\mathfrak{A}_b[\|\cdot\|_b]$ *is a Banach *-algebra, with respect to the multiplication* $\bullet$.

Proof (i) The fact that if $(\mathfrak{A}[\|\cdot\|], \mathfrak{A}_0)$ is normal, then $\mathfrak{A}_b$ is a *-algebra, with respect to $\blacktriangleright$, follows from the previous discussion. On the other hand, assume that $\mathfrak{A}_b$ is a *-algebra, with respect to $\blacktriangleright$; then, for every $a, b \in \mathfrak{A}_b$, $a{\blacktriangleright}b \in \mathfrak{A}_b$ and (see also (3*), before Lemma 4.1.6)

$$a{\blacktriangleleft}b = (b^*{\blacktriangleright}a^*)^* = a{\blacktriangleright}b.$$

(ii) It follows easily from Proposition 4.1.5 and the properties of involution. □

Example 4.1.10 Assume that for each $a \in \mathfrak{A}_b$ there exists a sequence $\{x_n\} \subset \mathfrak{A}_0$, such that

$$\sup_n \|x_n\|_0 < \infty \quad \text{and} \quad \lim_{n\to\infty} \|a - x_n\| = 0;$$

then $(\mathfrak{A}[\|\cdot\|], \mathfrak{A}_0)$ is normal. Indeed, in this case, (ii) or (iii) of Proposition 4.1.7 holds (see Proposition 4.1.19). Elements of the preceding type will be called *strongly bounded* and will be studied in Sect. 4.1.3.

Remark 4.1.11 If $(\mathfrak{A}[\|\cdot\|], \mathfrak{A}_0)$ is a commutative Banach quasi *-algebra, i.e., $ax = xa$, for every $a \in \mathfrak{A}$ and $x \in \mathfrak{A}_0$, then it is easily seen that each left bounded element a is also right bounded and one has $a{\blacktriangleright}b = b{\blacktriangleleft}a$, for every $a, b \in \mathfrak{A}_b$. Thus, if $a, b \in \mathfrak{A}_b$, then both $a{\blacktriangleright}b$ and $a{\blacktriangleleft}b$ are elements of $\mathfrak{A}_b$, but they need not be equal. In general, in such a case, $\mathfrak{A}_b$ is an algebra with respect to $\blacktriangleright$ (and also with respect to $\blacktriangleleft$). *Normality, in the commutative case, becomes equivalent to commutativity of the algebra* $\mathfrak{A}_b$.

Example 4.1.12 For the Banach quasi *-algebra $(L^p(I), C(I))$ considered in Example 3.1.6, one finds that $(L^p(I))_b = L^\infty(I)$ and the norm $\|\cdot\|_b$ is exactly the L^∞-norm. Since the multiplications $\blacktriangleright$ and $\blacktriangleleft$ both coincide with the ordinary multiplication of functions, $(L^p(I), C(I))$ is normal. This example also shows that, in general, $\mathfrak{A}_0$ is not dense in $\mathfrak{A}_b$, with respect to $\|\cdot\|_b$ since, as is well known, $C(I)$ is not dense in $L^\infty(I)$.

Similarly, $(L^p(\mathbb{R}), C_c^0(\mathbb{R}))$ is a Banach quasi *-algebra without unit. In this case, $(L^p(\mathbb{R}))_b = L^\infty(\mathbb{R}) \cap L^p(\mathbb{R})$ and $(L^p(\mathbb{R}), C_c^0(\mathbb{R}))$ is normal. The norm $\|\cdot\|_b$ is equivalent to $\|\cdot\|_p + \|\cdot\|_\infty$.

For the non-commutative L^p-spaces of Example 3.1.7 one finds that $(L^p(\varrho))_b = \mathcal{J}_p$, if ϱ is semifinite, or $(L^p(\varrho))_b = \mathfrak{M}$, if ϱ is finite. Normality follows from the fact that the multiplications ▸ and ◂ both coincide with the ordinary multiplication of bounded operators.

Example 4.1.13 In case of the Banach quasi *-algebra $(\mathcal{H}, \mathfrak{A}_0)$ constructed from a Hilbert algebra $\mathfrak{A}_0$, as in Example 3.1.8, the set $\mathcal{H}_b$ of bounded elements of $\mathcal{H}$ is the so-called *fulfillment* of $\mathfrak{A}_0$ ($\mathfrak{A}_0$ is called a *full* Hilbert algebra if $\mathcal{H}_b = \mathfrak{A}_0$). Moreover, $(\mathcal{H}, \mathfrak{A}_0)$ is normal. Indeed, let $a, b \in \mathcal{H}_b$, and let $\{x_n\}$, $\{y_n\}$ be sequences in $\mathfrak{A}_0$, $\|\cdot\|$-converging, respectively, to a and b. Then,

$$\langle a \blacktriangleright b|x\rangle = \lim_{n\to\infty} \langle a y_n|x\rangle = \lim_{n\to\infty} \langle y_n|a^* x\rangle = \langle b|a^* x\rangle, \quad \forall\, x \in \mathfrak{A}_0.$$

On the other hand,

$$\langle a \blacktriangleleft b|x\rangle = \lim_{n\to\infty} \langle x_n b|x\rangle = \lim_{n\to\infty} \langle b|x_n^* x\rangle = \langle b|a^* x\rangle, \quad \forall\, x \in \mathfrak{A}_0.$$

This implies that $a \blacktriangleright b = a \blacktriangleleft b$.

Note that, in general, a Banach quasi *-algebra $(\mathfrak{A}[\|\cdot\|], \mathfrak{A}_0)$ is said to be *full* if $\mathfrak{A}_b = \mathfrak{A}_0$.

Example 4.1.14 Let A be an unbounded densely defined selfadjoint operator in $L^2(\Omega)$, where Ω is an open subset of $\mathbb{R}^n$. Let $D(A)$ denote the domain of A. Then, $D(A)$ can be made into a Hilbert space under the inner product

$$\langle f|g\rangle_A = \langle f|g\rangle + \langle Af|Ag\rangle, \quad f, g \in D(A)$$

and norm $\|\cdot\|_A = \langle\cdot|\cdot\rangle_A^{1/2}$.

Let us assume that

(i) $C_c(\Omega) \subset D(A)$ and that it is a core for A, i.e., $\overline{A\lceil_{C_c(\Omega)}} = A$ [2, Definition 1.4.1], (here $C_c(\Omega)$ stands for the algebra of continuous functions on Ω, with compact support);
(ii) $f \in D(A)$ implies $f^* \in D(A)$, where $f^*(x) = \overline{f(x)}$;
(iii) $(Af)^* = Af^*$, for every $f \in D(A)$;
(iv) $f\phi \in D(A)$, for every $f \in D(A)$ and $\phi \in C_c(\Omega)$.

Then $(D(A), C_c(\Omega))$ is a Banach quasi *-algebra, when $D(A)$ is endowed with $\|\cdot\|_A$. The continuity of the multiplication follows from the closed graph theorem and the fact that, for each $\phi \in C_c(\Omega)$, the multiplication operator L_ϕ is closed in the Hilbert space $D(A)$. The space of bounded elements is given by the set

$$D_M(A) = \big\{f \in D(A) : \ fg \in D(A),\ \forall\, g \in D(A)\big\}.$$

Indeed, it is clear that for any bounded element f, the operator $\overline{L}_f$ maps $D(A)$ into $D(A)$. Conversely, if $f \in D_M(A)$, then the corresponding operator L_f is closed in the Hilbert space $D(A)$ and therefore bounded. Hence, $D(A)_{\mathrm{b}} = D_M(A)$. To conclude this example we notice that if $L_\phi A = AL_\phi$, for every $\phi \in C_c(\Omega)$, then $C_c(\Omega)$ is a Hilbert algebra with inner product $\langle\cdot|\cdot\rangle_A$. By Example 4.1.13 it follows that $(D(A), C_c(\Omega))$ is normal.

Lemma 4.1.15 *If* $(\mathfrak{A}[\|\cdot\|], \mathfrak{A}_0)$ *is a normal Banach quasi *-algebra, then,*

$$\overline{L}_a\overline{R}_b = \overline{R}_b\overline{L}_a, \quad \forall\, a, b \in \mathfrak{A}_{\mathrm{b}}. \tag{4.1.6}$$

Proof Indeed, let $a, b \in \mathfrak{A}_{\mathrm{b}}$, and $\{x_n\}$, $\{y_n\}$ be sequences in $\mathfrak{A}_0$, $\|\cdot\|$-converging, respectively, to a and b. Then, for every $x \in \mathfrak{A}_0$,

$$(\overline{L}_a\overline{R}_b)x = \overline{L}_a(\overline{R}_b x) = \lim_{m\to\infty} a(xy_m) = \lim_{m\to\infty}\lim_{n\to\infty} x_n(xy_m).$$

On the other hand,

$$(\overline{R}_b\overline{L}_a)x = \overline{R}_b(\overline{L}_a x) = \lim_{n\to\infty}(x_n x)b = \lim_{n\to\infty}\lim_{m\to\infty}(x_n x)y_m.$$

The statement then follows from (iv) of Proposition 4.1.7. □

Remark 4.1.16 If $(\mathfrak{A}[\|\cdot\|], \mathfrak{A}_0)$ has a unit, then (4.1.6) implies the normality of $(\mathfrak{A}[\|\cdot\|], \mathfrak{A}_0)$.

If $(\mathfrak{A}[\|\cdot\|], \mathfrak{A}_0)$ is a normal Banach quasi *-algebra the multiplications of an element of $\mathfrak{A}$ and an element of $\mathfrak{A}_{\mathrm{b}}$ are defined via (4.1.3) and (4.1.4). More precisely,

$$a \bullet b = \begin{cases} a \blacktriangleright b, & \text{if } a \in \mathfrak{A}_{\mathrm{b}} \\ a \blacktriangleleft b, & \text{if } b \in \mathfrak{A}_{\mathrm{b}}. \end{cases}$$

Proposition 4.1.17 *If* $(\mathfrak{A}[\|\cdot\|], \mathfrak{A}_0)$ *is a normal Banach quasi *-algebra, then* $(\mathfrak{A}, \mathfrak{A}_{\mathrm{b}})$ *is a BQ*-algebra.*

Proof Let $a, b \in \mathfrak{A}$, with $a \in \mathfrak{A}, b \in \mathfrak{A}_{\mathrm{b}}$. Then, if $\{x_n\}$ is a sequence in $\mathfrak{A}_0$ converging to a, we get

$$(a \bullet b)^* = (a \blacktriangleleft b)^* = (\overline{R}_b a)^* = \lim_{n\to\infty}(x_n b)^* = \lim_{n\to\infty} b^* x_n^* = L_{b^*}a^* = b^* \blacktriangleright a^* = b^* \bullet a^*.$$

The case $a \in \mathfrak{A}_{\mathrm{b}}, b \in \mathfrak{A}$ is similar. For $a \in \mathfrak{A}, b \in \mathfrak{A}_{\mathrm{b}}$, Lemma 4.1.6 implies that

$$\|a \bullet b\| \leq \|a\|\|b\|_{\mathrm{b}}.$$

It remains only to check the module associativity rules. Let $a \in \mathfrak{A}$ and $b_1, b_2 \in \mathfrak{A}_b$. Then,

$$\begin{aligned} a \bullet (b_1 \bullet b_2) &= a \blacktriangleleft (b_1 \bullet b_2) = a \blacktriangleleft (b_1 \blacktriangleleft b_2) \\ &= \overline{R}_{b_1 \blacktriangleleft b_2} a = (\overline{R}_{b_2} \overline{R}_{b_1}) a \\ &= \overline{R}_{b_2} (\overline{R}_{b_1} a) \\ &= (a \blacktriangleleft b_1) \blacktriangleleft b_2 = (a \bullet b_1) \bullet b_2. \end{aligned}$$

Using (4.1.6), we also have

$$\begin{aligned} (b_1 \bullet a) \bullet b_2 &= (b_1 \blacktriangleright a) \blacktriangleleft b_2 = \overline{R}_{b_2} (b_1 \blacktriangleright a) \\ &= \overline{R}_{b_2} (\overline{L}_{b_1} a) = (\overline{R}_{b_2} \overline{L}_{b_1}) a = (\overline{L}_{b_1} \overline{R}_{b_2}) a \\ &= \overline{L}_{b_1} (\overline{R}_{b_2} a) = \overline{L}_{b_1} (a \blacktriangleleft b_2) \\ &= b_1 \blacktriangleright (a \blacktriangleleft b_2) = b_1 \bullet (a \bullet b_2). \end{aligned}$$

The fact that, $\mathfrak{A}_b[\| \cdot \|_b]$ is a Banach algebra (see Corollary 4.1.9(ii)) completes the proof. □

4.1.3 Strongly Bounded Elements

Throughout this subsection $(\mathfrak{A}[\| \cdot \|], \mathfrak{A}_0)$ will be a Banach quasi *-algebra. In Example 4.1.10 we anticipated the definition of a strongly bounded element, which we want to discuss here in more details [81, 84]. We repeat, for convenience the definition.

Definition 4.1.18 An element $a \in \mathfrak{A}$ is called *strongly bounded* if there exists a sequence $\{x_n\} \subset \mathfrak{A}_0$, such that:

$$\sup_n \|x_n\|_0 < \infty \quad \text{and} \quad \lim_{n \to \infty} \|a - x_n\| = 0.$$

The set of strongly bounded elements is denoted with $\mathfrak{A}_{sb}$.

Proposition 4.1.19 *Each strongly bounded element in $\mathfrak{A}$ is bounded. If $a, b \in \mathfrak{A}_{sb}$, then $a \blacktriangleright b = a \blacktriangleleft b$.*

Proof Let a be strongly bounded and $\{x_n\} \subset \mathfrak{A}_0$ a sequence satisfying the conditions of Definition 4.1.18. Then, for every $x \in \mathfrak{A}_0$, we have

$$\|xa\| = \lim_{n \to \infty} \|x x_n\| \le \sup_n \|x_n\|_0 \, \|x\|, \quad \text{resp.} \quad \|ax\| \le \sup_n \|x_n\|_0 \, \|x\|.$$

Thus, since a is strongly bounded, Definition 4.1.1 is fulfilled, therefore a is bounded. Let $a, b \in \mathfrak{A}_{\rm sb}$ and $\{x_n\}$, $\{y_n\}$ $\|\cdot\|_0$-bounded sequences of elements from $\mathfrak{A}_0$, $\|\cdot\|$-converging, to the elements a and b of $\mathfrak{A}$, respectively. Then, we have

$$a \blacktriangleright b = \overline{L}_a b = \lim_{m\to\infty} L_a y_m = \lim_{m\to\infty}\lim_{n\to\infty} x_n y_m.$$

Similarly,

$$a \blacktriangleleft b = \overline{R}_b a = \lim_{m\to\infty} R_b x_m = \lim_{m\to\infty}\lim_{n\to\infty} x_m y_n.$$

Then,

$$\|a \blacktriangleright b - a \blacktriangleleft b\| = \lim_{m\to\infty}\lim_{n\to\infty} \|x_n y_m - x_m y_n\|.$$

But

$$\begin{aligned}\|x_n y_m - x_m y_n\| &\leq \|x_n y_m - x_m y_m\| + \|x_m y_m - x_m y_n\| \\ &\leq \sup_m \|y_m\|_0 \|x_n - x_m\| + \sup_m \|x_m\|_0 \|y_m - y_n\| \to 0.\end{aligned}$$

Therefore, $a \blacktriangleright b = a \blacktriangleleft b$. □

Example 4.1.20 An example of a Banach quasi *-algebra, where *every bounded element is strongly bounded* is provided by *Hilbert algebras* $(\mathcal{H}, \mathfrak{A}_0)$ (see Example 3.1.8). For this result we refer to [19, Proposition 11.7.9]. Of course, this statement applies also to the situation described at the end of Example 4.1.14.

Proposition 4.1.21 *The space* $\mathfrak{A}_{\rm sb}$*, endowed with the norm* $\|\cdot\|_{\rm b}$*, is a Banach *-algebra.*

Proof It is easily seen that $\mathfrak{A}_{\rm sb}$ is stable under addition and multiplication by scalars. Let $a, b \in \mathfrak{A}_{\rm sb}$. If $\{x_n\}$ and $\{y_n\}$ are $\|\cdot\|_0$-bounded sequences converging, respectively, to a and b in $\mathfrak{A}$, we have

$$\begin{aligned}\|a \bullet b - x_n y_n\| &\leq \|a \bullet b - a y_n\| + \|a y_n - x_m y_n\| + \|x_m y_n - x_n y_n\| \\ &\leq \|a \bullet b - a y_n\| + \sup_n \|y_n\|_0 \|a - x_m\| + \sup_n \|y_n\|_0 \|x_m - x_n\|,\end{aligned}$$

which goes to 0, for $n, m \to \infty$. It is clear that the sequence $\{x_n y_n\}$ is $\|\cdot\|_0$-bounded.

To prove the completeness, it is sufficient to show that $\mathfrak{A}_{\rm sb}$ is closed in $\mathfrak{A}_{\rm b}$. Let $a \in \overline{\mathfrak{A}_{\rm sb}}$. Then, for every $\varepsilon > 0$ there exists $a_\varepsilon \in \mathfrak{A}_{\rm sb}$ such that $\|a - a_\varepsilon\| < \varepsilon/2$. On the other hand, by Definition 4.1.18, we can find a sequence $\{x_n\} \subset \mathfrak{A}_0$, such that

$$\sup_n \|x_n\|_0 < \infty \quad \text{and} \quad \lim_{n\to\infty} \|a_\varepsilon - x_n\| = 0.$$

Hence, there exists $n_\varepsilon \in \mathbb{N}$, in such a way that, for every $n \geq n_\varepsilon$, $\|a_\varepsilon - x_n\| < \varepsilon/2$. Thus, in conclusion, for $n \geq n_\varepsilon$,

$$\|a - x_n\| \leq \|a - a_\varepsilon\| + \|a_\varepsilon - x_n\| < \varepsilon,$$

consequently, $a \in \mathfrak{A}_{\mathrm{sb}}$. □

Of course, if $\mathfrak{A}_{\mathrm{sb}} = \mathfrak{A}_{\mathrm{b}}$, then by Proposition 4.1.19, $(\mathfrak{A}[\|\cdot\|], \mathfrak{A}_0)$ is normal (see discussion before Proposition 4.1.5). In order to provide some sufficient condition for this equality to hold, we study some commutation properties of $(\mathfrak{A}[\|\cdot\|], \mathfrak{A}_0)$.

If $\mathcal{M} \subseteq \mathcal{B}(\mathfrak{A})$, we denote with $\mathcal{M}^c$ the commutant of $\mathcal{M}$, that is

$$\mathcal{M}^c = \{X \in \mathcal{B}(\mathfrak{A}) : XY = YX, \ \forall \, Y \in \mathcal{M}\}.$$

Let us consider the sets

$$\mathcal{L} = \{R_x : x \in \mathfrak{A}_0\}^c, \qquad \mathcal{R} = \{L_x : x \in \mathfrak{A}_0\}^c.$$

It is easily seen that both $\mathcal{L}$ and $\mathcal{R}$ are subalgebras of $\mathcal{B}(\mathfrak{A})$.

For every $x, y \in \mathfrak{A}_0$ one has $L_x R_y = R_y L_x$. Hence, $\{L_x : x \in \mathfrak{A}_0\} \subset \mathcal{L}$ and thus $\mathcal{L}^c \subseteq \mathcal{R}$. We shall prove that the last inclusion is, in fact, equality (see Proposition 4.1.26).

Lemma 4.1.22 *The following statements hold:*

(i) *if a is left bounded and $S \in \mathcal{L}$, then Sa is left bounded and $\overline{L}_{Sa} = S\overline{L}_a$;*
(ii) *if a is right bounded and $T \in \mathcal{R}$, then Ta is right bounded and $\overline{R}_{Ta} = T\overline{R}_a$.*

Proof We prove only (i); (ii) is similarly proved. For every $x \in \mathfrak{A}_0$ one has

$$L_{Sa}x = (Sa)x = R_x Sa = (R_x S)a = SR_x a = Sax.$$

This implies that

$$\|(Sa)x\| = \|Sax\| \leq \|S\| \, \|ax\| \leq \|S\| \, \gamma_a \, \|x\| \leq \gamma \|x\|,$$

for all $x \in \mathfrak{A}_0$ and $\gamma = \|S\|\gamma_a$ (see Definition 4.1.1). □

Lemma 4.1.23 *If a is left, respectively, right bounded, then $\overline{L}_a \in \mathcal{L}$, respectively, $\overline{R}_a \in \mathcal{R}$.*

Proof Using the module associativity one has, for every $x \in \mathfrak{A}_0$,

$$\overline{L}_a R_x y = L_a yx = a(yx) = R_x L_a y, \quad \forall \, y \in \mathfrak{A}_0.$$

Since $\mathfrak{A}_0$ is dense in $\mathfrak{A}$, the equality extends to $\mathfrak{A}$. □

Lemma 4.1.24 *Let $T \in \mathcal{R}$. Then, $T\overline{L}_a = \overline{L}_a T$, for every left bounded element $a \in \mathfrak{A}$.*

Proof Let a be left bounded and $\{x_n\}$ a sequence in $\mathfrak{A}_0$ converging to a. Then, for every $y \in \mathfrak{A}_0$, we have

$$\begin{aligned} T\overline{L}_a y = Tay &= T(\lim_{n\to\infty} x_n y) \\ &= \lim_{n\to\infty} Tx_n y = \lim_{n\to\infty} TL_{x_n} y = \lim_{n\to\infty} L_{x_n} Ty. \end{aligned}$$

On the other hand, since each $y \in \mathfrak{A}_0$ is right bounded (see discussion at the beginning of Sect. 3.3) and $T \in \mathcal{R}$, we can use (ii) of Lemma 4.1.22 and obtain

$$\overline{L}_a Ty = \overline{R}_{Ty} a = \lim_{n\to\infty} R_{Ty} x_n = \lim_{n\to\infty} L_{x_n} Ty.$$

□

In particular, we have

Corollary 4.1.25 *If a is left bounded and b is right bounded, then*

$$\overline{L}_a \overline{R}_b = \overline{R}_b \overline{L}_a.$$

Proof From Lemma 4.1.23, one has that $\overline{R}_b \in \mathcal{R}$; so the statement follows from Lemma 4.1.24. □

Proposition 4.1.26 *If either $(\mathfrak{A}[\|\cdot\|], \mathfrak{A}_0)$ has a unit e or $\mathfrak{A}_0^2$ is total in $\mathfrak{A}$, we have*

$$\mathcal{L}^c = \mathcal{R}, \quad \mathcal{R}^c = \mathcal{L}.$$

Proof It is sufficient to prove that each $S \in \mathcal{L}$ commutes with each $T \in \mathcal{R}$. This implies, in fact, that $\mathcal{L} \subseteq \mathcal{R}^c$. But, as noticed before, the converse inclusion always holds and so we obtain the equality. Let $x, y \in \mathfrak{A}_0$. Taking into account that each $x \in \mathfrak{A}_0$ is left bounded, Sx is left bounded (Lemma 4.1.22) and $\overline{L}_{Sx} = SL_x$; yet Lemma 4.1.24 can be applied to obtain

$$STxy = STL_x y = SL_x Ty = L_{Sx} Ty = TL_{Sx} y = TSxy.$$

Since either $\mathfrak{A}_0^2 = \mathfrak{A}_0$ or $\mathfrak{A}_0^2$ is total in $\mathfrak{A}$, the claim follows. □

The previous Proposition implies that $\mathcal{L}^{cc} = \mathcal{L}$ and $\mathcal{R}^{cc} = \mathcal{R}$.

Proposition 4.1.27 *Assume that $(\mathfrak{A}[\|\cdot\|], \mathfrak{A}_0)$ has a unit e. Then,*

$$\mathcal{L} = \left\{\overline{L}_a : a \in \mathfrak{A}_{\mathrm{lb}}\right\}, \quad \mathcal{R} = \left\{\overline{R}_a : a \in \mathfrak{A}_{\mathrm{rb}}\right\}.$$

Proof We know that $\overline{L}_a \in \mathcal{L}$, for every $a \in \mathfrak{A}_{\mathrm{lb}}$. Now assume that $X \in \mathcal{L}$ and put $a = Xe$. We prove that $X = L_{Xe}$. Indeed, for every $x, y \in \mathfrak{A}_0$, we obtain

$$Xxy = XR_y x = (XR_y)x = (R_y X)x = (Xx)y.$$

For $x = e$ we have $Xy = (Xe)y$, for every $y \in \mathfrak{A}_0$. Hence,

$$\|(Xe)y\| = \|Xy\| \leq \|X\| \, \|y\|, \quad \forall\, y \in \mathfrak{A}_0.$$

That is, Xe is left bounded. The equality concerning $\mathcal{R}$ is similarly proved. □

Before going forth we need some further notation and some definitions. In what follows, if $E[\|\cdot\|_E]$ is a normed space we denote with $\mathcal{U}(E)$ its unit ball, i.e., $\mathcal{U}(E) = \{a \in E : \|a\|_E \leq 1\}$. The strong topology in the algebra $\mathcal{B}(E)$ of bounded operators on E is defined by the seminorms

$$S \in \mathcal{B}(E) \to \|Sa\|_E \in \mathbb{R}_+, \ a \in E.$$

Proposition 4.1.28 *Let* $(\mathfrak{A}[\|\cdot\|], \mathfrak{A}_0)$ *have a unit* e. *Consider the following statements:*

(i) $\{L_x : x \in \mathfrak{A}_0\} \cap \mathcal{U}(\mathcal{B}(\mathfrak{A}))$ *is strongly dense in* $\mathcal{L} \cap \mathcal{U}(\mathcal{B}(\mathfrak{A}))$;
(ii) $\{R_x : x \in \mathfrak{A}_0\} \cap \mathcal{U}(\mathcal{B}(\mathfrak{A}))$ *is strongly dense in* $\mathcal{R} \cap \mathcal{U}(\mathcal{B}(\mathfrak{A}))$;
(iii) $\mathfrak{A}_{\mathrm{sb}} = \mathfrak{A}_{\mathrm{lb}} = \mathfrak{A}_{\mathrm{rb}} = \mathfrak{A}_{\mathrm{b}}$.

Then, (i) $\Leftrightarrow$ *(ii)* $\Rightarrow$ *(iii)*

Proof (i) $\Rightarrow$ (ii) By Proposition 4.1.27, if $S \in \mathcal{R} \cap \mathcal{U}(\mathcal{B}(\mathfrak{A}))$, then $S = \overline{R}_a$, for some $a \in \mathfrak{A}_{\mathrm{rb}}$, with $\|a\|_{\mathrm{rb}} \leq 1$. Since $a^* \in \mathfrak{A}_{\mathrm{lb}}$, then $\overline{L}_{a^*} \in \mathcal{L} \cap \mathcal{U}(\mathcal{B}(\mathfrak{A}))$. Therefore, there exists a sequence $\{x_n\} \subset \mathfrak{A}_0$, with $\|x_n\|_0 \leq 1$, such that

$$\|\overline{L}_{a^*}b - L_{x_n}b\| = \|a^* \blacktriangleright b - x_n b\| \underset{n\to\infty}{\longrightarrow} 0, \quad \forall\, b \in \mathfrak{A}.$$

The continuity of the involution in $\mathfrak{A}$ implies that

$$\|b^* \blacktriangleleft a - b^* x_n^*\| = \|\overline{R}_a b^* - R_{x_n^*} b^*\| \to 0, \quad \forall\, b \in \mathfrak{A}.$$

(ii) $\Rightarrow$ (i) can be proven in the very same way.

(i) $\Rightarrow$ (iii) Let $a \in \mathfrak{A}_{\mathrm{lb}}$. Put $b = a/\|a\|_{\mathrm{lb}}$. Then, $\overline{L}_b \in \mathcal{L} \cap \mathcal{U}(\mathcal{B}(\mathfrak{A}))$. Therefore, there exists a sequence $\{x_n\} \subset \mathfrak{A}_0$, with $\|x_n\|_0 \leq 1$, such that

$$\|\overline{L}_b c - L_{x_n} c\| \to 0, \quad \forall\, c \in \mathfrak{A}.$$

For $c = e$, $\|b - x_n\| \to 0$. Thus, the sequence $\{\|a\|_{\mathrm{lb}} x_n\}$ converges to a and $\|x_n\|_0 \leq \|a\|_{\mathrm{lb}}$, for every $n \in \mathbb{N}$. Therefore, $a \in \mathfrak{A}_{\mathrm{sb}}$. □

Remark 4.1.29 We note that if $(\mathfrak{A}[\|\cdot\|], \mathfrak{A}_0)$ has no unit, it can be proved, with a proof similar to that of Proposition 11.7.9 in [19], that (i) *and* (ii) imply (iii), but (i) *and* (ii) are no more equivalent, in general.

To conclude, we observe that (i) or (ii) closely reminds of Kaplansky's density theorem and this is the reason why, for the Banach quasi *-algebra $(\mathcal{H}, \mathfrak{A}_0)$ constructed from a Hilbert algebra $\mathfrak{A}_0$, as in Example 3.1.8, the set of bounded and strongly bounded elements coincide. In fact, in that case, $\mathcal{L}$ and $\mathcal{R}$ are both von Neumann algebras and Kaplansky's density theorem can be applied.

4.2 The Spectrum of an Element

In this section we will discuss how the notion of spectrum of an element, which is fundamental in the theory of Banach algebras, can be extended to Banach quasi *-algebras.

4.2.1 The Inverse of an Element

We first come back to a closer analysis of the linear maps L_a, R_a defined in (3.1.12). In general they are *unbounded* maps in the Banach space $\mathfrak{A}$. It is natural to deal with the problem of inverting an element $a \in \mathfrak{A}$ first by inverting them. As customary in the theory of unbounded operators, we will look for *bounded* inverses.

Definition 4.2.1 Let $(\mathfrak{A}[\|\cdot\|], \mathfrak{A}_0)$ be a Banach quasi *-algebra with unit e. An element $a \in \mathfrak{A}$ is called *closable* if the linear maps

$$L_a : x \in \mathfrak{A}_0 \to ax \in \mathfrak{A}, \qquad R_a : x \in \mathfrak{A}_0 \to xa \in \mathfrak{A}$$

are closable in $\mathfrak{A}$.

Remark 4.2.2 A Banach quasi *-algebra, such that every $a \in \mathfrak{A}$ is closable in $\mathfrak{A}$, was called *fully closable* in [49].

As we have done for bounded elements (see discussion before and after Remark 4.1.3), if $a \in \mathfrak{A}$ we denote by $\overline{L}_a$ the closure of L_a, i.e., the linear operator defined on,

$$D(\overline{L}_a) := \big\{b \in \mathfrak{A} : \exists\, \{x_n\} \subset \mathfrak{A}_0 : \|b - x_n\| \to 0 \text{ and } \{ax_n\} \text{ is Cauchy}\big\}.$$

by

$$\overline{L}_a b = \lim_{n\to\infty} ax_n.$$

Similarly, $\overline{R}_a$ will denote the closure of R_a and $D(\overline{R}_a)$ its domain. The definitions are obvious modifications of the previous ones.

Lemma 4.2.3 *Let a be a closable element of* $(\mathfrak{A}[\|\cdot\|], \mathfrak{A}_0)$. *The following statements hold:*

(i) *if* $y \in D(\overline{L}_a)$ *and* $x \in \mathfrak{A}_0$, *then* $yx \in D(\overline{L}_a)$ *and*

$$\overline{L}_a yx = (\overline{L}_a y)x;$$

(ii) *if* $y \in D(\overline{R}_a)$ *and* $x \in \mathfrak{A}_0$, *then* $xy \in D(\overline{R}_a)$ *and*

$$\overline{R}_a xy = x(\overline{R}_a y);$$

(iii) $D(\overline{R}_a) = D(\overline{L}_{a^*})^*$ *and* $\overline{L}_{a^*} y^* = (\overline{R}_a y)^*$, *for every* $y \in D(\overline{R}_a)$;

(iv) *if* $\overline{L}_a$ *has an everywhere defined inverse, then*

$$\big((\overline{L}_a)^{-1} c\big)x = (\overline{L}_a)^{-1} cx, \quad \forall c \in \mathfrak{A},\ x \in \mathfrak{A}_0;$$

(v) *if* $\overline{R}_a$ *has an everywhere defined inverse, then*

$$x\big((\overline{R}_a)^{-1} c\big) = (\overline{R}_a)^{-1} xc, \quad \forall c \in \mathfrak{A},\ x \in \mathfrak{A}_0.$$

Proof We begin with proving (i). Let $y \in D(\overline{L}_a)$. Then, there exists a sequence $\{y_n\} \subset \mathfrak{A}_0$ such that $\|y - y_n\| \to 0$ and $\{ay_n\}$ is Cauchy. Clearly, if $x \in \mathfrak{A}_0$, $\|y_n x - yx\| \to 0$ and, since

$$\|a(y_n x) - a(y_m x)\| = \|a(y_n - y_m)x\| \le \|a(y_n - y_m)\| \|x\|_0,$$

the sequence $\{a(y_n x)\}$ is Cauchy too. Thus, $yx \in D(\overline{L}_a)$ and

$$\overline{L}_a(yx) = \lim_{n\to\infty} a(y_n x) = \lim_{n\to\infty} (ay_n)x = (\overline{L}_a y)x.$$

The proof of (ii) is similar and we omit it.

As for (iii), let $y \in D(\overline{R}_a)$. Then there exists a sequence $\{y_n\} \subset \mathfrak{A}_0$ such that $y = \lim_{n\to\infty} y_n$ and $\lim_{n\to\infty} y_n a = z =: \overline{R}_a y$. This implies that $y^* = \lim_{n\to\infty} y_n^*$ and $\lim_{n\to\infty} a^* y_n^* = z^* =: \overline{L}_{a^*} y^*$. Hence $y^* \in D(\overline{L}_{a^*})$ and $\overline{L}_{a^*} y^* = (\overline{R}_a y)^*$. The inclusion $D(\overline{L}_{a^*})^* \subset D(\overline{R}_a)$ can be proved in analogous way.

Now we prove (iv). If $c \in \mathfrak{A}$, there exists $z \in D(\overline{L}_a)$ such that $(\overline{L}_a)^{-1} c = z$. Then by (i), we get that, for every $x \in \mathfrak{A}_0$, $\big((\overline{L}_a)^{-1} c\big)x = zx \in D(\overline{L}_a)$. On the other hand, since $zx \in D(\overline{L}_a)$ there exists a unique $w \in \mathfrak{A}$, such that $zx = ((\overline{L}_a)^{-1} c)x = (\overline{L}_a)^{-1} w$. Now, by applying $\overline{L}_a$ to both sides of this equality we obtain

$$\overline{L}_a(zx) = \overline{L}_a\big((\overline{L}_a)^{-1} c\big)x = \overline{L}_a(\overline{L}_a)^{-1} cx = cx.$$

Hence,

$$((\overline{L}_a)^{-1}c)x = (\overline{L}_a)^{-1}cx, \quad \forall\, c \in \mathfrak{A},\ x \in \mathfrak{A}_0.$$

The proof of (v) is similar and we omit it. □

Definition 4.2.4 Let $(\mathfrak{A}[\|\cdot\|], \mathfrak{A}_0)$ be a Banach quasi *-algebra with unit e and $a \in \mathfrak{A}$ a closable element. We say that a has a *bounded inverse* if there exists $b \in \mathfrak{A}_b \cap D(\overline{L}_a) \cap D(\overline{R}_a)$, such that $\overline{R}_b a = \overline{L}_b a = e$. If $(\mathfrak{A}[\|\cdot\|], \mathfrak{A}_0)$ is normal, then this element b, if any, is unique by (4.1.6). In this case, we denote the bounded inverse of a by a^{-1}.

For a Banach (*-)algebra, the existence of the inverse of an element a can be characterized through the invertibility of the corresponding maps L_a, R_a.

A similar result holds also in our case.

Proposition 4.2.5 *Let $(\mathfrak{A}[\|\cdot\|], \mathfrak{A}_0)$ be a normal Banach quasi *-algebra with unit e and $a \in \mathfrak{A}$ a closable element. The following statements are equivalent:*

(i) *the element a has a bounded inverse;*
(ii) *both $\overline{L}_a$ and $\overline{R}_a$ possess everywhere defined* (hence, bounded) *inverses.*

Proof (i) ⇒ (ii) Suppose that there exists $b \in \mathfrak{A}_b \cap D(\overline{L}_a) \cap D(\overline{R}_a)$, such that $\overline{R}_b a = \overline{L}_b a = e$. We prove that $\overline{L}_a$ is injective and surjective. Let $y \in D(\overline{L}_a)$ with $\overline{L}_a y = 0$. By definition there exists a sequence $\{y_n\} \subset \mathfrak{A}_0$ such that $y_n \to y$ and $ay_n \to \overline{L}_a y = 0$. Since b is bounded, we have $\overline{L}_b a y_n \to 0$. By (i) of Lemma 4.2.3, we get

$$y = \lim_{n\to\infty} y_n = \lim_{n\to\infty} (\overline{L}_b a) y_n = \lim_{n\to\infty} \overline{L}_b a y_n = 0.$$

Therefore $\overline{L}_a$ is injective.

Let now $c \in \mathfrak{A}$. Then, c is the limit of a sequence $\{z_n\} \subset \mathfrak{A}_0$. Let $y_n := bz_n$ and $y = \lim_{n\to\infty} y_n$. Again by (i) of Lemma 4.2.3 it follows that $bz_n \in D(\overline{L}_a)$, for every $n \in \mathbb{N}$. Taking into account the assumption of normality we have $\overline{L}_a b z_n = (\overline{L}_a b) z_n = (\overline{R}_b a) z_n = z_n \to c$. Hence, $y \in D(\overline{L}_a)$ and $\overline{L}_a y = c$. This proves that $\overline{L}_a$ is surjective. Therefore $(\overline{L}_a)^{-1}$ exists, it is everywhere defined and closed, hence bounded.

With a similar argument, taking into account that $b \in D(\overline{R}_a)$, one proves that $\overline{R}_a$ has a bounded inverse.

(ii) ⇒ (i) Let us now assume that $\overline{L}_a$ and $\overline{R}_a$ have bounded inverses $(\overline{L}_a)^{-1}$ and $(\overline{R}_a)^{-1}$, respectively. Let us define $b = (\overline{L}_a)^{-1}e$ and $b' = (\overline{R}_a)^{-1}e$. Then $b \in D(\overline{L}_a)$ and $b' \in D(\overline{R}_a)$ and, clearly, $b' = (\overline{R}_a)^{-1}\overline{L}_a b$. The element b is left bounded and b' is right bounded,

$$\|bx\| = \|((\overline{L}_a)^{-1}e)x\| = \|(\overline{L}_a)^{-1}x\| \le \|(\overline{L}_a)^{-1}\|\,\|x\|, \quad \forall\, x \in \mathfrak{A}_0.$$

$$\|xb'\| = \|x((\overline{R}_a)^{-1}e)\| = \|(\overline{R}_a)^{-1}x\| \le \|(\overline{R}_a)^{-1}\|\,\|x\|, \quad \forall\, x \in \mathfrak{A}_0.$$

Let $\{x_n\} \subset \mathfrak{A}_0$ be a sequence converging to a.Then, by (v) of Lemma 4.2.3, we obtain

$$\overline{L_b}a = \lim_{n\to\infty} \left((\overline{L_a})^{-1}e\right)x_n = \lim_{n\to\infty} (\overline{L_a})^{-1}x_n = (\overline{L_a})^{-1}a = (\overline{L_a})^{-1}L_a e = e.$$

Similarly one shows that $\overline{R_{b'}}a = e$. Using Corollary 4.1.25, we get

$$b' = \overline{R_{b'}}\,\overline{L_b}a = \overline{L_b}\,\overline{R_{b'}}a = b.$$

Hence, $b \in \mathfrak{A}_\mathrm{b} \cap D(\overline{L_a}) \cap D(\overline{R_a})$ and $\overline{L_b}a = \overline{R_b}a = e$. This completes the proof. □

We denote by $\mathfrak{S}_\mathrm{b}(\mathfrak{A})$ the set of closable elements of $\mathfrak{A}$ having a bounded inverse.

Remark 4.2.6 In contrast with the case of Banach algebras, $\mathfrak{S}_\mathrm{b}(\mathfrak{A})$ need not be an open subset of $\mathfrak{A}$. For instance, let us consider again the Banach quasi *-algebra $(L^p(I), C(I))$ with $I = [0, 1]$. In this case, as seen in Example 4.1.12, $(L^p(I))_\mathrm{b} = L^\infty(I)$ and the norm $\|\cdot\|_\mathrm{b}$ is exactly the L^∞-norm. Every neighborhood of the unit function $u(x) = 1$, for every $x \in [0, 1]$, contains noninvertible elements: the function $u_\varepsilon(x) = 0$, for $x \in [0, \frac{\varepsilon}{2})$ and $u_\varepsilon(x) = 1$, for $x \in (\frac{\varepsilon}{2}, 1]$ is in the ball $B(u, \varepsilon)$ of $L^p(I)$, but it is not invertible, since it is zero on a set of positive measure.

Proposition 4.2.7 *Let $(\mathfrak{A}[\|\cdot\|], \mathfrak{A}_0)$ be a normal Banach quasi *-algebra with unit e and $a \in \mathfrak{A}$ an element having a bounded inverse a^{-1}. If $b \in \mathfrak{A}$ and $h := a - b \in \mathfrak{A}_\mathrm{b}$, with $\|h\|_\mathrm{b} < \|a^{-1}\|_\mathrm{b}^{-1}$, then b has a bounded inverse.*

Proof We have $b = a \bullet (e - a^{-1} \bullet h)$. By the assumptions, the element $e - a^{-1} \bullet h$ is invertible in $\mathfrak{A}_\mathrm{b}$, because $\|a^{-1} \bullet h\|_\mathrm{b} < 1$ and $\mathfrak{A}_\mathrm{b}[\|\cdot\|_\mathrm{b}]$ is a Banach algebra (Corollary 4.1.9(ii)). Hence, b has a bounded inverse. □

4.2.2 The Spectrum

Let $(\mathfrak{A}[\|\cdot\|], \mathfrak{A}_0)$ be a Banach quasi *-algebra with unit e.

Definition 4.2.8 The *resolvent* $\rho(a)$ of $a \in \mathfrak{A}$ is the set

$$\rho(a) := \left\{\lambda \in \mathbb{C} : a - \lambda e \in \mathfrak{S}_\mathrm{b}(\mathfrak{A})\right\}.$$

The set $\sigma(a) := \mathbb{C} \setminus \rho(a)$ is called the *spectrum* of a.

Proposition 4.2.9 *Let $(\mathfrak{A}[\|\cdot\|], \mathfrak{A}_0)$ be a normal Banach quasi *-algebra with unit. Let $a \in \mathfrak{A}$, then the following statements hold:*

(i) *the resolvent $\rho(a)$ is an open subset of the complex plane;*
(ii) *the resolvent function $R_\lambda(a) := (a - \lambda e)^{-1} \in \mathfrak{A}_\mathrm{b}$, $\lambda \in \rho(a)$, is $\|\cdot\|_\mathrm{b}$-analytic on each connected component of $\rho(a)$;*

(iii) *for any two points* $\lambda, \mu \in \rho(a)$, $R_\lambda(a)$ *and* $R_\mu(a)$ *commute and*

$$R_\lambda(a) - R_\mu(a) = (\mu - \lambda) R_\mu(a) \bullet R_\lambda(a).$$

Proof

(i) Let $\lambda_0 \in \rho(a)$ and $\lambda \in \mathbb{C}$, such that $|\lambda - \lambda_0| < \|R_{\lambda_0}(a)\|_{\mathrm{b}}^{-1}$. Then, the series

$$\sum_{n=1}^{\infty} (\lambda_0 - \lambda)^n R_{\lambda_0}(a)^n$$

converges to an element $S_{\lambda,a}$ in $\mathfrak{A}_{\mathrm{b}}$, with respect to $\| \cdot \|_{\mathrm{b}}$.

Let now $T_{\lambda,a} \equiv R_{\lambda_0}(a)(e + S_{\lambda,a})$. It is easily checked, using the $\| \cdot \|$-convergence for the product $T_{\lambda,a}(a - \lambda e)$, that $T_{\lambda,a}$ is a bounded inverse of $a - \lambda e$. Hence, $\lambda \in \rho(a)$.

(ii) It follows immediately from the proof of (i).

The proof of (iii) is straightforward. □

The classical argument (see Theorem A.3.14) based on Liouville's theorem can be applied to prove the following

Proposition 4.2.10 *Let* $a \in \mathfrak{A}$. *Then,* $\sigma(a)$ *is non-empty.*

Example 4.2.11 Let us consider again the Banach quasi *-algebra $(L^p(I), C(I))$ and let $f \in L^p(I)$. Then, it is easily seen that the spectrum $\sigma(f)$ of f coincides with its essential range; that is the set of all $\lambda \in \mathbb{C}$, such that the set

$$\{x \in I : |f(x) - \lambda| < \varepsilon\}$$

has positive Lebesgue measure, for every $\varepsilon > 0$.

Definition 4.2.12 Let $a \in \mathfrak{A}$. The non-negative number $r(a) = \sup\{|\lambda|, \lambda \in \sigma(a)\}$ is called *spectral radius* of a.

Remark 4.2.13 Of course, if $a \in \mathfrak{A}_{\mathrm{b}}$, then $\sigma(a)$ is the same set as that obtained regarding it as an element of the Banach *-algebra $\mathfrak{A}_{\mathrm{b}}$. For an arbitrary element $a \in \mathfrak{A}$, $\sigma(a)$, which is a closed set, could be an unbounded subset of $\mathbb{C}$. The next proposition shows that, if $a \in \mathfrak{A} \setminus \mathfrak{A}_{\mathrm{b}}$, then $\sigma(a)$ is necessarily unbounded.

Proposition 4.2.14 *Let* $a \in \mathfrak{A}$. *Then,* $r(a) < \infty$, *if and only if,* $a \in \mathfrak{A}_{\mathrm{b}}$.

Proof The "if" part has been discussed in the previous remark.

Assume, now that $r(a) < \infty$. Then, the function $\lambda \to (a - \lambda e)^{-1}$ is $\|\cdot\|_b$-analytic in the region $|\lambda| > r(a)$. Therefore, it has there a $\|\cdot\|_b$-convergent Laurent expansion

$$(a - \lambda e)^{-1} = \sum_{k=1}^{\infty} \frac{x_k}{\lambda^k}, \quad |\lambda| > r(a),$$

with $x_k \in \mathfrak{A}_b$, for each $k \in \mathbb{N}$. As usually,

$$x_k = \frac{1}{2\pi i} \int_\gamma \frac{(a - \lambda e)^{-1}}{\lambda^{-k+1}} d\lambda, \quad k \in \mathbb{N},$$

where γ is a circle centered in 0, with radius $R > r(a)$. The integral on the right hand side converges with respect to $\|\cdot\|_b$. The $\|\cdot\|$-continuity of the multiplication implies, as in the ordinary case, that

$$ax_k = \frac{1}{2\pi i} \int_\gamma \frac{a(a - \lambda e)^{-1}}{\lambda^{-k+1}} d\lambda = \frac{1}{2\pi i} \int_\gamma \frac{(a - \lambda e)^{-1}}{\lambda^{-k}} d\lambda = x_{k+1}.$$

In particular, using Cauchy's integral formula, we find $ax_1 = a$. This implies that $a \in \mathfrak{A}_b$. □

Remark 4.2.15 If $\lambda \in \rho(a)$ then all powers $(a - \lambda e)^{-n}$ do exist in $\mathfrak{A}_b$, for every $n \in \mathbb{N}$. This does not imply the existence of $(a - \lambda e)^n$, for $n > 1$. As an example, let us consider the Banach quasi *-algebra $(L^2(I), C(I))$, where $I = [0, 1]$ (cf. Example 3.1.6). The function $v(x) = x^{-\frac{1}{4}}$ (we put $v(0) = 0$) is in $L^2(I)$; obviously, $0 \in \rho(v)$, since $v^{-1}(x) = x^{\frac{1}{4}} \in L^\infty(I)$. We have $v^{-n}(x) = x^{\frac{n}{4}} \in L^2(I)$, for all $n \in \mathbb{N}$, but $v^2(x) = x^{-\frac{1}{2}} \notin L^2(I)$.

4.2.3 The *-Semisimple Case

Proposition 4.2.16 *Let* $(\mathfrak{A}[\|\cdot\|], \mathfrak{A}_0)$ *be a *-semisimple Banach quasi *-algebra. Then,* $(\mathfrak{A}[\|\cdot\|], \mathfrak{A}_0)$ *is normal.*

Proof Assume that $a, b \in \mathfrak{A}_b$. Then, for every $\varphi \in \mathcal{S}_{\mathfrak{A}_0}(\mathfrak{A})$ and $z \in \mathfrak{A}_0$, we have

$$\varphi\big((a \blacktriangleright b)z, z\big) = \varphi\big((\overline{L}_a b)z, z\big) = \varphi\big((\overline{R}_b a)z, z\big) = \varphi\big((a \blacktriangleleft b)z, z\big).$$

Therefore,

$$\varphi\big((a \blacktriangleright b - a \blacktriangleleft b)z, z\big) = 0, \quad \forall\, \varphi \in \mathcal{S}_{\mathfrak{A}_0}(\mathfrak{A}),\ z \in \mathfrak{A}_0.$$

Hence, by *-semisimplicity $a \blacktriangleright b = a \blacktriangleleft b$, for all a, b in $\mathfrak{A}_b$. This completes the proof. □

Definition 4.2.17 A Banach quasi *-algebra $(\mathfrak{A}[\|\cdot\|], \mathfrak{A}_0)$ is called *quasi regular*, respectively *regular*, if it is *-semisimple and

$$\|a\| = \max\{\mathfrak{p}(a), \mathfrak{p}(a^*)\}, \ \text{resp.,} \ \mathfrak{p}(a) = \|a\|, \quad \forall\, a \in \mathfrak{A}.$$

We recall that the seminorm $\mathfrak{p}$ was defined in (3.1.9) by

$$\mathfrak{p}(a) := \sup_{\varphi \in \mathcal{S}_{\mathfrak{A}_0}(\mathfrak{A})} \varphi(a, a)^{1/2}, \ a \in \mathfrak{A}.$$

It is clear that every regular Banach quasi *-algebra is quasi regular. Moreover, if $(\mathfrak{A}[\|\cdot\|], \mathfrak{A}_0)$ is commutative and quasi regular, then it is regular. In fact, in this case, if $\varphi \in \mathcal{S}_{\mathfrak{A}_0}(\mathfrak{A})$ then $\varphi^* \in \mathcal{S}_{\mathfrak{A}_0}(\mathfrak{A})$, where $\varphi^*(a, b) = \overline{\varphi(b^*, a^*)}$, $a, b \in \mathfrak{A}$. This easily implies that $\mathfrak{p}(a) = \mathfrak{p}(a^*)$, for every $a \in \mathfrak{A}$. Examples will be given after Proposition 4.2.24.

Remark 4.2.18 Taking into account Proposition 3.1.18, if $(\mathfrak{A}[\|\cdot\|], \mathfrak{A}_0)$ is a regular Banach quasi *-algebra, then for each element $a \in \mathfrak{A}$, there exists $\varphi \in \mathcal{S}_{\mathfrak{A}_0}(\mathfrak{A})$, such that $\varphi(a, a) = \|a\|^2$, a property, which is closely reminiscent of the behaviour of unital C*-algebras (see, Proposition A.5.4).

It is not expected that every *-semisimple Banach quasi *-algebra $(\mathfrak{A}[\|\cdot\|], \mathfrak{A}_0)$ is regular, because this is not true in the case, where $\mathfrak{A} = \mathfrak{A}_0$ and $\|\cdot\| = \|\cdot\|_0$, that is when $\mathfrak{A}[\|\cdot\|]$ is a *-semisimple Banach *-algebra. In that situation, the C*-seminorm p defined in Proposition A.5.1, is a norm which, in general, does not coincide with the initial norm $\|\cdot\|_0$. Every *-semisimple Banach *-algebra is, in fact, an A*-algebra, but not necessarily a C*-algebra. A well-known example is provided by the convolution algebra $L^1(\mathbb{R})$, which is an A*-algebra but not a C*-algebra (see [21, Corollary (4.6.10); Example A.3.1]).

Our next goal is to show that, for regular Banach quasi *-algebras the set of elements having finite spectral radius can also be described in terms of the seminorm $\mathfrak{q}$ defined in (3.1.13).

We begin with the following

Proposition 4.2.19 *Let $(\mathfrak{A}[\|\cdot\|], \mathfrak{A}_0)$ be a *-semisimple Banach quasi *-algebra. Then, for every $a \in \mathfrak{A}$, the maps*

$$L_a : x \in \mathfrak{A}_0 \to ax \in \mathfrak{A}, \qquad R_a : x \in \mathfrak{A}_0 \to xa \in \mathfrak{A}$$

are closable in $\mathfrak{A}$.

Proof Let $a \in \mathfrak{A}$ and $\{x_n\}$ a sequence in $\mathfrak{A}_0$, $\|\cdot\|$-converging to zero and such that $ax_n \to b$, with respect to $\|\cdot\|$. Then, if $\varphi \in \mathcal{S}_{\mathfrak{A}_0}(\mathfrak{A})$ and $y_1, y_2 \in \mathfrak{A}_0$, we get

$$\begin{aligned} |\varphi(by_1, y_2)| &\le |\varphi\big((b - ax_n)y_1, y_2\big)| + |\varphi(x_n y_1, a^* y_2)| \\ &\le \|b - ax_n\|\, \|y_1\|_0 \|y_2\|_0 + \|x_n\|\, \|y_1\|_0 \|a^* y_2\| \to 0. \end{aligned}$$

Therefore, $\varphi(by_1, y_2) = 0$, for every $\varphi \in \mathcal{S}_{\mathfrak{A}_0}(\mathfrak{A})$ and for every $y_1, y_2 \in \mathfrak{A}_0$. From this we obtain that $\varphi(b, b) = 0$, for every $\varphi \in \mathcal{S}_{\mathfrak{A}_0}(\mathfrak{A})$. Therefore, $b = 0$, i.e., L_a is closable (see Proposition B.4.5 in Appendix B). The proof for R_a is similar. □

As we have seen, it is possible to introduce in a *-semisimple Banach quasi *-algebra $(\mathfrak{A}[\|\cdot\|], \mathfrak{A}_0)$ a weak multiplication □ which makes it into a partial *-algebra. As a consequence of this, for every $a \in \mathfrak{A}$, the operators L_a, R_a have closed extensions (see Proposition 3.1.34) defined, respectively, in $R_w(a)$, $L_w(a)$ and denoted, respectively by $\widehat{L}_a$, $\widehat{R}_a$. It is clear that (for the notation of the left parts of the inclusions, see beginning of Sect. 4.1.1) $\overline{L}_a \subset \widehat{L}_a$ and $\overline{R}_a \subset \widehat{R}_a$.

The following proposition gives an interesting variant of the associativity law.

Proposition 4.2.20 *Let $(\mathfrak{A}[\|\cdot\|], \mathfrak{A}_0)$ be a *-semisimple Banach quasi *-algebra. Suppose that a, b, c are elements of $\mathfrak{A}$, such that $(a \square b) \square c$ is well defined. If $c \in D(\overline{L}_b)$, then $a \square (b \square c)$ is well defined and*

$$(a \square b) \square c = a \square (b \square c).$$

Proof Since $c \in D(\overline{L}_b)$, there exists a sequence $\{z_n\} \subset \mathfrak{A}_0$ such that $\|c - z_n\| \to 0$ and $\|\overline{L}_b c - b z_n\| \to 0$ (see discussion before Lemma 4.2.3). Then, for every $\varphi \in \mathcal{S}_{\mathfrak{A}_0}(\mathfrak{A})$ and $x, y \in \mathfrak{A}_0$, we have

$$\begin{aligned}
\varphi\big(((a \square b) \square c)x, y\big) &= \varphi\big(cx, (a \square b)^* y\big) = \varphi\big(cx, (b^* \square a^*)y\big) \\
&= \lim_{n\to\infty} \varphi(z_n x, (b^* \square a^*)y) = \lim_{n\to\infty} \varphi(b(z_n x), a^* y) \\
&= \lim_{n\to\infty} \varphi\big((b z_n)x, a^* y\big) = \varphi\big((\overline{L}_b c)x, a^* y\big) \\
&= \varphi\big((b \square c)x, a^* y\big).
\end{aligned}$$

From these equalities it follows that $a \square (b \square c)$ is well defined and the required equality, indeed holds. □

The next Proposition 4.2.21 provides a handy criterion for the existence of a bounded inverse of an element.

Proposition 4.2.21 *Let $(\mathfrak{A}[\|\cdot\|], \mathfrak{A}_0)$ be a *-semisimple Banach quasi *-algebra with unit e. Then, $a \in \mathfrak{A}$ has a bounded inverse, if and only if, both of the following conditions hold:*

(i) *there exists $\gamma > 0$, such that*

$$\min\big\{\|ax\|, \|xa\|\big\} \geq \gamma \|x\|, \quad \forall\, x \in \mathfrak{A}_0;$$

(ii) *the sets $\big\{ax,\ x \in \mathfrak{A}_0\big\}$ and $\big\{xa,\ x \in \mathfrak{A}_0\big\}$ are both dense in $\mathfrak{A}$.*

Proof Suppose that (i) and (ii) hold. Using the definitions of $\overline{L}_a$ and $\overline{R}_a$, it is easy to prove that (i) extends as follows:

$$\|\overline{L}_a y\| \geq \gamma \|y\|, \quad \forall\, y \in D(\overline{L}_a) \text{ and } \|\overline{R}_a z\| \geq \gamma \|z\|, \quad \forall\, z \in D(\overline{R}_a). \tag{4.2.7}$$

Hence, both $\overline{L}_a$ and $\overline{R}_a$ are injective maps and their ranges, $\overline{L}_a D(\overline{L}_a)$, $\overline{R}_a D(\overline{R}_a)$, are dense by assumption. The inequalities (4.2.7) imply also that $\overline{L}_a D(\overline{L}_a)$ and $\overline{R}_a D(\overline{R}_a)$ are closed in $\mathfrak{A}$. Indeed, let $w = \lim_{n\to\infty} w_n$ with $w_n \in \overline{L}_a D(\overline{L}_a)$. Then, for every $n \in \mathbb{N}$ there is a unique $y_n \in D(\overline{L}_a)$ such that $w_n = \overline{L}_a y_n$. It follows from (4.2.7) that the sequence $\{y_n\}$ is Cauchy. Then, it converges to an element $y \in \mathfrak{A}$. The closedness of $\overline{L}_a$ implies that $y \in D(\overline{L}_a)$ and $\overline{L}_a y = \lim_{n\to\infty} \overline{L}_a y_n = w$. This proves the closedness of $\overline{L}_a D(\overline{L}_a)$. Since it is also dense, it follows that $\overline{L}_a D(\overline{L}_a) = \mathfrak{A}$. Similarly, one shows that $\overline{R}_a D(\overline{R}_a) = \mathfrak{A}$. Hence, both $\overline{L}_a$ and $\overline{R}_a$ have bounded inverses, $(\overline{L}_a)^{-1}$ and $(\overline{R}_a)^{-1}$, everywhere defined in $\mathfrak{A}$. Taking into account Propositions 4.2.16 and 4.2.5, we conclude that a has a bounded inverse. The converse is easily seen. □

We consider now the seminorm $\mathfrak{q}$ defined in (3.1.13). As we have seen, in general, this seminorm is defined on a domain $\mathcal{D}(\mathfrak{q})$, which is smaller than $\mathfrak{A}$ and it has the following properties:

(a) $\mathfrak{q}(a^*) = \mathfrak{q}(a), \quad \forall\, a \in \mathfrak{A}$;
(b) $\mathfrak{q}(x^*x) = \mathfrak{q}(x)^2, \quad \forall\, x \in \mathfrak{A}_0$;

i.e., it is an *extended C*-seminorm*, in the sense of [77]. If $(\mathfrak{A}, \mathfrak{A}_0)$ has a unit e and $\mathcal{S}_{\mathfrak{A}_0}(\mathfrak{A})$ is sufficient, then $\mathfrak{q}$ is a C*-norm. One has

$$\mathfrak{p}(a) \leq \mathfrak{q}(a), \quad \forall\, a \in \mathcal{D}(\mathfrak{q}). \tag{4.2.8}$$

The space $\mathcal{D}(\mathfrak{q})$ endowed with the topology defined by $\mathfrak{q}$ is denoted by $\mathfrak{A}_\mathfrak{q}$. Then, we have the following

Proposition 4.2.22 *Let $(\mathfrak{A}[\|\cdot\|], \mathfrak{A}_0)$ be a *-semisimple Banach quasi *-algebra with a unit e. Then, $\mathfrak{A}_\mathfrak{q}$ is a normed space containing $\mathfrak{A}_0$ as a subspace. Moreover, if $(\mathfrak{A}[\|\cdot\|], \mathfrak{A}_0)$ is quasi regular, then $\mathfrak{A}_\mathfrak{q}$ is a Banach space.*

Proof The first part of the statement follows from (i) of Proposition 3.1.40. To prove that, $\mathfrak{A}_\mathfrak{q}$ is a Banach space, when $(\mathfrak{A}[\|\cdot\|], \mathfrak{A}_0)$ is quasi regular, we only have to show its completeness. Let $\{a_n\}$ be a $\mathfrak{q}$-Cauchy sequence in $\mathfrak{A}_\mathfrak{q}$.

The inequality (4.2.8), in the quasi regular case, becomes $\|a\| \leq \mathfrak{q}(a)$, for all $a \in \mathfrak{A}_\mathfrak{q}$. Therefore, $\{a_n\}$ is also $\|\cdot\|$-Cauchy. Using the $\|\cdot\|$-completeness of $\mathfrak{A}$, we conclude that there exists an element $a \in \mathfrak{A}$, which is the $\|\cdot\|$-limit of a_n.

Let $\varphi \in \mathcal{P}_{\mathfrak{A}_0}(\mathfrak{A})$. Then, $\varphi(a, a) = \lim_{n\to\infty} \varphi(a_n, a_n)$. The sequence $\mathfrak{q}(a_n)$ is bounded, because $\{a_n\}$ is $\mathfrak{q}$-Cauchy. If M is its supremum, we have

$$\varphi(a_n, a_n)^{1/2} \leq \mathfrak{q}(a_n)\varphi(e, e) \leq M\varphi(e, e).$$

In conclusion,

$$\varphi(a,a)^{1/2} \leq M\varphi(e,e).$$

Thus, clearly $\mathfrak{q}(a) < \infty$, i.e., $a \in \mathfrak{A}_{\mathfrak{q}}$.

Finally, using the uniqueness of the limit in the completion of $\mathfrak{A}_{\mathfrak{q}}$, we conclude that $a = \mathfrak{q} - \lim_{n\to\infty} a_n$. Therefore, $\mathfrak{A}_{\mathfrak{q}}$ is complete. □

We observe that, in general, the inclusion $\mathfrak{A}_0 \subseteq \mathfrak{A}_{\mathfrak{q}}$ is proper. For instance, in $(L^p(I), C(I))$ any step function s defined on $[0, 1]$ belongs to $L^p(I)$, but not to $C(I)$. It is immediate to verify that $s \in (L^p(I))_{\mathfrak{q}}$.

Proposition 4.2.23 *Let $(\mathfrak{A}[\|\cdot\|], \mathfrak{A}_0)$ be a regular Banach quasi *-algebra with unit e. Let $a \in \mathfrak{A}_{\mathfrak{q}}$ and $\lambda \in \mathbb{C}$, such that $|\lambda| > \mathfrak{q}(a)$. Then, $a - \lambda e$ has a bounded inverse $(a-\lambda e)^{-1}$ in $\mathfrak{A}_{\mathrm{b}}$. Thus,*

$$\{\lambda \in \mathbb{C} : |\lambda| > \mathfrak{q}(a)\} \subseteq \rho(a).$$

Proof If $a \in \mathcal{D}(\mathfrak{q})$, by our assumption and definition of $\mathfrak{q}(a)$, one has that

$$|\lambda| > \mathfrak{q}(a) \geq \varphi(ay, ay), \quad \forall\, y \in \mathfrak{A}_0, \text{ such that } \varphi(y,y) = 1.$$

Therefore, for every $x \in \mathfrak{A}_0$,

$$\begin{aligned}\varphi\big((a-\lambda e)x, (a-\lambda e)x\big)^{1/2} &\geq \big(|\lambda|\varphi(x,x) - \varphi(ax, ax)\big)^{1/2} \\ &\geq \big(|\lambda|\varphi(x,x) - \mathfrak{q}(a)\big)^{1/2}.\end{aligned}$$

Taking the supremum over the family $\mathcal{S}_{\mathfrak{A}_0}(\mathfrak{A})$ we obtain

$$\mathfrak{p}\big((a-\lambda e)x\big) \geq \mathfrak{p}(x)\big(|\lambda| - \mathfrak{q}(a)\big), \quad \forall\, x \in \mathfrak{A}_0.$$

Taking into account the regularity of $(\mathfrak{A}[\|\cdot\|], \mathfrak{A}_0)$, we finally obtain

$$\|(a-\lambda e)x\| \geq \|x\|(|\lambda| - \mathfrak{q}(a)), \quad \forall\, x \in \mathfrak{A}_0. \tag{4.2.9}$$

Now we prove that if $\mathfrak{q}(a) < \infty$ and $|\lambda| > \mathfrak{q}(a)$ the sets

$$\operatorname{Ran}(L_{a-\lambda e}) := \big\{(a-\lambda e)y : y \in \mathfrak{A}_0\big\}, \quad \operatorname{Ran}(R_{a-\lambda e}) := \big\{y(a-\lambda e) : y \in \mathfrak{A}_0\big\}$$

are $\|\cdot\|$-dense in $\mathfrak{A}$.

If it is not so, there would exist a non zero $\|\cdot\|$-continuous functional f on $\mathfrak{A}$, such that $f((a-\lambda e)y) = 0$, for every $y \in \mathfrak{A}_0$. Therefore, we should have $f(ay) = \lambda f(y)$, for every $y \in \mathfrak{A}_0$. From the $\|\cdot\|$-continuity of f we obtain $|f(ay)| \leq \|f\|^\star \|ay\|$, for every $y \in \mathfrak{A}_0$, where $\|f\|^\star$ is the norm of f in the dual Banach space $\mathfrak{A}^\star$ of $\mathfrak{A}[\|\cdot\|]$.

From Proposition 3.1.40(iii), we obtain $\mathfrak{p}(ay) \le \mathfrak{q}(a)\mathfrak{p}(y)$, for all $a \in \mathcal{D}(\mathfrak{q})$ and $y \in \mathfrak{A}_0$. Thus, by the regularity of $(\mathfrak{A}[\|\cdot\|], \mathfrak{A}_0)$,

$$|f(ay)| \le \|f\|^\star \|ay\| = \|f\|^\star \mathfrak{p}(ay) \le \|f\|^\star \mathfrak{q}(a)\mathfrak{p}(y) = \|f\|^\star \mathfrak{q}(a)\|y\|, \qquad (4.2.10)$$

for all $y \in \mathfrak{A}_0$. The functional f_a, defined by $f_a(y) := f(ay)$, $y \in \mathfrak{A}_0$, is $\|\cdot\|$-continuous, since

$$|f_a(y)| = |\lambda f(y)\| \le |\lambda| \, \|f\|^\star \|y\|, \quad \forall\, y \in \mathfrak{A}_0. \qquad (4.2.11)$$

An easy computation shows that $\|f_a\|^\star = |\lambda|\|f\|^\star$. From this and the inequalities (4.2.10), (4.2.11) we deduce $|\lambda| \le \mathfrak{q}(a)$, which is a contradiction. A similar argument shows the corresponding statement for Ran $(R_{a-\lambda e})$. By Proposition 4.2.21 it finally follows that $a - \lambda e$ has a bounded inverse □

Let $(\mathfrak{A}[\|\cdot\|], \mathfrak{A}_0)$ be *-semisimple normed quasi *-algebra. Then, the seminorm $\mathfrak{p}$ on $\mathfrak{A}$ becomes a norm as follows from the very definitions. Denote by $\mathfrak{A}_\mathcal{S}$ the completion of $\mathfrak{A}_0[\mathfrak{p}]$. In this regard, we have the following

Proposition 4.2.24 *Let* $(\mathfrak{A}[\|\cdot\|], \mathfrak{A}_0)$ *be a *-semisimple Banach quasi *-algebra and assume that* $\mathfrak{p}(a^*) = \mathfrak{p}(a)$, *for every* $a \in \mathfrak{A}$. *Then, there exists a regular Banach quasi *-algebra* $(\mathfrak{A}_\mathcal{S}, \mathfrak{A}_0)$, *such that* $\mathfrak{A}_\mathcal{S}$ *contains* $\mathfrak{A}$, *as a dense subspace.*

Proof Note that $(\mathfrak{A}_\mathcal{S}, \mathfrak{A}_0)$ is a Banach quasi *-algebra, because of Proposition 3.1.40(iv) and the assumption that $\mathfrak{p}(a^*) = \mathfrak{p}(a)$, for every $a \in \mathfrak{A}$. We now prove that $\mathfrak{A}$ can be identified with a subspace of $\mathfrak{A}_\mathcal{S}$. Indeed, if $a \in \mathfrak{A}$ then there exists a sequence $\{x_n\} \subset \mathfrak{A}_0$, such that

$$\|x_n - a\| \underset{n\to\infty}{\to} 0.$$

Since $\mathfrak{p}(a) \le \|a\|$, for every $a \in \mathfrak{A}$, $\{x_n\}$ is a Cauchy sequence with respect to $\mathfrak{p}$, too. Thus, there exists an element $\overline{a} \in \mathfrak{A}_\mathcal{S}$, such that

$$\mathfrak{p}(x_n - \overline{a}) \underset{n\to\infty}{\to} 0.$$

The element $\overline{a}$ does not depend on the particular sequence $\{x_n\}$ used to approximate a in $\mathfrak{A}$. Indeed, if $\{x'_n\}$ is another sequence with the same property, then

$$\mathfrak{p}(x_n - x'_n) \le \|x_n - x'_n\| \to 0, \quad \text{as } n \to \infty.$$

We have defined in this way a map $i : a \in \mathfrak{A} \to \overline{a} \in \mathfrak{A}_\mathcal{S}$; we shall prove that i is injective.

Assume that $\overline{a} = 0$, for some $a \in \mathfrak{A}$ and let $\{x_n\}$ be a sequence in $\mathfrak{A}_0$ approximating a in the norm of $\mathfrak{A}$ and such that $\mathfrak{p}(x_n) \to 0$; this implies that $\varphi(x_n, x_n) \to 0$, for each $\varphi \in \mathcal{S}_{\mathfrak{A}_0}(\mathfrak{A})$. Therefore,

$$\varphi(a, a) = \lim_{n\to\infty} \varphi(x_n, x_n) = 0.$$

From the sufficiency of $\mathcal{S}_{\mathfrak{A}_0}(\mathfrak{A})$, we obtain $a = 0$ (see Definition 3.1.17 and Corollary 3.1.24). To conclude the proof, we need to show that $\mathcal{S}_{\mathfrak{A}_0}(\mathfrak{A}_{\mathcal{S}})$ is sufficient and that $(\mathfrak{A}_{\mathcal{S}}, \mathfrak{A}_0)$ is regular.

First, we prove that the two families of sesquilinear forms $\mathcal{S}_{\mathfrak{A}_0}(\mathfrak{A})$ and $\mathcal{S}_{\mathfrak{A}_0}(\mathfrak{A}_{\mathcal{S}})$ can be identified. Indeed, assume that $\Phi \in \mathcal{S}_{\mathfrak{A}_0}(\mathfrak{A}_{\mathcal{S}})$, then its restriction $\Phi_{\mathfrak{A}}$ to $\mathfrak{A}$ belongs, as it is easily seen, to $\mathcal{S}_{\mathfrak{A}_0}(\mathfrak{A})$. Conversely, if $\Phi_0 \in \mathcal{S}_{\mathfrak{A}_0}(\mathfrak{A})$, making use of the Cauchy–Schwarz inequality, we obtain

$$|\Phi_0(a, b) \leq \mathfrak{p}(a)\mathfrak{p}(b), \quad \forall\, a, b \in \mathfrak{A}.$$

Therefore, Φ_0 has a unique extension Φ to $\mathfrak{A}_{\mathcal{S}}$ and $\Phi \in \mathcal{S}_{\mathfrak{A}_0}(\mathfrak{A}_{\mathcal{S}})$. Taking this fact into account, the sufficiency of $\mathcal{S}_{\mathfrak{A}_0}(\mathfrak{A}_{\mathcal{S}})$ follows by the definition of $\mathfrak{A}_{\mathcal{S}}$. Thus, $(\mathfrak{A}_{\mathcal{S}}, \mathfrak{A}_0)$ is *-semisimple and the extension, say $\mathfrak{p}_{\mathcal{S}}$, of $\mathfrak{p}$ on $\mathfrak{A}_{\mathcal{S}}$ is also a norm. The regularity is now a simple consequence of the definition of $\mathfrak{p}_{\mathcal{S}}$ □

Example 4.2.25 The BQ*-algebra (see Definition 2.1.4) $(L^p(I), C(I))$ is regular [50], if and only if, $p \geq 2$. For $1 \leq p < 2$, $\mathcal{S}_{C(I)}(L^p(I)) = \{0\}$. In the case of the non-commutative L^p in Example 3.1.7, it has been proved in [52] that, for finite ϱ, $(L^p(\varrho), \mathfrak{M})$ is regular if $p \geq 2$.

Example 4.2.26 The Banach quasi *-algebra $(\mathcal{H}, \mathfrak{A}_0)$ of Example 3.1.8 is *-semisimple, since $\mathcal{S}_{\mathfrak{A}_0}(\mathfrak{A})$ contains the inner product $\langle\cdot|\cdot\rangle$. For the same reason, $(\mathcal{H}, \mathfrak{A}_0)$ is regular.

Finally, using the uniqueness of the limit in the completion of $\mathfrak{A}_{\mathfrak{q}}$, we conclude that $a = \mathfrak{q} - \lim_{n\to\infty} a_n$. Therefore, $\mathfrak{A}_{\mathfrak{q}}$ is complete.

We can now prove the following, where for the term *A*-algebra*, the reader is referred to [21, p. 181].

Theorem 4.2.27 *Let $(\mathfrak{A}[\|\cdot\|], \mathfrak{A}_0)$ be a regular Banach quasi *-algebra with unit e. Then, $\mathcal{D}(\mathfrak{q})$ coincides with the set $\mathfrak{A}_{\mathrm{b}}$ of all bounded elements of $\mathfrak{A}$. Moreover,*

$$\mathfrak{q}(a) = \|a\|_{\mathrm{b}}, \quad \forall\, a \in \mathfrak{A}_{\mathrm{b}}.$$

Therefore, $\mathfrak{A}_{\mathrm{b}}[\|\cdot\|_{\mathrm{b}}]$ is a C-algebra.*

Proof Proposition 4.2.23 shows that $\mathcal{D}(\mathfrak{q}) \subseteq \mathfrak{A}_{\mathrm{b}}$. On the other hand, let us consider, for each $\varphi \in \mathcal{P}_{\mathfrak{A}_0}(\mathfrak{A})$, the linear functional ω_φ defined by

$$\omega_\varphi(a) := \varphi(a, e), \;\; a \in \mathfrak{A}_{\mathrm{b}}.$$

A simple limit argument shows that ω_φ is positive (i.e., $\omega(a^* \bullet a) \geq 0$, for each $a \in \mathfrak{A}_\flat$), so the corresponding GNS representation π_φ is bounded and $\|\overline{\pi_\varphi(a)}\| \leq \|a\|_\flat$. Then, if $a \in \mathfrak{A}_\flat$,

$$\begin{aligned}\mathfrak{q}(a) &= \sup\left\{\varphi(ax, ax)^{1/2} : \varphi \in \mathcal{P}_{\mathfrak{A}_0}(\mathfrak{A}),\ x \in \mathfrak{A}_0,\ \varphi(x, x) = 1\right\}\\ &= \sup_{\varphi \in \mathcal{P}_{\mathfrak{A}_0}(\mathfrak{A})} \|\overline{\pi_\varphi(a)}\| \leq \|a\|_\flat.\end{aligned}$$

On the other hand, from Proposition 3.1.40(iii) follows that

$$\|xa\| = \mathfrak{p}(xa) \leq \mathfrak{q}(a)\mathfrak{p}(x) = \mathfrak{q}(a)\|x\|, \quad \forall\, a \in \mathcal{D}(q),\ x \in \mathfrak{A}_0.$$

This implies that $\|a\|_\flat \leq \mathfrak{q}(a)$, for every a in $\mathcal{D}(q)$ (see (4.1.2)). Thus, in conclusion, $\|\cdot\|_\flat$ is a C*-norm. □

A further characterization of the set of bounded elements of $(\mathfrak{A}[\|\cdot\|], \mathfrak{A}_0)$, in the case where $\mathcal{S}_{\mathfrak{A}_0}(\mathfrak{A})$ is sufficient, can be obtained in terms of *-representations. In particular, under the latter assumption, $(\mathfrak{A}[\|\cdot\|], \mathfrak{A}_0)$ accepts a faithful *-representation.

Theorem 4.2.28 *Let $(\mathfrak{A}[\|\cdot\|], \mathfrak{A}_0)$ be a *-semisimple Banach quasi *-algebra with unit e. Then, $(\mathfrak{A}[\|\cdot\|], \mathfrak{A}_0)$ admits a faithful *-representation π in a Hilbert space $\mathcal{H}$. Moreover,*

$$\mathfrak{A}_\flat = \left\{a \in \mathfrak{A} : \overline{\pi(a)} \in \mathcal{B}(\mathcal{H})\right\} \text{ and } \|\overline{\pi(a)}\| = \mathfrak{q}(a), \quad \forall\, a \in \mathfrak{A}_\flat.$$

Proof Let $\varphi \in \mathcal{S}_{\mathfrak{A}_0}(\mathfrak{A})$ and let π_φ be the corresponding GNS representation with dense domain $\mathcal{D}_\varphi \subseteq \mathcal{H}_\varphi$. Put

$$\mathcal{H} = \bigoplus_{\varphi \in \mathcal{S}_{\mathfrak{A}_0}(\mathfrak{A})} \mathcal{H}_\varphi = \left\{(\xi_\varphi)_{\varphi \in \mathcal{S}_{\mathfrak{A}_0}(\mathfrak{A})} : \xi_\varphi \in \mathcal{H}_\varphi,\ \sum_{\varphi \in \mathcal{S}_{\mathfrak{A}_0}(\mathfrak{A})} \|\xi_\varphi\|^2 < \infty\right\},$$

with the usual inner product

$$\langle(\xi_\varphi)|(\eta_\varphi)\rangle = \sum_{\varphi \in \mathcal{S}_{\mathfrak{A}_0}(\mathfrak{A})} \langle\xi_\varphi|\eta_\varphi\rangle, \quad (\xi_\varphi), (\eta_\varphi) \in \mathcal{H}.$$

Let

$$\mathcal{D} = \left\{(\xi_\varphi)_{\varphi \in \mathcal{S}_{\mathfrak{A}_0}(\mathfrak{A})} \in \mathcal{H} : \xi_\varphi \in \mathcal{D}_\varphi, \varphi \in \mathcal{S}_{\mathfrak{A}_0}(\mathfrak{A}) : \sum_{\varphi \in \mathcal{S}_{\mathfrak{A}_0}(\mathfrak{A})} \|\pi_\varphi(a)\xi_\varphi\|^2 < \infty,\ \forall\, a \in \mathfrak{A}\right\}.$$

Then, $\mathcal{D}$ is a dense domain in $\mathcal{H}$ and so we can define

$$\pi(a)(\xi_\varphi) = (\pi_\varphi(a)\xi_\varphi), \quad \forall\, a \in \mathfrak{A},\ (\xi_\varphi)_{\varphi \in \mathcal{S}_{\mathfrak{A}_0}(\mathfrak{A})} \in \mathcal{D}.$$

Thus, $\pi(a) \in \mathcal{L}^\dagger(\mathcal{D}, \mathcal{H})$, for each $a \in \mathfrak{A}$ and $\pi : a \in \mathfrak{A} \to \pi(a) \in \mathcal{L}^\dagger(\mathcal{D}, \mathcal{H})$ is a *-representation of $(\mathfrak{A}[\|\cdot\|], \mathfrak{A}_0)$. Moreover, π is faithful, since

$$\pi(a) = 0 \Leftrightarrow \pi_\varphi(a) = 0,\ \forall\, \varphi \in \mathcal{S}_{\mathfrak{A}_0}(\mathfrak{A}) \Leftrightarrow \varphi(a, a) = 0,\ \forall\, \varphi \in \mathcal{S}_{\mathfrak{A}_0}(\mathfrak{A}).$$

The sufficiency of $\mathcal{S}_{\mathfrak{A}_0}(\mathfrak{A})$ (see Corollary 3.1.24) now implies that $a = 0$.

Finally, $\pi(a)$ is bounded, if and only if, each π_φ, $\varphi \in \mathcal{S}_{\mathfrak{A}_0}(\mathfrak{A})$, is bounded and

$$\sup_{\varphi \in \mathcal{S}_{\mathfrak{A}_0}(\mathfrak{A})} \|\overline{\pi_\varphi(a)}\| < \infty;$$

in this case,

$$\|\overline{\pi(a)}\| = \sup_{\varphi \in \mathcal{S}_{\mathfrak{A}_0}(\mathfrak{A})} \|\overline{\pi_\varphi(a)}\| = \sup_{\varphi \in \mathcal{P}_{\mathfrak{A}_0}(\mathfrak{A})} \|\overline{\pi_\varphi(a)}\|, \quad a \in \mathfrak{A}.$$

The latter equality follows from the fact that for every $\varphi \in \mathcal{P}_{\mathfrak{A}_0}(\mathfrak{A})$, there exists $\psi \in \mathcal{S}_{\mathfrak{A}_0}(\mathfrak{A})$, such that $\|\overline{\pi_\varphi(a)}\| = \|\overline{\pi_\psi(a)}\|$, a in $\mathfrak{A}$. For this it is enough to choose $\psi = \frac{\varphi}{n(\varphi)}$, φ as before, where $n(\varphi)^2 = \sup_{a \in \mathfrak{A}} \frac{\varphi(a,a)}{\|a\|^2}$. Then, $\mathcal{P}_{\mathfrak{A}_0}(\mathfrak{A}) = \mathcal{S}_{\mathfrak{A}_0}(\mathfrak{A})$, therefore we obtain

$$\sup_{\varphi \in \mathcal{P}_{\mathfrak{A}_0}(\mathfrak{A})} \|\overline{\pi_\varphi(a)}\| = \sup\{\varphi(ax, ax)^{1/2} : \varphi \in \mathcal{P}_{\mathfrak{A}_0}(\mathfrak{A}),\ x \in \mathfrak{A}_0,\ \varphi(x, x) = 1\} = \mathsf{q}(a).$$

Hence, $\|\overline{\pi(a)}\| = \mathsf{q}(a)$, for every $a \in \mathfrak{A}$ and this concludes the proof. □

If $(\mathfrak{A}[\|\cdot\|], \mathfrak{A}_0)$ has a unit, the *-representation π constructed in Theorem 4.2.28 enjoys the following property: *for every* $\varphi \in \mathcal{S}_{\mathfrak{A}_0}(\mathfrak{A})$, *there exists a vector* $\xi \in \mathcal{D}$, *with* $\|\xi\| = 1$, *such that*

$$\varphi(a, b) = \langle \pi(a)\xi | \pi(b)\xi \rangle, \quad \forall\, a, b \in \mathfrak{A}.$$

Borrowing the terminology from the theory of *-algebras (see, e.g., [7, Definition 3.1.16] and also Remark A.6.16), we will call a *-representation π with this property *universal*. Then, we can restate Theorem 4.2.28 in the following terms, having thus a *Gelfand–Naimark type theorem for *-semisimple Banach quasi *-algebras with unit*.

Theorem 4.2.29 *Let* $(\mathfrak{A}[\|\cdot\|], \mathfrak{A}_0)$ *be a Banach quasi *-algebra with unit. The following statements are equivalent:*

(i) *there exists a faithful universal *-representation* π *of* $(\mathfrak{A}[\|\cdot\|], \mathfrak{A}_0)$*;*
(ii) $(\mathfrak{A}[\|\cdot\|], \mathfrak{A}_0)$ *is *-semisimple.*

Chapter 5
CQ*-Algebras

This chapter is devoted to a special class of Banach quasi *-algebras, the so-called *CQ*-algebras*. Their essential feature consists of the fact that they contain a C*-algebra as dense subspace.

5.1 Basic Aspects

Definition 5.1.1 Let $(\mathfrak{A}[\|\cdot\|], \mathfrak{A}_0)$ be a Banach quasi *-algebra. We say that $(\mathfrak{A}[\|\cdot\|], \mathfrak{A}_0)$ is a *proper CQ*-algebra* if

(i) $\mathfrak{A}_0$ is a C*-algebra with norm $\|\cdot\|_0$ and involution * inherited by that of $\mathfrak{A}$;
(ii) $\mathfrak{A}_0$ is dense in $\mathfrak{A}[\|\cdot\|]$;
(iii) $\|x\|_0 = \sup_{a\in\mathfrak{A}, \|a\|\leq 1} \|ax\|, \; x \in \mathfrak{A}_0.$

We have defined the norm $\|\cdot\|_0$ for a Banach quasi *-algebra $(\mathfrak{A}[\|\cdot\|], \mathfrak{A}_0)$ by (3.1.4) (see also (3.1.1)). It is worth mentioning that condition (iii) of Definition 5.1.1, which is seemingly imposing a stronger requirement, is exactly equivalent to the definition given in Chap. 3, due to the fact that $\mathfrak{A}_0$ is supposed to be a C*-algebra. This is a consequence of the following lemma which can be applied, for $x \in \mathfrak{A}_0$, to $\|x\|_0$ as defined in (3.1.4) and to $\|x\|_1 := \sup_{a\in\mathfrak{A}, \|a\|\leq 1} \|ax\|$.

Lemma 5.1.2 *Let $\mathfrak{A}_0[\|\cdot\|_0]$ be a C*-algebra and $\|\cdot\|_1$ another norm on $\mathfrak{A}_0$, which makes of it a normed algebra. Suppose that $\|x\|_1 \leq \|x\|_0$, for every $x \in \mathfrak{A}_0$. Then,*

$$\|x\|_1 = \|x\|_0, \quad \forall\, x \in \mathfrak{A}_0.$$

Proof Let $x = x^* \in \mathfrak{A}_0$ and let $\mathfrak{M}(x)$ denote the abelian C*-algebra generated by x. Since every norm that makes an abelian C*-algebra into a normed algebra is

M. Fragoulopoulou, C. Trapani, *Locally Convex Quasi *-Algebras and their Representations*, Lecture Notes in Mathematics 2257,
https://doi.org/10.1007/978-3-030-37705-2_5

necessarily stronger than the C*-norm [22, Theorem 1.2.4], we obtain the equality $\|x\|_1 = \|x\|_0$, $\forall\, x = x^* \in \mathfrak{A}_0$. For an arbitrary element $y \in \mathfrak{A}_0$, we have

$$\|y\|_0^2 = \|y^* y\|_0 = \|y^* y\|_1 \leq \|y^*\|_1 \, \|y\|_1.$$

But $\|y^*\|_1 \leq \|y^*\|_0$ and so $\|y\|_0^2 \leq \|y\|_0 \, \|y\|_1$ and this implies that $\|y\|_0 \leq \|y\|_1$. □

A proper CQ*-algebra can be viewed as the completion of an arbitrary C*-algebra $\mathfrak{A}_0[\|\cdot\|]_0$ with respect to a weaker norm. Indeed, we have

Proposition 5.1.3 *Let $\mathfrak{A}_0$ be a C*-algebra, with norm $\|\cdot\|_0$ and involution *. Let $\|\cdot\|$ be another norm on $\mathfrak{A}_0$, weaker than $\|\cdot\|_0$, i.e.,*

$$\|x\| \leq \|x\|_0, \quad \forall\, x \in \mathfrak{A}_0$$

and such that the following conditions are satisfied:

(i) $\|xy\| \leq \|x\| \, \|y\|_0, \quad \forall\, x, y \in \mathfrak{A}_0$;
(ii) $\|x^*\| = \|x\|, \quad \forall\, x \in \mathfrak{A}_0$.

Let $\mathfrak{A}$ denotes the $\|\cdot\|$-completion of $\mathfrak{A}_0$. Then, $(\mathfrak{A}[\|\cdot\|], \mathfrak{A}_0)$ is a proper CQ-algebra.*

Proof Let $\mathfrak{A}$ be the Banach space completion of $\mathfrak{A}_0[\|\cdot\|]$. For $x \in \mathfrak{A}_0$, put

$$\|x\|_1 := \sup_{\|a\| \leq 1} \|ax\|, \; a \in \mathfrak{A}, \tag{5.1.1}$$

It follows that the operator $R_x : \mathfrak{A}[\|\cdot\|] \to \mathfrak{A}[\|\cdot\|]$, such that $R_x(a) := ax$, for all $a \in \mathfrak{A}$, is bounded and $\|xy\|_1 \leq \|x\|_1 \|y\|_1$, for every $x, y \in \mathfrak{A}_0$. Moreover, from (i), we obtain

$$\|x\|_1 \leq \|x\|_0, \quad \forall\, x \in \mathfrak{A}_0.$$

Hence, the statement follows by Lemma 5.1.2 and Definition 5.1.1. □

Corollary 5.1.4 *Let $(\mathfrak{A}[\|\cdot\|], \mathfrak{A}_0)$ be a proper CQ*-algebra and $\mathfrak{B}_0$ any *-subalgebra of $\mathfrak{A}_0$. Let $\mathfrak{M}_0(\mathfrak{B}_0)$ be the closure of $\mathfrak{B}_0$ in $\mathfrak{A}_0$ and $\mathfrak{M}(\mathfrak{B}_0)$ the closure of $\mathfrak{B}_0$ in $\mathfrak{A}$. Then, $(\mathfrak{M}(\mathfrak{B}_0), \mathfrak{M}_0(\mathfrak{B}_0))$ is a proper CQ*-algebra.*

5.1.1 Commutative CQ*-Algebras

We begin with giving some examples of commutative CQ*-algebras.

Example 5.1.5 (CQ-Algebras of Functions)* Let μ be a measure in a non-empty point set X. Let M^+ be the collection of all μ-measurable nonnegative functions on

X. We assume that to each $f \in M^+$ it corresponds a number $\rho(f) \in [0, \infty]$, such that:

(i) $\rho(f) = 0$, if and only if, $f = 0$, a.e. in X;
(ii) $\rho(f_1 + f_2) \leq \rho(f_1) + \rho(f_2), \quad \forall\, f_1, f_2 \in M^+$;
(iii) $\rho(\lambda f) = \lambda \rho(f), \quad \forall\, \lambda \in \mathbb{R}^+, f \in M^+$;
(iv) let $\{f_n\} \subset M^+$ and $f_n \uparrow f$ a.e. in X. Then, $\rho(f_n) \uparrow \rho(f)$.

Following [30] we call ρ a *function norm*. Let us define

$$L_\rho := \left\{ f \in M^+ : \rho(f) < \infty \right\}.$$

With this definition it has been proved in [30] that the space L_ρ is a Banach space, that is complete, with respect to the norm $\|f\| \equiv \rho(|f|)$.

Some L_ρ spaces generate examples of abelian proper CQ*-algebras.

(A) Let (X, μ) be a measure space with μ a regular Borel measure on a compact Hausdorff space X. As usual, we denote by $L^p(X, \mu)$ the Banach space of all (equivalence classes of) measurable functions $f : X \longrightarrow \mathbb{C}$, such that

$$\|f\|_p := \left(\int_X |f|^p \, d\mu \right)^{1/p} < \infty$$

On $L^p(X, \mu)$ we consider the natural involution $f \in L^p(X, \mu) \mapsto f^* \in L^p(X, \mu)$ with $f^*(x) = \overline{f(x)}$. Clearly $L^p(X, \mu)$ is an L_ρ space (with $\|\cdot\| \equiv \|\cdot\|_p$). We denote by $C(X)$ the C*-algebra of continuous functions defined on X. The pair $(L^p(X, \mu), C(X))$ provides the basic commutative example of a Banach quasi *-algebra. It turns also out that $(L^p(X, \mu), C(X))$ is a proper abelian CQ*-algebra, for any $p \geq 1$, since the p-norm satisfies all the conditions of Proposition 5.1.3. These spaces have been analyzed with a certain care in [50].

(B) Let X be a compact Hausdorff space and $\mathcal{M} = \{\mu_\alpha, \alpha \in I\}$ a family of Borel measures on X, for which there exists a constant $c > 0$, such that $\mu_\alpha(X) \leq c$, for all $\alpha \in I$. Let $\|\cdot\|_{p,\alpha}$ be the norm on $L^p(X, \mu_\alpha)$. Of course, each norm is related to a particular function norm $\rho_{p,\alpha}(\cdot)$. Let us define, for $f \in C(X)$, the function

$$\|f\|_{p,I} := \sup_{\alpha \in I} \|f\|_{p,\alpha}.$$

In [30] it is shown that the map $\rho_{p,I}$ related to this norm still satisfies all the requirements of a function norm, so that the completion $L^p_I(X, \mathcal{M})$ of $C(X)$ with respect to $\|\cdot\|_{p,I}$, is a Banach space. The norm $\|\cdot\|_{p,I}$ also satisfies the conditions of Proposition 5.1.3. Indeed, since $\|f\|_{p,I} \leq c\|f\|_\infty$, for all

$f \in C(X)$, it follows that $\|\cdot\|_{p,I}$ is finite on $C(X)$ and really defines a norm on $C(X)$ satisfying

(i) $\|f^*\|_{p,I} = \|f\|_{p,I}, \quad \forall\, f \in C(X)$;
(ii) $\|fg\|_{p,I} \le \|f\|_{p,I}\|g\|_\infty, \quad \forall\, f, g \in C(X)$.

Therefore, $(L^p_I(X, \mathcal{M}), C(X))$ is a commutative proper CQ*-algebra. It is clear that $L^p_I(X, \mathcal{M})$ can be identified with a subspace of $L^p(X, \mu_\alpha)$, for all $\alpha \in I$. It is obvious that $L^p_I(X, \mathcal{M})$ may contain also noncontinuous functions (depending on the set X and on the family $\mathcal{M}$ of measures).

(C) Let X, $\mathcal{M}$ and $\rho_{p,\alpha}$ be as above. For a set $\{c_\alpha\}$ of positive constants, we define

$$\rho_p(f) := \sum_{\alpha\in I} c_\alpha \rho_{p,\alpha}(f).$$

Then, the space $L_p(X, \mathcal{M})$ (the completion of $C(X)$ with respect to the norm $\|\cdot\|_p$ generated by ρ_p) is a Banach space, which contains the space $L^p_I(X, \mathcal{M})$ of the previous example, if the set of numbers $\{c_\alpha\}$ is summable. Again, $(L_p(X, \mathcal{M}), C(X))$ is an abelian proper CQ*-algebra.

(D) Interesting examples of commutative proper CQ*-algebras are also provided by Nachbin spaces (see [43]). Let X be a locally compact Hausdorff space and v an upper semicontinuous nonnegative function, such that $\inf_{t\in X} v(t) > 0$. As usual, $C(X)$ and $C_0(X)$ stand, respectively, for the set of continuous functions on X and for the set of continuous functions on X vanishing at infinity. Let us consider the spaces

$$C^v_b(X) := \{f \in C(X) : vf \text{ is bounded on } X\}$$

and

$$C^v_0(X) := \{f \in C(X) : vf \text{ vanishes at infinity on } X\}.$$

The *weighted uniform topology* is defined by the norm

$$\|f\|_v := \sup_{t\in X} v(t)|f(t)|, \quad f \in C^v_b(X).$$

It turns out that both $C^v_b(X)$ and $C^v_0(X)$, when equipped with the weighted uniform topology, are Banach modules over $C_0(X)$. Actually, $(C^v_b(X), C_0(X))$ and $(C^v_0(X), C_0(X))$ are commutative proper CQ*-algebras.

In this subsection we shall describe the structure of commutative CQ*-algebras.

As is known, for C*-algebras the situation is completely clear: a commutative C*-algebra with unit is isometrically *-isomorphic to the C*-algebra $C(X)$ of all $\mathbb{C}$-valued continuous functions on the compact space X of all characters of $C(X)$. The respective correspondence in this case is the so-called *Gelfand map*. CQ*-algebras do not behave so nicely: the first reason is that Proposition 5.1.3 allows the existence of non isomorphic CQ*-algebras over $C(X)$; indeed, in the case considered in Example 5.1.5 (A), it is clear that $L^p(X)$ is not isomorphic to $L^q(X)$, if $p \neq q$, in a usual situation; the second reason is that, as is known [4, 8, 18] already for Banach *-algebras the Gelfand map is not, in general, an isometric *-isomorphism.

Proposition 5.1.6 *Let* $(\mathfrak{A}[\|\cdot\|], \mathfrak{A}_0)$ *be a commutative proper CQ*-algebra and X the space of all characters of* $\mathfrak{A}_0$. *Then, the following statements hold:*

(i) *if* $\mathfrak{A}_0$ *admits a faithful state* ω, *which is continuous with respect to the norm* $\|\cdot\|$ *of* $\mathfrak{A}$, *then there exists a linear map* ϕ *from* $\mathfrak{A}$ *into* $C(X)^\star$ (the dual of the Banach space $C(X)$), *whose restriction to* $\mathfrak{A}_0$ *is a linear isomorphism of* $\mathfrak{A}_0$ *onto* $C(X)$;

(ii) *if, in addition, for some positive constant* $\gamma > 0$

$$|\omega(y^*x)| \leq \gamma \|x\| \|y\|, \quad \forall\, x, y \in \mathfrak{A}_0,$$

then there exists a regular Baire measure μ *on X and a linear map* ϕ *from* $\mathfrak{A}$ *into* $L^2(X, d\mu)$, *whose restriction to* $\mathfrak{A}_0$ *is a linear isomorphism of* $\mathfrak{A}_0$ *onto* $C(X)$.

In both cases ϕ *preserves involution.*

Proof

(i) Let ω be a faithful state, which is continuous with respect to the norm $\|\cdot\|$ of $\mathfrak{A}$; then ω has a continuous extension (denoted by the same symbol) to the whole of $\mathfrak{A}$; for each $a \in \mathfrak{A}$, the linear functional ω_a defined by $\omega_a(x) := \omega(ax)$, $x \in \mathfrak{A}_0$, is therefore, bounded on $\mathfrak{A}_0$. Let us now consider the linear functional $\widehat{\omega}_a$ on $C(X)$ defined by $\widehat{\omega}_a(\widehat{x}) := \omega_a(x)$, $x \in \mathfrak{A}_0$ and let

$$f : a \in \mathfrak{A} \to f(a) := \widehat{\omega}_a \in C(X)^\star.$$

It is easily shown that f is a linear map of $\mathfrak{A}$ into $C(X)^\star$. We define $\phi := f \restriction_{\mathfrak{A}_0}$. Then, ϕ is a linear isomorphism: indeed, if $x \in \mathfrak{A}_0$ and $\widehat{\omega}_x(\widehat{y}) = 0$, for every $y \in \mathfrak{A}_0$, then in particular, $\widehat{\omega}_x(\widehat{x^*}) = \omega(xx^*) = 0$ and then, by the faithfulness of ω, $x = 0$.

(ii) Set

$$\varphi_0(x, y) := \omega(y^*x), \ x, y \in \mathfrak{A}_0.$$

By the assumption for ω in (ii), φ_0 can be extended, by continuity, to a positive sesquilinear form φ on $\mathfrak{A} \times \mathfrak{A}$.

Since ω is linear and continuous on $\mathfrak{A}_0$, $\widehat{\omega}$ (defined exactly as $\widehat{\omega}_a$, above) is continuous on $C(X)$, so that by the Riesz representation theorem [14], there exists a unique Borel measure μ_ω on X, such that

$$\omega(x) = \widehat{\omega}(\widehat{x}) = \int_X \widehat{x}(\rho) d\mu_\omega(\rho), \quad \forall\, x \in \mathfrak{A}_0.$$

□

For $a \in \mathfrak{A}$ the conjugate linear form F_a on $C(X)$ defined by $F_a(\widehat{x}) = \varphi(a, x)$, $x \in \mathfrak{A}_0$, is bounded in $L^2(X, d\mu_\omega)$, due to the Schwarz inequality. Therefore, there exists a function $\widehat{a} \in L^2(X, d\mu_\omega)$, such that

$$F_a(\widehat{x}) = \int_X \widehat{a}(\rho)\overline{\widehat{x}(\rho)} d\mu_\omega(\rho), \; x \in \mathfrak{A}_0.$$

The map $\phi : a \in \mathfrak{A} \to \phi(a) = \widehat{a} \in L^2(X, d\mu_\omega)$ is, as it is readily checked, a linear map of $\mathfrak{A}$ into $L^2(X, d\mu_\omega)$, preserving the involution * of $\mathfrak{A}_0$. It is easy to see that ϕ satisfies the requirements of our proposition. The fact that $\phi \upharpoonright_{\mathfrak{A}_0}$ is a linear isomorphism of $\mathfrak{A}_0$ onto $C(X)$ follows, as before, from the faithfulness of ω.

Remark 5.1.7 It is clear that the assumption $|\omega(y^*x)| \leq \gamma \|x\|\|y\|$, for all $x, y \in \mathfrak{A}_0$, made in Proposition 5.1.6(ii), implies the continuity of ω required in (i). The converse is, however, not true in general. This could appear in contradiction to the fact that any separately continuous sesquilinear form on a Banach space, as $\mathfrak{A}$, is necessarily jointly continuous. But, as a matter of fact, the continuity of ω does not imply the separate continuity of φ_0. There is, however, one relevant exception: if the state ω of Proposition 5.1.6(i) can be taken to be *pure*, then it is multiplicative, therefore

$$\varphi_0(x, y) = \omega(x)\overline{\omega(y)}, \quad \forall\, x, y \in \mathfrak{A}_0.$$

Hence, φ_0 is separately continuous with respect to the norm $\|\cdot\|$ of $\mathfrak{A}$ and the same holds for its extension φ to the whole $\mathfrak{A}$.

Now we show that the *-semisimple and commutative case is completely understood: in fact, as we shall see below, any CQ*-algebra with these two properties can be thought as a CQ*-algebra of functions. We remind the reader that, in the case of the L^p-spaces, *-semisimplicity occurs, if and only if, $p \geq 2$ (see Example 3.1.29).

Let X be a compact Hausdorff space and $\mathcal{M} = \{\mu_\alpha : \alpha \in \mathrm{I}\}$ a family of Borel measures on X.

The CQ*-algebra $(L^p_{\mathrm{I}}(X, \mathcal{M}), C(X))$ constructed in Example 5.1.5(B) is *-semisimple, for $p \geq 2$. This depends on the fact that, for each α, an element of $\mathcal{S}_{C(X)}(L^p(X, \mu_\alpha))$ gives rise, by restriction, to an element of $\mathcal{S}_{C(X)}(L^p_{\mathrm{I}}(X, \mathcal{M}))$ (for the latter notation, see Definition 3.1.14).

Proposition 5.1.8 *Let* $(\mathfrak{A}[\|\cdot\|], \mathfrak{A}_0)$ *be a *-semisimple commutative CQ*-algebra with unit e. Then, there exists a family* $\mathcal{M}$ *of Borel measures on the Hausdorff* *compact space X of the characters of $\mathfrak{A}_0$ and a map $\Phi : a \in \mathfrak{A} \mapsto \Phi(a) \equiv \widehat{a} \in L^2_{\mathrm{I}}(X, \mathcal{M})$ with the following properties:*

(i) *Φ extends the Gelfand transform of elements of $\mathfrak{A}_0$ and $\Phi(\mathfrak{A}) \supset C(X)$;*
(ii) *Φ is linear and injective;*
(iii) $\Phi(ax) = \Phi(a)\Phi(x), \ \forall\, a \in \mathfrak{A},\ x \in \mathfrak{A}_0$;
(iv) $\Phi(a^*) = \Phi(a)^*, \quad \forall\, a \in \mathfrak{A}$.

Thus, $\mathfrak{A}$ can be identified with a subspace of $L^2_{\mathrm{I}}(X, \mathcal{M})$.

If $\mathfrak{A}$ is regular (cf. Definition 4.2.17), *that is, if*

$$\|a\|^2 = \sup_{\varphi \in \mathcal{S}_{\mathfrak{A}_0}(\mathfrak{A})} \varphi(a, a),\ a \in \mathfrak{A},$$

*then Φ is an isometric *-isomorphism of $\mathfrak{A}$ onto $L^2_{\mathrm{I}}(X, \mathcal{M})$.*

Proof Define first Φ on $\mathfrak{A}_0$ as the usual Gelfand map

$$\Phi : x \in \mathfrak{A}_0 \mapsto \widehat{x} \in C(X).$$

As it is well-known, the Gelfand map is an isometric *-isomorphism of $\mathfrak{A}_0$ onto $C(X)$. Let $\varphi \in \mathcal{S}_{\mathfrak{A}_0}(\mathfrak{A})$. Define the linear functional ω on $C(X)$ by

$$\omega(\widehat{x}) := \varphi(x, e),\ \ x \in \mathfrak{A}_0.$$

It is easy to check that ω is bounded on $C(X)$; then, by the Riesz representation theorem, there exists a unique regular positive Borel measure μ_φ on X, such that

$$\omega(\widehat{x}) = \varphi(x, e) = \int_X \widehat{x}(\eta) d\mu_\varphi(\eta), \quad \forall\, x \in \mathfrak{A}_0.$$

We have $\mu_\varphi(X) \leq \|e\|^2$, for all $\varphi \in \mathcal{S}_{\mathfrak{A}_0}(\mathfrak{A})$.

Let $\mathcal{M} \equiv \{\mu_\varphi : \varphi \in \mathcal{S}_{\mathfrak{A}_0}(\mathfrak{A})\}$ and let $L^2_{\mathcal{S}_{\mathfrak{A}_0}(\mathfrak{A})}(X, \mathcal{M})$ be the CQ*-algebra constructed as above. Now, if $a \in \mathfrak{A}$, there exists a sequence $\{x_n\} \subset \mathfrak{A}_0$ converging to a in the norm of $\mathfrak{A}$. We have then

$$\|\widehat{x_n} - \widehat{x_m}\|^2_{2,\mathcal{S}_{\mathfrak{A}_0}(\mathfrak{A})} = \sup_{\varphi \in \mathcal{S}_{\mathfrak{A}_0}(\mathfrak{A})} \varphi(x_n - x_m, x_n - x_m) \leq \|x_n - x_m\|^2 \to 0\,.$$

Let $\widehat{a}$ be the $\|\cdot\|_{2,\mathcal{S}_{\mathfrak{A}_0}(\mathfrak{A})}$-limit of $\{\widehat{x_n}\}$ in $L^2_{\mathcal{S}_{\mathfrak{A}_0}(\mathfrak{A})}(X, \mathcal{M})$. Define

$$\Phi(a) := \widehat{a},\ a \in \mathfrak{A}.$$

Evidently, $\|\widehat{a}\|_{2,\mathcal{S}_{\mathfrak{A}_0}(\mathfrak{A})} = \sup_{\varphi \in \mathcal{S}_{\mathfrak{A}_0}(\mathfrak{A})} \varphi(a, a)$, $a \in \mathfrak{A}$. This implies that if $\widehat{a} = 0$, then $\varphi(a, a) = 0$, for all $\varphi \in \mathcal{S}_{\mathfrak{A}_0}(\mathfrak{A})$ and thus $a = 0$, since $\mathfrak{A}$ is *-semisimple. The proof of (ii), (iii) and (iv) is straightforward.

Now, if $\mathfrak{A}$ is regular, it follows immediately from the preceding discussion that Φ is an isometry. We conclude by proving, in this case, that Φ is onto. Let h be an element of $L^2_{\mathcal{S}_{\mathfrak{A}_0}(\mathfrak{A})}(X, \mathcal{M})$. Then, there exists a sequence $\{\widehat{a}_n\} \subset C(X)$ converging to h with respect to $\|\cdot\|_{2,\mathcal{S}_{\mathfrak{A}_0}(\mathfrak{A})}$. The sequence $\{a_n\}$ converges to some $a \in \mathfrak{A}$. It is easily seen that $h = \widehat{a}$. Hence, Φ is onto. □

5.2 General CQ*-Algebras

In the previous sections we have considered *proper* CQ*-algebras, only in the commutative case. However, there are also noncommutative examples, where just here we give an easy one. Other examples are provided by noncommutative L^p-spaces in Sect. 5.6.2.

Example 5.2.1 Let $\mathfrak{A}_0[\|\cdot\|_0]$ be a C*-algebra with unit e. Let Φ be a linear map of $\mathfrak{A}_0$ into itself with $\Phi(x^*) = \Phi(x)^*$, for every $x \in \mathfrak{A}_0$. Suppose that the following inequality is fulfilled

$$\|\Phi(xy)\|_0 \leq \|\Phi(x)\|_0 \|y\|_0, \quad \forall\, x, y \in \mathfrak{A}_0. \tag{5.2.2}$$

Let us assume that $\|\Phi(e)\|_0 = 1$ and define a new norm on $\mathfrak{A}_0$ by

$$\|x\| := \|\Phi(x)\|_0, \quad x \in \mathfrak{A}_0.$$

It is easy to verify that this norm satisfies the conditions of Proposition 5.1.3. Therefore, the $\|\cdot\|$-completion $\mathfrak{A}$ of $\mathfrak{A}_0$ is a proper CQ*-algebra over $\mathfrak{A}_0$.

Of course, the inequality (5.2.2) automatically holds if Φ is a *-homomorphism, [22, Corollary 1.2.6]. However, in this case the two norms coincide, as always, when $\|\cdot\|$ is a Banach algebra norm on $\mathfrak{A}_0$ (see also Lemma 5.1.2).

Originally, the proper CQ*-algebras were introduced as a subcase of a richer structure, where three different involutions were involved. Now we shall discuss the essential features of CQ*-algebras in their original formulation.

Definition 5.2.2 Let $\mathfrak{A}_\#$ be a C*-algebra, with norm $\|\cdot\|_\#$ and involution $\#$. Let $\mathfrak{A}[\|\cdot\|]$ be a left Banach module over the C*-algebra $\mathfrak{A}_\#$, with isometric involution $*$, such that $\mathfrak{A}_\# \subset \mathfrak{A}$. Set $\mathfrak{A}_\flat = (\mathfrak{A}_\#)^*$. We say that $\{\mathfrak{A}, *, \mathfrak{A}_\#, \#\}$ is a *CQ*-algebra* if

(i) $\mathfrak{A}_\#$ is dense in $\mathfrak{A}$ with respect to its norm $\|\cdot\|$;
(ii) $\mathfrak{A}_0 \equiv \mathfrak{A}_\# \cap \mathfrak{A}_\flat$ is dense in $\mathfrak{A}_\#$ with respect to its norm $\|\cdot\|_\#$;
(iii) $(xy)^* = y^* x^*, \quad \forall\, x, y \in \mathfrak{A}_0$;
(iv) $\|x\|_\# = \sup_{a \in \mathfrak{A}, \|a\| \leq 1} \|xa\|, \; x \in \mathfrak{A}_\#$.

Since $*$ is isometric, it is easy to see that the space $\mathfrak{A}_\flat$ itself is a C*-algebra with respect to the norm $\|x\|_\flat := \|x^*\|_\#$ and the involution $x^\flat := ((x^*)^\#)^*$. Moreover, notice that for every CQ*-algebra $\{\mathfrak{A}, *, \mathfrak{A}_\#, \#\}$, the pair $(\mathfrak{A}[\|\cdot\|], \mathfrak{A}_0)$ is a Banach quasi *-algebra. This follows by the very definitions after some technical calculations.

Remark 5.2.3 It is quite clear that we can restate the previous definition starting from a C*-algebra $\mathfrak{A}_\flat$ and a right module $\mathfrak{A}$ over $\mathfrak{A}_\flat$, with $\mathfrak{A}_\flat \subset \mathfrak{A}$, satisfying the following properties:

(i′) $\mathfrak{A}_\flat$ is dense in $\mathfrak{A}$ with respect to its norm $\|\cdot\|$;
(ii′) $\mathfrak{A}_0 \equiv \mathfrak{A}_\flat \cap \mathfrak{A}_\#$ is dense in $\mathfrak{A}_\flat$ with respect to its norm $\|\cdot\|_\flat$;
(iii′) $(xy)^* = y^*x^*, \quad \forall\, x, y \in \mathfrak{A}_0$;
(iv′) $\|x\|_\flat = \sup_{a\in\mathfrak{A}, \|a\|\leq 1} \|ax\|, \quad x \in \mathfrak{A}_\flat$.

It is then also natural to adopt the notation $\{\mathfrak{A}, *, \mathfrak{A}_\flat, \flat\}$ for indicating a CQ*-algebra as it has been done in many papers on this subject.

Remark 5.2.4 Let $\{\mathfrak{A}, *, \mathfrak{A}_\#, \#\}$ be a CQ*-algebra. If $\mathfrak{A}_\#$ has a unit $e_\#$, then $e_\flat := e_\#^*$ is a unit for $\mathfrak{A}_\flat$. *We say that* $\{\mathfrak{A}, *, \mathfrak{A}_\#, \#\}$ *has a unit* if the quasi *-algebra $(\mathfrak{A}, \mathfrak{A}_0)$ has a unit e. In this case, e is a unit for both $\mathfrak{A}_\#$ and $\mathfrak{A}_\flat$.

The situation for the structure of a CQ*-algebra is illustrated in the diagram of Fig. 5.1, where each arrow denotes a continuous embedding.

According to Definition 5.1.1, a proper CQ*-algebra is then a CQ*-algebra, with $\mathfrak{A}_\# = \mathfrak{A}_\flat = \mathfrak{A}_0$ and $* = \#$ on $\mathfrak{A}_0$.

There are several situations where a scheme like that of Fig. 5.1 below, can be constructed. However, the density conditions required in Definition 5.2.2 are not always fulfilled. We illustrate this point with the two following examples.

Example 5.2.5 (Operators on Scales of Hilbert Spaces) Let $\mathcal{H}$ be a Hilbert space with scalar product $\langle\cdot|\cdot\rangle$ and $\lambda(.,.)$ a positive sesquilinear closed form defined on a dense domain $\mathcal{D}_\lambda \subset \mathcal{H}$. Then, $\mathcal{D}_\lambda$ becomes a Hilbert space, that we denote by $\mathcal{H}_{+1}$,

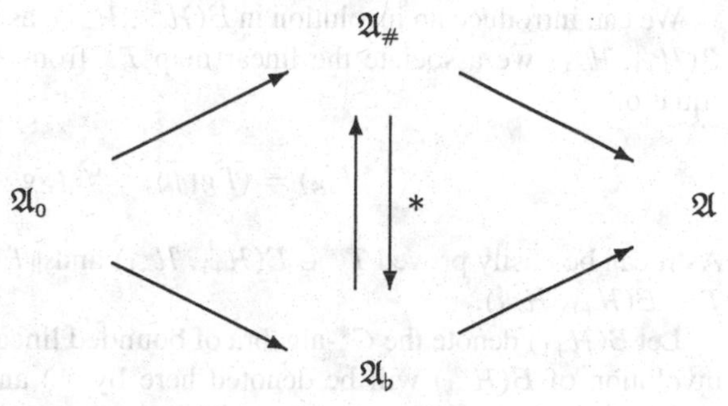

Fig. 5.1 Structure of a CQ*-algebra

with respect to the scalar product

$$\langle f|g\rangle_{+1} = \langle f|g\rangle + \lambda(f,g), \quad \forall\, f, g \in \mathcal{D}_\lambda. \tag{5.2.3}$$

The representation theorem for sesquilinear forms implies [16, Ch. VI, Sect. 2] the existence of a positive selfadjoint operator H such that $D((\mathbb{I}+H)^{1/2}) = \mathcal{D}_\lambda = \mathcal{H}_{+1} \subseteq \mathcal{H}$ and

$$\langle f|g\rangle_{+1} = \langle (\mathbb{I}+H)^{1/2} f|(\mathbb{I}+H)^{1/2} g\rangle, \quad \forall\, f, g \in \mathcal{D}_\lambda. \tag{5.2.4}$$

Let $\mathcal{H}_{-1}$ be the Hilbert space of conjugate linear functionals on $\mathcal{H}_{+1}$. Its natural norm is denoted by $\|\cdot\|_{-1}$. The operator $S = (1+H)^{1/2}$ has a bounded inverse S^{-1}, which maps $\mathcal{H}$ into $\mathcal{H}_\lambda$. As a result, we can write:

$$\langle f|g\rangle_{+1} = \langle Sf|Sg\rangle_{+1} = \langle Uf|Ug\rangle_{-1} \quad \forall\, f, g \in \mathcal{H}_{+1}$$

Here U is the operator from $\mathcal{H}_{+1}$ onto $\mathcal{H}_{-1}$, whose existence is ensured by the Riesz lemma.

This is the canonical way to get a scale of Hilbert spaces [20, VIII.6]

$$\mathcal{H}_{+1} \xrightarrow{i} \mathcal{H} \xrightarrow{j} \mathcal{H}_{-1}, \tag{5.2.5}$$

where $\mathcal{H}_{-1}$ is the conjugate dual space of $\mathcal{H}_{+1}$ and i, j are continuous embeddings with dense range. In fact, the identity map i embeds $\mathcal{H}_{+1}$ in $\mathcal{H}$ and the map $j : \psi \in \mathcal{H} \to j(\psi) \in \mathcal{H}_{-1}$, where $j(\psi)(\phi) := \langle \phi|\psi\rangle$ for all $\phi \in \mathcal{H}_{+1}$, is a linear embedding of $\mathcal{H}$ into $\mathcal{H}_{-1}$. Identifying $\mathcal{H}_{+1}$ and $\mathcal{H}$ with their respective images in $\mathcal{H}_{-1}$ we can read (5.2.5) as a chain of topological inclusions

$$\mathcal{H}_{+1} \subset \mathcal{H} \subset \mathcal{H}_{-1}$$

Let $\mathcal{B}(\mathcal{H}_{+1}, \mathcal{H}_{-1})$ be the Banach space of bounded operators from $\mathcal{H}_{+1}$ into $\mathcal{H}_{-1}$ and let us denote with $\|T\|_{+1,-1}$ the natural norm of T in $\mathcal{B}(\mathcal{H}_{+1}, \mathcal{H}_{-1})$.

We can introduce an involution in $\mathcal{B}(\mathcal{H}_{+1}, \mathcal{H}_{-1})$, as follows: to each element $T \in \mathcal{B}(\mathcal{H}_{+1}, \mathcal{H}_{-1})$ we associate the linear map T^* from $\mathcal{H}_{+1}$ into $\mathcal{H}_{-1}$ defined by the equation

$$\langle T^* f|g\rangle = \overline{\langle Tg|f\rangle}, \quad \forall\, f, g \in \mathcal{H}_{+1}$$

As it can be easily proved $T^* \in \mathcal{B}(\mathcal{H}_{+1}, \mathcal{H}_{-1})$ and $\|T^*\|_{+1,-1} = \|T\|_{+1,-1}$, for every $T \in \mathcal{B}(\mathcal{H}_{+1}, \mathcal{H}_{-1})$.

Let $\mathcal{B}(\mathcal{H}_{+1})$ denote the C^*-algebra of bounded linear operators on $\mathcal{H}_{+1}$ (the usual involution of $\mathcal{B}(\mathcal{H}_{+1})$ will be denoted here by $\flat$) and $\mathcal{B}(\mathcal{H}_{-1})$ the C^*-algebra of bounded operators on $\mathcal{H}_{-1}$ (the natural involution of $\mathcal{B}(\mathcal{H}_{-1})$ is denoted by $\#$). Then, both $\mathcal{B}(\mathcal{H}_{+1})$ and $\mathcal{B}(\mathcal{H}_{-1})$ are contained in $\mathcal{B}(\mathcal{H}_{+1}, \mathcal{H}_{-1})$ and $T \in \mathcal{B}(\mathcal{H}_{+1})$, if and only if, $T^* \in \mathcal{B}(\mathcal{H}_{-1})$. Moreover $(S^\flat)^* = (S^*)^\#$, for every $S \in \mathcal{B}(\mathcal{H}_{+1})$. The following subspace of $\mathcal{B}(\mathcal{H}_{+1}, \mathcal{H}_{-1})$

$$\mathcal{B}^+(\mathcal{H}_{+1}) := \{T \in \mathcal{B}(\mathcal{H}_{+1}, \mathcal{H}_{-1}) : T,\ T^* \in \mathcal{B}(\mathcal{H}_{+1})\}$$

is a *-algebra, as it is easily seen.

Then, one can show that $\mathcal{B}(\mathcal{H}_{+1}, \mathcal{H}_{-1})$ is a Banach module over the normed algebra $\mathcal{B}^+(\mathcal{H}_{+1})$, that can be viewed as a locally convex (Banach) partial *-algebra. This family of spaces realizes the situation depicted in Fig. 5.1, when one takes

- $\mathfrak{A} = \mathcal{B}(\mathcal{H}_{+1}, \mathcal{H}_{-1})$, a Banach space;
- $\mathfrak{A}_\flat = \mathcal{B}(\mathcal{H}_{+1})$, a C*-algebra;
- $\mathfrak{A}_\# = \mathcal{B}(\mathcal{H}_{-1})$, also a C*-algebra;
- $\mathfrak{A}_0 = \mathcal{B}^+(\mathcal{H}_{+1})$.

However, as proved in [2] it is not a CQ*-algebra, because the density conditions (i) and (ii) of Definition 5.2.2 are never satisfied. It is only possible to prove the existence of a maximal CQ*-algebra contained in it. This is obtained by completing $\mathcal{B}^+(\mathcal{H}_{+1})$ with respect to the norm $\|\cdot\|_{+1,-1}$ and using Proposition 5.3.1 of the next subsection.

Example 5.2.6 (Left Hilbert Algebras) For reader's convenience, we briefly review here the definition and the basic properties of left Hilbert algebras. A *-algebra $\mathfrak{A}_0$ with involution $\#$ is said to be a *left Hilbert algebra* [25, Section 10.1] if it is a dense subspace in a Hilbert space $\mathcal{H}$ with inner product $\langle\cdot|\cdot\rangle$ satisfying the following conditions:

(i) for any $x \in \mathfrak{A}_0$, the map $y \in \mathfrak{A}_0 \to xy \in \mathfrak{A}_0$ is continuous;
(ii) $\langle xy|z\rangle = \langle y|x^\# z\rangle, \quad \forall\, x, y, z \in \mathfrak{A}_0$;
(iii) $\mathfrak{A}_0^2 \equiv \{xy : x, y \in \mathfrak{A}_0\}$ is total in $\mathcal{H}$;
(iv) The involution $x \to x^\#$ is closable in $\mathcal{H}$.

By (i), for any $x \in \mathfrak{A}_0$, denote by L_x, the unique continuous linear extension to $\mathcal{H}$ of the map $y \in \mathfrak{A}_0 \to xy \in \mathfrak{A}_0$; then, using (ii), it is easy to see that the map

$$L : x \in \mathfrak{A}_0 \to L_x \in \mathcal{B}(\mathcal{H})$$

is a bounded *-representation of $\mathfrak{A}_0$ on $\mathcal{H}$.

Denote by S the closure of the operator S_0 defined on the dense domain $\mathfrak{A}_0^2$ by

$$y \in \mathfrak{A}_0^2 \mapsto y^\# \in \mathcal{H}.$$

Let $S = J\Delta^{\frac{1}{2}}$ be the polar decomposition of S. Then, J is an isometric involution on $\mathcal{H}$ and Δ is a non-singular positive selfadjoint operator in $\mathcal{H}$, such that $S = J\Delta^{\frac{1}{2}} = \Delta^{-\frac{1}{2}}J$ and $S^* = J\Delta^{-\frac{1}{2}} = \Delta^{\frac{1}{2}}J$; J is called the *modular conjugation operator* of $\mathfrak{A}_0$ and Δ is called the *modular operator* of $\mathfrak{A}_0$.

We introduce now the *commutant* $\mathfrak{A}_0'$ of $\mathfrak{A}_0$ as follows: for any $y \in D(S^*)$ we define a linear map $R_y : D(S^*) \to \mathcal{H}$ by $R_y x := L_x y$, $x \in \mathfrak{A}_0$, and we put

$$\mathfrak{A}_0' := \{y \in D(S^*) : R_y \text{ is bounded}\}.$$

Then, $\mathfrak{A}_0'$ is a *right Hilbert algebra* in $\mathcal{H}$ with involution $y \to y^\flat := S^* y$ and multiplication $y_1 y_2 \equiv R_{y_2} y_1$, $y_1, y_2 \in \mathfrak{A}_0'$.

➽ *We* want to *warn the reader* that *we use here the word commutant* (resp., *bicommutant*) and *the symbol* $'$ (resp., $''$) to call and denote both the (von Neumann) *commutant* of a family of bounded operators (see Example A.1.7) and the "commutant" of a *left Hilbert algebra* as defined above: we follow, in fact, an established tradition.

We recall that a *right Hilbert algebra* $\mathfrak{B}_0$ is defined similarly to the left one: $\mathfrak{B}_0$ is a *-algebra with involution $\flat$, which is a dense subspace in a Hilbert space $\mathcal{H}$, with inner product $\langle\cdot|\cdot\rangle$ satisfying the following conditions:

(i′) for any $y \in \mathfrak{B}_0$, the map $x \in \mathfrak{B}_0 \to xy \in \mathfrak{B}_0$ is continuous;
(ii′) $\langle xy|z\rangle = \langle x|zy^\flat\rangle, \quad \forall\, x, y, z \in \mathfrak{B}_0$;
(iii′) $\mathfrak{B}_0^2 \equiv \{xy : x, y \in \mathfrak{A}\}$ is total in $\mathcal{H}$;
(iv′) the involution $y \to y^\flat$ is closable in $\mathcal{H}$.

Now, the *bicommutant* $\mathfrak{A}_0''$ of $\mathfrak{A}_0$ is defined by

$$\mathfrak{A}_0'' = \{x \in D(S) : y \in \mathfrak{A}_0' \mapsto xy \in \mathcal{H} \text{ is continuous}\}.$$

For any $x \in \mathfrak{A}_0''$ we denote by L_x the unique continuous linear operator on $\mathcal{H}$, such that $L_x y := R_y x$, $y \in \mathfrak{A}_0'$. Then, $\mathfrak{A}_0''$ is a left Hilbert algebra in $\mathcal{H}$ with involution S and multiplication $x_1 x_2 \equiv L_{x_1} x_2$, containing $\mathfrak{A}_0$. A left Hilbert algebra $\mathfrak{A}_0$ is said to be *full* or *achieved* if $\mathfrak{A}_0 = \mathfrak{A}_0''$. It is well-known, as the Tomita fundamental theorem says, that $JL(\mathfrak{A}_0)''J = L(\mathfrak{A}_0)'$ and $\Delta^{it} L(\mathfrak{A}_0)'' \Delta^{-it} = L(\mathfrak{A}_0)''$, for every $t \in \mathbb{R}$; here $L(\mathfrak{A}_0) = \{L_x, x \in \mathfrak{A}_0\}$, with L as in Example 5.2.6. Let $\mathfrak{A}_0$ be a full left Hilbert algebra in $\mathcal{H}$, and

$$\mathfrak{A}_{00} \equiv \left\{x \in \bigcap_{\alpha \in \mathbb{C}} D(\Delta^\alpha) : \Delta^\alpha x \in \mathfrak{A}_0, \forall\, \alpha \in \mathbb{C}\right\}.$$

Then, $\mathfrak{A}_{00}$ is a left Hilbert subalgebra in $\mathcal{H}$, such that $\mathfrak{A}_{00}'' = \mathfrak{A}_0$, $J\mathfrak{A}_{00} = \mathfrak{A}_{00}$ [27, VI, p. 22, Theorem 2.2]; $\{\Delta^\alpha : \alpha \in \mathbb{C}\}$ is a complex one-parameter group of automorphisms of $\mathfrak{A}_{00}$, such that

$$(\Delta^\alpha x)^\# = \Delta^{-\overline{\alpha}} x^\# \text{ and } (\Delta^\alpha x)^* = \Delta^{-\overline{\alpha}} x^*, \quad \forall\, \alpha \in \mathbb{C} \text{ and } x \in \mathfrak{A}_{00}.$$

The left Hilbert subalgebra $\mathfrak{A}_{00}$ is called the *maximal Tomita algebra* of $\mathfrak{A}_0$. For more details, we refer to [25–27, 90].

Let now $\mathfrak{A}_0$ be a full left Hilbert algebra with unit e and involution $\#$ in $\mathcal{H}$. Then, as seen above, the commutant $\mathfrak{A}_0'$ of $\mathfrak{A}_0$ is a full right Hilbert algebra in $\mathcal{H}$ with (the same) unit and involution $\flat$. The involution in $\mathcal{H}$ is defined by the modular conjugation operator J. For short we put $\mathcal{H}_\flat = \mathfrak{A}_0'$ and $\mathcal{H}_\# = \mathfrak{A}_0$. Consider now the system $\{\mathcal{H}, J, \mathcal{H}_\flat, \flat\}$ and introduce a topological structure in it. We start by defining for $y \in \mathcal{H}_\flat$,

$$\|y\|_\flat \equiv \|R_y\| = \sup_{\|x\|\leq 1} \|xy\|,$$

where R denotes the regular *-representation of $\mathfrak{A}_0'$ in $\mathcal{B}(\mathcal{H})$. We also define $\|x\|_\# := \|Jx\|_\flat$, for every $x \in \mathcal{H}_\#$.

Is $\{\mathcal{H}, J, \mathcal{H}_\#, \#\}$ a CQ*-algebra? First of all, we observe that conditions (i) and (iv) of Definition 5.2.2 are obviously fulfilled, whereas condition (iii) follows from the known equality $(Jx)^\flat = Jx^\#$, for every $x \in \mathcal{H}_\#$. As for (ii), the C*-property for the norm $\|\cdot\|_\flat$ is easily obtained from the fact that the linear map $y \mapsto R_y$ is a *-representation of $\mathcal{H}_\flat$ into $\mathcal{B}(\mathcal{H})$.

To show the completeness of $\mathcal{H}_\flat = \mathfrak{A}_0'$, one has to take into account the equality:

$$\mathfrak{A}_0' = \{y \in D(S^*) : R_y \text{ is bounded}\}.$$

Now, if $\{y_k\}$ is a $\|\cdot\|_\flat$-Cauchy sequence in $\mathcal{H}_\flat$, since $e \in \mathfrak{A}_0'$, one can find an element $y \in \mathcal{H}$, such that y_k converges to y with respect to the Hilbert norm; moreover, since as is known, for each $y \in \mathfrak{A}_0'$, $y^\flat = S^*y$, the sequence $\{S^*y_k\}$ is also convergent. Therefore, $y \in D(S^*)$. The fact that R_y is bounded follows easily from the norm completeness of $\mathcal{B}(\mathcal{H})$.

To conclude that $\{\mathcal{H}, J, \mathcal{H}_\#, \#\}$ is a CQ*-algebra, we should prove the density of $\mathcal{H}_\flat \cap \mathcal{H}_\#$ in $\mathcal{H}_\flat$ with respect to $\|\cdot\|_\flat$. This question is still open; however in [25, Section 10.19] it is shown that the set

$$\{f_r(\Delta) f_s(\Delta^{-1})y : y \in \mathcal{H}_\flat,\ r, s > 0\},$$

where $f_m(x) = \exp(-mx)$ and Δ is the modular operator, is contained in $\mathcal{H}_\flat \cap \mathcal{H}_\#$. This set is, in a sense, quite rich; indeed, a simple application of the spectral theorem for the operator Δ and the Lebesgue dominated convergence theorem shows that $f_r(\Delta) f_s(\Delta^{-1})y$ converges to y with respect to the Hilbert norm, for each $y \in \mathcal{H}_\flat$. A deeper analysis of these points can be found in Sect. 5.5.

Remark 5.2.7 In Chap. 3, given a Banach quasi *-algebra $(\mathfrak{A}[\|\cdot\|], \mathfrak{A}_0)$, an important role has been played by the family of sesquilinear forms $\mathcal{S}_{\mathfrak{A}_0}(\mathfrak{A})$ of Definition 3.1.14. As we have noticed, from the very definitions follows that, if $\{\mathfrak{A}, *, \mathfrak{A}_\#, \#\}$ is a CQ*-algebra, then $(\mathfrak{A}[\|\cdot\|], \mathfrak{A}_0)$ is a Banach quasi *-algebra. Hence, it makes sense to consider the set $\mathcal{S}_{\mathfrak{A}_0}(\mathfrak{A})$ also in this case. The density properties required in Definition 5.2.2 make stronger the condition (ii) of Definition 3.1.14. In fact, the following condition holds: for every $\varphi \in \mathcal{S}_{\mathfrak{A}_0}(\mathfrak{A})$,

$$\varphi(xa, y) = \varphi(x, ya^*), \quad \forall\, a \in \mathfrak{A},\ x, y \in \mathfrak{A}_\#; \tag{5.2.6}$$

$$\varphi(ax, y) = \varphi(x, a^* y), \quad \forall\, a \in \mathfrak{A},\ x, y \in \mathfrak{A}_\flat. \tag{5.2.7}$$

For this reason, when dealing with a CQ*-algebra $\{\mathfrak{A}, *, \mathfrak{A}_\#, \#\}$ we shall not distinguish with a different notation the family of sesquilinear forms satisfying (5.2.6) or (5.2.7), which are equivalent.

5.3 Construction of CQ*-Algebras

5.3.1 Constructive Method

The next proposition extends the constructive Proposition 5.1.3.

Proposition 5.3.1 *Let $\mathfrak{A}_\#$ be a C*-algebra, with norm $\|\cdot\|_\#$ and involution $\#$; let $\|\cdot\|$ be another norm on $\mathfrak{A}_\#$, weaker than $\|\cdot\|_\#$ and such that*

(i) $\|xy\| \le \|x\|_\# \|y\|, \quad \forall\, x, y \in \mathfrak{A}_\#$;
(ii) *there exists a $\|\cdot\|_\#$-dense subalgebra $\mathfrak{A}_0$ of $\mathfrak{A}_\#$, where an involution* $*$ (which makes $\mathfrak{A}_0$ into a *-algebra) *is defined with the property*

$$\|x^*\| = \|x\|, \quad \forall\, x \in \mathfrak{A}_0.$$

Let $\mathfrak{A}$ be the $\|\cdot\|$-completion of $\mathfrak{A}_\#$. Then, $\{\mathfrak{A}, *, \mathfrak{A}_\#, \#\}$ is a CQ-algebra.*

Proof Since $\mathfrak{A}_0$ is $\|\cdot\|_\#$-dense in $\mathfrak{A}_\#$, then the $\|\cdot\|$-completions of $\mathfrak{A}_0$ and $\mathfrak{A}_\#$ can be identified with the same Banach space $\mathfrak{A}$. Now, for $x \in \mathfrak{A}_\#$, put

$$\|x\|_{\widetilde{\#}} = \sup_{\|a\| \le 1} \|xa\|. \tag{5.3.8}$$

Because of (i) we have

$$\|x\|_{\widetilde{\#}} \le \|x\|_\#, \quad \forall\, x \in \mathfrak{A}_\#.$$

The converse inequality follows from Lemma 5.1.2. □

Corollary 5.3.2 *Let $\mathfrak{A}_\#$ be a C*-algebra and $\mathfrak{A}$ a left Banach $\mathfrak{A}_\#$-module with involution $*$ and $\mathfrak{B}_0$ any *-subalgebra of $\mathfrak{A}_\#^* \cap \mathfrak{A}_\#$, which is also $\#$-invariant. Let $\mathcal{M}(\mathfrak{B}_0)_\#$ be the closure of $\mathfrak{B}_0$ in $\mathfrak{A}_\#$ and $\mathcal{M}(\mathfrak{B}_0)$ the closure of $\mathfrak{B}_0$ in $\mathfrak{A}$. Then, $\{\mathcal{M}(\mathfrak{B}_0), *, \mathcal{M}(\mathfrak{B}_0)_\#, \#\}$ is a CQ*-algebra.*

Proof We notice that $\mathcal{M}(\mathfrak{B}_0)_\#$ is a C*-algebra, with respect to the involution # and the norm $\|\cdot\|_\#$, since $\mathfrak{B}_0$ is an involutive algebra also with respect to #. The statement then follows from the previous proposition. □

Let us now give some explicit applications of Proposition 5.3.1.

Example 5.3.3 Let S be an unbounded selfadjoint operator in a Hilbert space $\mathcal{H}$ with domain $D(S)$ and with bounded inverse S^{-1} in $\mathcal{B}(\mathcal{H})$, such that $\|S^{-1}\| \leq 1$. We define the commutant of the operator S^{-1}, as follows

$$(S^{-1})' := \left\{T \in \mathcal{B}(\mathcal{H}) : TS^{-1} = S^{-1}T\right\}. \tag{5.3.9}$$

It is straightforward to check that $(S^{-1})'$ is a C*-algebra (indeed, a von Neumann algebra), being a norm closed *-subalgebra of $\mathcal{B}(\mathcal{H})$. Moreover, if S has not simple spectrum, $(S^{-1})'$ is not abelian [1, II, Ch.VI, n.69]. Define a norm $\|\cdot\|_{\mathfrak{C}}$ on $(S^{-1})'$, weaker than the operator norm on $\mathcal{B}(\mathcal{H})$, in the following way

$$\|T\|_{\mathfrak{C}} := \|S^{-1}TS^{-1}\| = \|S^{-2}T\| = \|TS^{-2}\|. \tag{5.3.10}$$

If we call $\mathfrak{C}$ the $\|\cdot\|_{\mathfrak{C}}$-completion of $(S^{-1})'$, Proposition 5.3.1 ensures us that $(\mathfrak{C}[\|\cdot\|_{\mathfrak{C}}], (S^{-1})')$ is a proper nonabelian CQ*-algebra. The non triviality of the construction follows from the fact that $\|\cdot\|_{\mathfrak{C}}$ is not equivalent to $\|\cdot\|$, as it is easily checked. We now prove the *-semisimplicity of $(\mathfrak{C}[\|\cdot\|_{\mathfrak{C}}], (S^{-1})')$.

First we observe that any sesquilinear form of the following type

$$\varphi_\xi(A, B) = \langle S^{-1}AS^{-1}\xi | S^{-1}BS^{-1}\xi\rangle, \ A, B \in \mathfrak{C}[\|\cdot\|_{\mathfrak{C}}],$$

belongs to $\mathcal{S}_{(S^{-1})'}(\mathfrak{C})$, for any $\xi \in \mathcal{H}$, with $\|\xi\| \leq 1$. Therefore, if $\varphi(A, A) = 0$, for all $\varphi \in \mathcal{S}_{(S^{-1})'}(\mathfrak{C})$, it follows, in particular, that $\varphi_\xi(A, A) = 0$, for every $\xi \in \mathcal{H}$. This implies that $S^{-1}AS^{-1} = 0$ and therefore that $A = 0$.

Example 5.3.4 As before, let $\mathcal{H}$ be a Hilbert space with scalar product $\langle\cdot|\cdot\rangle$ and consider the triplet of Hilbert spaces $\mathcal{H}_{+1} \subset \mathcal{H} \subset \mathcal{H}_{-1}$ as generated by a positive selfadjoint operator S, with dense domain $D(S)$ and $S \geq I$, as in Example 5.2.5. In that example it was claimed that $\{\mathcal{B}(\mathcal{H}_{+1}, \mathcal{H}_{-1}), *, \mathcal{B}(\mathcal{H}_{-1}), \#\}$ is never a CQ*-algebra. Nevertheless, Proposition 5.3.1 provides a canonical way of constructing a CQ*-algebra of operators acting in the given scale of Hilbert spaces.

Indeed, since $\mathcal{B}^+(\mathcal{H}_{+1}) \subset \mathcal{B}(\mathcal{H}_{-1})$, we may consider the largest *-subalgebra $\mathcal{B}_0$ of $\mathcal{B}^+(\mathcal{H}_{-1})$, which is also invariant with respect to the involution # and define $\mathcal{B}_c(\mathcal{H}_{+1})$ as the C*-subalgebra of $\mathcal{B}(\mathcal{H}_{-1})$ generated by $\mathcal{B}_0$. The non triviality of this set is discussed in [2, Section 10.4], where a characterization of $\mathcal{B}_0$ is also given. Then, the conditions of Proposition 5.3.1 are fulfilled, by choosing the weaker norm on $\mathcal{B}_c(\mathcal{H}_{+1})$ as equal to $\|\cdot\|_{+1,-1}$. Therefore, if we denote with $\mathcal{B}_c(\mathcal{H}_{+1}, \mathcal{H}_{-1})$ the subspace of $\mathcal{B}(\mathcal{H}_{+1}, \mathcal{H}_{-1})$ obtained by completing $\mathcal{B}_c(\mathcal{H}_{-1})$ with respect to the norm $\|\cdot\|_{+1,-1}$ we obtain the CQ*-algebra $\{\mathcal{B}_c(\mathcal{H}_{+1}, \mathcal{H}_{+1}), *, \mathcal{B}_c(\mathcal{H}_{+1}), \#\}$.

Example 5.3.5 (CQ-Algebras of Compact Operators)* The same approach can be repeated starting from any C*-subalgebra $\mathcal{Q}$ of $\mathcal{B}_c(\mathcal{H}_{+1})$, since conditions (i) and (ii) of Proposition 5.3.1 are satisfied, whenever the weaker norm $\|\cdot\|$ is just $\|\cdot\|_{+1,-1}$ and the adjoint is the one in $\mathcal{B}(\mathcal{H}_{+1},\mathcal{H}_{-1})$. In particular, we give now an example, with $\mathcal{H}$ separable, where all the spaces involved are explicitly identified.

We start introducing the following sets of operators

$$\mathfrak{A} = \left\{ T \in \mathcal{B}(\mathcal{H}_{+1},\mathcal{H}_{-1}) : S^{-1}TS^{-1} \text{ is compact in } \mathcal{H} \right\},$$

$$\mathfrak{A}_\flat = \left\{ T \in \mathcal{B}(\mathcal{H}_{+1}) : STS^{-1} \text{ is compact in } \mathcal{H} \right\},$$

$$\mathfrak{A}_\# = \left\{ T \in \mathcal{B}(\mathcal{H}_{-1}) : S^{-1}TS \text{ is compact in } \mathcal{H} \right\}.$$

These sets are non empty: for instance, $\mathfrak{A}$ contains any operator of the form SZS, with Z compact in $\mathcal{H}$. As in the previous example, we indicate with the same symbol, S, both the operator from $\mathcal{H}_{+1}$ into $\mathcal{H}$ and its extension from $\mathcal{H}$ into $\mathcal{H}_{-1}$. The sets above coincide with the following ones:

$$\mathfrak{A} = \left\{ T \in \mathcal{B}(\mathcal{H}_{+1},\mathcal{H}_{-1}) : T \text{ is compact from } \mathcal{H}_{+1} \text{ into } \mathcal{H}_{-1} \right\},$$

$$\mathfrak{A}_\flat = \left\{ T \in \mathcal{B}(\mathcal{H}_{+1}) : T \text{ is compact in } \mathcal{H}_{+1} \right\},$$

$$\mathfrak{A}_\# = \left\{ T \in \mathcal{B}(\mathcal{H}_{-1}) : T \text{ is compact in } \mathcal{H}_{-1} \right\}.$$

It is easy to check that $\mathfrak{A}_\flat$ is a C*-algebra with respect to the involution $\flat$ and to the norm $\|\cdot\|_\flat = \|S\cdot S^{-1}\|$. Analogously $\mathfrak{A}_\#$ is a C*-algebra with respect to the involution $\# = *\flat*$ and to the norm $\|\cdot\|_\# = \|S^{-1}\cdot S\|$, while $\mathfrak{A}$ is a Banach space with respect to the involution $*$ and to the norm $\|\cdot\| = \|S^{-1}\cdot S^{-1}\|$. The norms $\|\cdot\|_\flat$, $\|\cdot\|_\#$ and $\|\cdot\|$ coincide with those defined in $\{\mathcal{B}(\mathcal{H}_{+1},\mathcal{H}_{-1}), *, \mathcal{B}(\mathcal{H}_{-1}), \#\}$ and the involutions $\#$ and $*$ are the ones defined respectively in $\mathcal{B}(\mathcal{H}_{-1})$ and $\mathcal{B}(\mathcal{H}_{+1},\mathcal{H}_{-1})$.

In order to prove the density conditions, let us consider the family of projection operators P_ξ, $\xi \in \mathcal{H}_{+1}$ with $\|\xi\| = 1$ (the norm in $\mathcal{H}$) defined by

$$P_\xi \eta = \langle \eta | \xi \rangle \xi, \quad \eta \in \mathcal{H}.$$

Each operator P_ξ has an obvious extension to $\mathcal{H}_{-1}$, which we still call P_ξ. It is straightforward to prove that $P_\xi \in \mathcal{B}(\mathcal{H}_{+1})$, $P_\xi \in \mathcal{B}(\mathcal{H}_{-1})$ and $P_\xi \in \mathfrak{A}_\flat$. Let $\mathfrak{A}_0$ be the subalgebra of $\mathfrak{A}_\flat$ generated by all the operators P_ξ, $\xi \in \mathcal{H}_{+1}$. This is closed with respect to the adjoint $*$ and it is also $\|\cdot\|_\flat$-dense in $\mathfrak{A}_\flat$, since any compact operator is the norm limit of operators of finite rank. Moreover, it is also $\|\cdot\|$-dense in $\mathfrak{A}$. Applying Proposition 5.3.1, we conclude that $\{\mathfrak{A}, *, \mathfrak{A}_\flat, \flat\}$ is a CQ*-algebra of operators.

5.3.2 *Starting from a *-Algebra*

Let $\mathfrak{A}_0[\|\cdot\|]$ be a *-algebra endowed with a norm $\|\cdot\|$. Suppose that the involution $*$ is isometric with respect to $\|\cdot\|$ and the multiplication is separately (but not jointly) continuous. Then, as we have seen in the previous sections, the pair $(\mathfrak{A}[\|\cdot\|], \mathfrak{A}_0)$, where $\mathfrak{A}[\|\cdot\|]$ is the completion $\widetilde{\mathfrak{A}_0}[\|\cdot\|]$ of $\mathfrak{A}_0[\|\cdot\|]$, is a Banach quasi *-algebra. As before, for any $a \in \mathfrak{A}$, we put

$$L_a x = ax \text{ and } R_a x = xa, \quad x \in \mathfrak{A}_0.$$

Then, L_a and R_a are linear maps of $\mathfrak{A}_0$ into $\mathfrak{A}[\|\cdot\|]$. In particular, if $z \in \mathfrak{A}_0$, then L_z and R_z can be extended to bounded linear operators on the Banach space $\mathfrak{A}[\|\cdot\|]$ and they are denoted by the same symbols L_z and R_z.

Let $(\mathfrak{A}[\|\cdot\|], \mathfrak{A}_0)$ be a Banach quasi *-algebra and assume that the *-algebra $\mathfrak{A}_0$ has another norm $\|\cdot\|_\#$ and another involution $\#$ satisfying the following conditions:

(a.1) $\|x^\# x\|_\# = \|x\|_\#^2, \quad \forall\, x \in \mathfrak{A}_0$;
(a.2) $\|x\| \leq \|x\|_\#, \quad \forall\, x \in \mathfrak{A}_0$;
(a.3) $\|xy\| \leq \|x\|_\# \|y\|, \quad \forall\, x, y \in \mathfrak{A}_0$.

Then, by (a.2), the identity map $i : \mathfrak{A}_0[\|\cdot\|_\#] \to \mathfrak{A}_0[\|\cdot\|]$ has a continuous extension $\widehat{i}$ from the completion $\mathfrak{A}_\#$ of $\mathfrak{A}[\|\cdot\|_\#]$ ($\mathfrak{A}_\#$ is, of course, a C*-algebra) into $\mathfrak{A}[\|\cdot\|]$. If $\widehat{i}$ is injective, then $\mathfrak{A}_\#$ is (identified with) a dense subspace of $\mathfrak{A}$. This happens, if and only if,

(a.4) the two norms $\|\cdot\|$ and $\|\cdot\|_\#$ are compatible in the following sense [11]: for any sequence $\{x_n\} \subset \mathfrak{A}_0$, such that $\|x_n\| \to 0$ and $x_n \to x$ in $\mathfrak{A}_\#[\|\cdot\|_\#]$, one obtains $x = 0$; i.e., this happens when $\widehat{i}^{-1} : \mathfrak{A}_0[\|\cdot\|] \to \mathfrak{A}_\#[\|\cdot\|_\#]$ is closable.

Definition 5.3.6 A Banach quasi *-algebra $(\mathfrak{A}[\|\cdot\|], \mathfrak{A}_0)$ is said to be a *pseudo CQ*-algebra*, if the *-algebra $\mathfrak{A}_0$ has another norm $\|\cdot\|_\#$ and another involution $\#$ satisfying the conditions (a.1)–(a.4) above. A pseudo CQ*-algebra $(\mathfrak{A}[\|\cdot\|], \mathfrak{A}_0)$ is said to be a *strict CQ*-algebra*, if

$$\|x\|_\# = \|L_x\| \equiv \sup\{\|xy\| : y \in \mathfrak{A}_0 \text{ with } \|y\| \leq 1\}, \quad \forall\, x \in \mathfrak{A}_0. \tag{5.3.11}$$

Proposition 5.3.7 *Let $(\mathfrak{A}[\|\cdot\|], \mathfrak{A}_0)$ be a strict CQ*-algebra and $\mathfrak{A}_\#$ the C*-algebra obtained by completing $\mathfrak{A}_0$ with respect to $\|\cdot\|_\#$. Then, $\{\mathfrak{A}, *, \mathfrak{A}_\#, \#\}$ is a CQ*-algebra. Conversely, if $\{\mathfrak{A}, *, \mathfrak{A}_\#, \#\}$ is a CQ*-algebra and $\mathfrak{A}_0 := \mathfrak{A}_\# \cap \mathfrak{A}_\flat$, then $(\mathfrak{A}[\|\cdot\|], \mathfrak{A}_0)$ is a strict CQ*-algebra.*

Proof As discussed before, if $(\mathfrak{A}[\|\cdot\|], \mathfrak{A}_0)$ is a pseudo CQ*-algebra, then the completion $\mathfrak{A}_\#$ of $\mathfrak{A}_0[\|\cdot\|_\#]$ is a C*-algebra identified with a subspace of $\mathfrak{A}$ (which we call $\mathfrak{A}_\#$ too) and the same holds, of course, for $\mathfrak{A}_\flat := \mathfrak{A}_\#^*$, with respect to the norm $\|x\|_\flat := \|x^*\|_\#$. The density properties required in Definition 5.2.2 can be derived from the construction itself. It remains to show that (iv) of Definition 5.2.2 holds, that is

$$\|x\|_\# = \sup_{a\in\mathfrak{A},\|a\|\le 1} \|xa\|, \quad \forall\, x \in \mathfrak{A}_\#. \tag{5.3.12}$$

But this follows easily from the density properties of $\mathfrak{A}_0$ and from (5.3.11).

Conversely, if $\{\mathfrak{A}, *, \mathfrak{A}_\#, \#\}$ is a CQ*-algebra and $\mathfrak{A}_0 := \mathfrak{A}_\# \cap \mathfrak{A}_\flat$, then the pair $(\mathfrak{A}[\|\cdot\|], \mathfrak{A}_0)$ is a Banach quasi *-algebra. It is clear that the norm $\|\cdot\|_\#$ and the involution $\#$ of $\mathfrak{A}_\#$, restricted to $\mathfrak{A}_0$, satisfy the conditions (a.1)–(a.3) above. As for (a.4), we just notice that, for any sequence $\{x_n\} \subset \mathfrak{A}_0$, such that $\|x_n\| \to 0$ and $\|x_n - x\|_\# \to 0$, for some $x \in \mathfrak{A}_\#$, we necessarily have $x = 0$, since $\|y\| \le \|y\|_\#$, for every $y \in \mathfrak{A}_\#$. □

- A *pseudo* CQ*-algebra $(\mathfrak{A}[\|\cdot\|], \mathfrak{A}_0)$ *is fully determined* by the involution $\#$ and the C*-norm $\|\cdot\|_\#$. For this reason, *it will be often denoted* by $(\mathfrak{A}[\|\cdot\|], \#, \mathfrak{A}_0, \|\cdot\|_\#)$, in the case when it will be necessary to indicate explicitly the different norms.
- A *strict* CQ*-algebra *is fully determined*, when the new involution $\#$ is known; so *it can be simply denoted* by $(\mathfrak{A}[\|\cdot\|], \mathfrak{A}_0, \#)$.

Let $(\mathfrak{A}[\|\cdot\|], \#, \mathfrak{A}_0, \|\cdot\|_\#)$ be a pseudo CQ*-algebra and, as above, $\mathfrak{A}_\#$ the C*-algebra obtained by completing the #-algebra $\mathfrak{A}_0$ with respect to the C*-norm $\|\cdot\|_\#$. Let $J_\mathfrak{A}$ be the involution $*$ of the Banach quasi *-algebra $(\mathfrak{A}[\|\cdot\|], \mathfrak{A}_0)$. Then, as noted before, $\mathfrak{A}_\flat \equiv J_\mathfrak{A}\mathfrak{A}_\#$ is a C*-algebra with involution $x^\flat \equiv x^{*\#*}$ $\big($where $x^{*\#*} = \big((x^*)^\#\big)^*\big)$ and C*-norm $\|x^*\|_\flat \equiv \|x\|_\#$, for all $x \in \mathfrak{A}_\flat$.

If $\mathfrak{M} \subset \mathfrak{A}_\#$ let $L_\mathfrak{M} = \{L_x : x \in \mathfrak{M}\}$ and if $\mathfrak{N} \subset \mathfrak{A}_\flat$ let $R_\mathfrak{N} = \{R_y : y \in \mathfrak{N}\}$.

Proposition 5.3.8 *A pseudo CQ*-algebra $(\mathfrak{A}[\|\cdot\|], \#, \mathfrak{A}_0, \|\cdot\|_\#)$ contains two C*-algebras $\mathfrak{A}_\#$ and $\mathfrak{A}_\flat \equiv J_\mathfrak{A}\mathfrak{A}_\#$ with different involutions $\#$ and $\flat$, respectively, as dense subalgebras. In particular, if $(\mathfrak{A}[\|\cdot\|], \mathfrak{A}_0, \#)$ is a strict CQ*-algebra, then $L_{\mathfrak{A}_\#}$ and $R_{\mathfrak{A}_\flat}$ are C*-algebras, $L_xR_y = R_yL_x$, for each $x \in \mathfrak{A}_\#$, $y \in \mathfrak{A}_\flat$ and $R_{\mathfrak{A}_\flat} = J_\mathfrak{A} L_{\mathfrak{A}_\#} J_\mathfrak{A}$.*

We summarize the situation with the following scheme

$$\begin{array}{ccccc} & \subset & \mathfrak{A}_\# & \subset & \\ \mathfrak{A}_0[\|\cdot\|] & & \updownarrow J & & \mathfrak{A}[\|\cdot\|] \\ & \subset & \mathfrak{A}_\flat & \subset & \\ \text{*-algebra, with norm} & & \text{C*-algebras} & & \text{pseudo CQ*-algebra,} \end{array}$$

i.e., the *-algebra $\mathfrak{A}_0[\|\cdot\|]$ is contained in its closure $\mathfrak{A}_\flat = \widetilde{\mathfrak{A}_0}[\|\cdot\|_\flat]$ and $\mathfrak{A}_\# = \widetilde{\mathfrak{A}_0}[\|\cdot\|_\#] = J\mathfrak{A}_\flat$. Moreover, these C*-algebras are both contained in $\mathfrak{A}[\|\cdot\|]$.

We shall come back to pseudo CQ*-algebras and strict CQ*-algebras in Sect. 5.5.

5.3.3 Construction of CQ*-Algebras through Families of Forms

In Sect. 3.1.3 we discussed *-semisimple Banach quasi *-algebras and we obtained several consequences of this definition; for *-semisimplicity of CQ*-algebras, see discussion before Proposition 5.4.7. Here we introduce a different definition of semisimplicity, the #-semisimplicity, that allows to define a new norm satisfying the assumptions of Proposition 5.3.1. The application of this proposition leads to the construction of a new CQ*-algebra, whose norm closely reminds the characterization of a C*-norm in terms of states given by Gelfand.

Definition 5.3.9 Let $\{\mathfrak{A}, *, \mathfrak{A}_\#, \#\}$ be a CQ*-algebra. We denote as $\mathcal{T}_\#(\mathfrak{A})$ the set of sesquilinear forms φ on $\mathfrak{A} \times \mathfrak{A}$ with the following properties:

(i) $\varphi(a, a) \geq 0, \quad \forall\, a \in \mathfrak{A}$;
(ii) $\varphi(xa, b) = \varphi(a, x^\# b), \quad \forall\, a, b \in \mathfrak{A}, \forall\, x \in \mathfrak{A}_\#$;
(iii) $|\varphi(a, b)| \leq \|a\|\,\|b\|, \quad \forall\, a, b \in \mathfrak{A}$;
(iv) $\varphi(a, b) = \varphi(b^*, a^*), \quad \forall\, a, b \in \mathfrak{A}$.

The CQ*-algebra $\{\mathfrak{A}, *, \mathfrak{A}_\#, \#\}$ is called #-semisimple, if $a \in \mathfrak{A}$ with $\varphi(a, a) = 0$, for all $\varphi \in \mathcal{T}_\#(\mathfrak{A})$, implies $a = 0$.

It is worth remarking that while conditions (i) and (iii) were already present in the definition of the family $\mathcal{S}_{\mathfrak{A}_0}(\mathfrak{A})$, condition (iv) is peculiar for this family of forms, and (ii) is a natural modification of that for $\mathcal{S}_{\mathfrak{A}_0}(\mathfrak{A})$. The non triviality of this definition will appear clearer in Sect. 5.5, where it is shown how the Tomita–Takesaki theory naturally provides examples of sesquilinear forms of this kind.

In a very similar way, we could speak of $\flat$-semisimplicity, simply starting with a family of sesquilinear forms $\mathcal{T}_\flat(\mathfrak{A})$, where (ii) is replaced by the specular condition

(ii′) $\varphi(ay, b) = \varphi(a, by^\flat), \quad \forall\, a, b \in \mathfrak{A}$ and $y \in \mathfrak{A}_\flat$,

while the other conditions are kept fixed. However, *due to condition* (iv) *and* to *the equality* $(x^\flat)^* = (x^*)^\#$, for each $x \in \mathfrak{A}_\flat$, it is easily seen that *the two families* $\mathcal{T}_\flat(\mathfrak{A})$ and $\mathcal{T}_\#(\mathfrak{A})$ *coincide*. By the way, *it is also interesting to remark that* without condition (iv) in the definition of the two preceding sets, *their equality is replaced by a weaker, but still interesting, result*: there is a *one-to-one correspondence between their sesquilinear forms*, given by

$$\mathcal{T}_\flat(\mathfrak{A}) \ni \varphi \mapsto \varphi^* \in \mathcal{T}_\#(\mathfrak{A}) \text{ with } \varphi^*(a, b) \equiv \overline{\varphi(b, a)}, \quad \forall\, a, b \in \mathfrak{A}.$$

It is easy to check that $\varphi \in \mathcal{T}_\flat(\mathfrak{A})$, if and only if, $\varphi^* \in \mathcal{T}_\#(\mathfrak{A})$. It is evident that, either if condition (iv) is assumed or not, $\flat$-semisimplicity and #-semisimplicity are equivalent.

One of the main reasons for the introduction of the set $\mathcal{T}_\#(\mathfrak{A})$ is that it allows, by means of the constructive Proposition 5.3.1, to build up examples of CQ*-algebras starting with a given #-semisimple CQ*-algebra $\{\mathfrak{A}, *, \mathfrak{A}_\#, \#\}$. First, we introduce on $\mathfrak{A}$ a new norm

$$\|a\|_{\mathcal{T}} := \sup_{\varphi \in \mathcal{T}_\#(\mathfrak{A})} \varphi(a, a)^{1/2}, \ a \in \mathfrak{A}. \tag{5.3.13}$$

It is not difficult to see that this is really a norm. In particular, the #-semisimplicity implies that $\|a\|_{\mathcal{T}} = 0$, if and only if, $a = 0$. Property (iv) of the family $\mathcal{T}_\#(\mathfrak{A})$ implies that $\|a\|_{\mathcal{T}} = \|a^*\|_{\mathcal{T}}$, $a \in \mathfrak{A}$. Furthermore, condition (iii) implies that $\|a\|_{\mathcal{T}} \leq \|a\|$, for any $a \in \mathfrak{A}$. In order to apply Proposition 5.3.1 we still have to check that the following inequality holds

$$\|xy\|_{\mathcal{T}} \leq \|x\|_\# \|y\|_{\mathcal{T}}, \quad \forall\, x, y \in \mathfrak{A}_\#. \tag{5.3.14}$$

Indeed, defining $\omega(x) := \varphi(x, e)$, for $x \in \mathfrak{A}_\#$, we get $\omega(x^\# x) \geq 0$. This implies that ω is also continuous and that the following inequality holds,

$$|\omega(y^\# xy)| \leq \omega(y^\# y)\|x\|_\#, \quad \forall\, x, y \in \mathfrak{A}_\#.$$

Now, inequality (5.3.14) is an immediate consequence of the definition of $\|\cdot\|_{\mathcal{T}}$.

Applying Proposition 5.3.1 we can conclude that $\{\mathfrak{A}_{\mathcal{T}}, *, \mathfrak{A}_\#, \#\}$, where $\mathfrak{A}_{\mathcal{T}}$ is the completion of $\mathfrak{A}_\#$ under the norm $\|\cdot\|_{\mathcal{T}}$, is a CQ*-algebra, *containing* $\{\mathfrak{A}, *, \mathfrak{A}_\#, \#\}$. Indeed, $\mathfrak{A} \subset \mathfrak{A}_{\mathcal{T}}$, since both $\mathfrak{A}$ and $\mathfrak{A}_{\mathcal{T}}$ are completions of the same C*-algebra $\mathfrak{A}_\#$ with respect to the norms $\|\cdot\|$ and $\|\cdot\|_{\mathcal{T}}$ that satisfy the condition $\|\cdot\|_{\mathcal{T}} \leq \|\cdot\|$ and are compatible in the sense of [11]; this means, that it is possible to extend by continuity the identity map $i : \mathfrak{A}_\flat[\|\cdot\|] \to \mathfrak{A}_\flat[\|\cdot\|_{\mathcal{T}}]$ to the respective completions of the normed algebras involved: i.e., $\widehat{i} : \mathfrak{A}[\|\cdot\|] \to \mathfrak{A}_{\mathcal{T}}[\|\cdot\|_{\mathcal{T}}]$, preserving its injectivity.

Let us now prove the following

Lemma 5.3.10 *Let* $\{\mathfrak{A}, *, \mathfrak{A}_\#, \#\}$ *be a #-semisimple CQ*-algebra. Then,* $\mathcal{T}_\#(\mathfrak{A}) = \mathcal{T}_\#(\mathfrak{A}_{\mathcal{T}})$.

Proof Let $\varphi \in \mathcal{T}_\#(\mathfrak{A}_{\mathcal{T}})$. We call φ_r the restriction of φ to $\mathfrak{A} \times \mathfrak{A}$. Since $\|\cdot\|_{\mathcal{T}}$ is weaker than $\|\cdot\|$, then φ_r belongs to $\mathcal{T}_\#(\mathfrak{A})$.

Conversely, if $\varphi \in \mathcal{T}_\#(\mathfrak{A})$, then because of the following bound,

$$|\varphi(a, b)| \leq \varphi(a, a)^{1/2} \varphi(b, b)^{1/2} \leq \|a\|_{\mathcal{T}} \|b\|_{\mathcal{T}},$$

for all $a, b \in \mathfrak{A}$, φ can be extended to $\mathfrak{A}_{\mathcal{T}} \times \mathfrak{A}_{\mathcal{T}}$ and still satisfies all the required properties. □

It is evident from the proof that the equality of the two sets must be understood as the possibility of associating a form of $\mathcal{T}_\#(\mathfrak{A}_{\mathcal{T}})$ to a form of $\mathcal{T}_\#(\mathfrak{A})$ and vice versa.

Proposition 5.3.11 *The CQ*-algebra* $\{\mathfrak{A}_{\mathcal{T}}, *, \mathfrak{A}_\#, \#\}$ *is #-semisimple.*

Proof Let $a \in \mathfrak{A}$ with $\varphi(a, a) = 0$, for every $\varphi \in \mathcal{T}_\#(\mathfrak{A}_\mathcal{T})$. Then, due to the above Lemma and to the #-semisimplicity of $\{\mathfrak{A}, *, \mathfrak{A}_\#, \#\}$, we conclude that $a = 0$. □

With the previous construction, one can always construct, starting from a #-semisimple CQ*-algebra $\{\mathfrak{A}, *, \mathfrak{A}_\#, \#\}$, a new CQ*-algebra, whose norm has the form (5.3.13), a property that closely reminds what happens for C*-algebras.

We now give another similar construction, starting this time from a family of sesquilinear forms on the C*-algebra $\mathfrak{A}_\#$. Let $\{\mathfrak{A}, *, \mathfrak{A}_\#, \#\}$ be a CQ*-algebra with unit e. We denote with $\Sigma_\#$ the family of all sesquilinear forms φ of the C*-algebra $\mathfrak{A}_\#$ satisfying the following properties:

(c.1) $\varphi(x, x) \geq 0, \ \forall\, x \in \mathfrak{A}_\#$;
(c.2) $\varphi(xy, z) = \varphi(y, x^\# z), \ \forall\, x, y, z \in \mathfrak{A}_\#$;
(c.3) $\varphi(e, e) \leq 1$;
(c.4) $\varphi(x, y) = \varphi(y^*, x^*), \ \forall\, x, y \in \mathfrak{A}_\# \cap \mathfrak{A}_\flat$.

Remark 5.3.12 If $\{\mathfrak{A}, *, \mathfrak{A}_\#, \#\}$ is a CQ*-algebra with unit e, then there is a one-to-one correspondence between the sesquilinear forms φ of $\mathfrak{A}_\#$ satisfying the conditions (c.1)–(c.3) (denote the respective family with $\mathcal{F}$) and the positive linear functionals ω on $\mathfrak{A}_\#$ fulfilling the property $\|\omega\| \leq 1$. Then, if we define

$$\|x\|_\mathcal{F} := \sup_{\varphi \in \mathcal{F}} \varphi(x, x)^{1/2}, \quad x \in \mathfrak{A}_\#,$$

which is analogous to $\|x\|_\Sigma$ below, we get exactly $\|\cdot\|_\#$ (in this respect, see also Theorem A.5.5). Considering the adjoint forms φ^*, we get a result for $\mathfrak{A}_\flat$, similar to the preceding one. So what makes the difference is condition (c.4).

Furthermore, assume that $\{\mathfrak{A}, *, \mathfrak{A}_\#, \#\}$ satisfies the following property:

$$\text{If } x \in \mathfrak{A}_\# \text{ and } \varphi(x, x) = 0, \quad \forall\, \varphi \in \Sigma_\#, \text{ then } x = 0. \tag{5.3.15}$$

Since for each $x \in \mathfrak{A}_\#$ the linear functional

$$\omega_y(x) = \varphi(xy, y), \quad y \in \mathfrak{A}_\#,$$

is positive, one can easily prove the inequality

$$|\varphi(xy, y)| \leq \varphi(y, y)\|x\|_\#, \quad \forall\, x, y \in \mathfrak{A}_\#,$$

and then that

$$|\varphi(x, y)| = |\varphi(y^\# x, e)| \leq \|x\|_\# \|y\|_\#, \quad \forall\, x, y \in \mathfrak{A}_\#.$$

We now can define a new norm on $\mathfrak{A}_\#$ by

$$\|x\|_\Sigma := \sup_{\varphi \in \Sigma_\#} \varphi(x,x)^{1/2}, \quad x \in \mathfrak{A}_\#.$$

This norm defines on $\mathfrak{A}_\#$ a topology coarser than that of the original norm. With the help of the previous inequalities and of (c.4), we can prove that

$$\|xy\|_\Sigma \le \|x\|_\# \|y\|_\Sigma, \quad \forall\, x, y \in \mathfrak{A}_\#,$$

and

$$\|x^*\|_\Sigma = \|x\|_\Sigma, \quad \forall\, x \in \mathfrak{A}_\# \cap \mathfrak{A}_\flat.$$

Thus Proposition 5.3.1 applies and if $\widehat{\mathfrak{A}}$ denotes the completion of $\mathfrak{A}_\#[\|\cdot\|_\Sigma]$, then $\{\widehat{\mathfrak{A}}, *, \mathfrak{A}_\#, \#\}$ is a CQ*-algebra.

Now, it is easy to see that if $\varphi \in \mathcal{T}_\#(\widehat{\mathfrak{A}})$, then $\varphi\restriction_{\mathfrak{A}_\#}$ is an element of $\Sigma_\#$.

Conversely, if $\varphi_0 \in \Sigma_\#$, since

$$\|\varphi_0(x,y)\| \le \|x\|_\Sigma \|y\|_\Sigma, \quad \forall\, x, y \in \mathfrak{A}_\#,$$

then φ_0 has a continuous extension to $\widehat{\mathfrak{A}}$ denoted by φ. It is easy to check that $\varphi \in \mathcal{T}_\#(\widehat{\mathfrak{A}})$. However, in spite of this very close relation between the sets $\Sigma_\#$ and $\mathcal{T}_\#(\widehat{\mathfrak{A}})$, $\{\widehat{\mathfrak{A}}, *, \mathfrak{A}_\#, \#\}$ need not be #-semisimple.

Remark 5.3.13 Every C*-algebra with unit possesses sesquilinear forms satisfying the properties (c.1)–(c.3), but not necessarily the property (c.4); there are in fact C*-algebras, which do not have any nonzero trace. On the other hand, if (c.1)–(c.4) and condition (5.3.15) hold, the previous construction shows that it is possible to define a *new* CQ*-algebra, where each element of $\Sigma_\#$ is bounded.

5.4 *-Homomorphisms of CQ*-Algebras

The automatic continuity of *-homomorphisms and the uniqueness of the C*-norm are certainly among the most important features of the theory of C*-algebras (see, e.g., Sect. A.6.2). These properties, however, are not preserved in the framework of CQ*-algebras. In particular, as we shall see below, there exist non equivalent norms on a quasi *-algebra $(\mathfrak{A}, \mathfrak{A}_0)$, with $\mathfrak{A}_0$ a C*-algebra, that make it into topologically different CQ*-algebras. To address this question, we begin with introducing a convenient notion of a *-homomorphism for CQ*-algebras [51].

Definition 5.4.1 Let $\{\mathfrak{A}, *, \mathfrak{A}_\#, \#\}$, $\{\mathfrak{B}, *, \mathfrak{B}_\#, \#\}$ be two CQ*-algebras. A linear map $\Phi : \mathfrak{A} \mapsto \mathfrak{B}$ is said to be a **-homomorphism* of $\{\mathfrak{A}, *, \mathfrak{A}_\#, \#\}$ into $\{\mathfrak{B}, *, \mathfrak{B}_\#, \#\}$ if

(i) $\Phi(a^*) = \Phi(a)^*, \quad \forall\, a \in \mathfrak{A}$;
(ii) $\Phi_\# := \Phi\restriction_{\mathfrak{A}_\#}$ maps $\mathfrak{A}_\#$ into $\mathfrak{B}_\#$ and is a *-homomorphism of C*-algebras;
(iii) $\Phi(xa) = \Phi(x)\Phi(a), \quad \forall\, a \in \mathfrak{A},\ x \in \mathfrak{A}_\#$.

The *-homomorphism $\Phi_\#$ from $\mathfrak{A}_\#$ into $\mathfrak{B}_\#$ defines a *-homomorphism $\Phi_\flat$ from the C*-algebra $\mathfrak{A}_\flat$ into the C*-algebra $\mathfrak{B}_\flat$ by $\Phi_\flat(y) := \Phi_\#(y^*)^*$, $y \in \mathfrak{A}_\flat$. Clearly $\Phi_\flat$ is a *-isomorphism of C*-algebras, if and only if, $\Phi_\#$ is a *-isomorphism.

➠ Observe that, in what follows, *for simplicity's sake*, we use the same symbol $\|\cdot\|$ *for the norms of both Banach spaces* $\mathfrak{A}$ and $\mathfrak{B}$.

A bijective *-homomorphism Φ, such that $\Phi(\mathfrak{A}_\#) = \mathfrak{B}_\#$, is called a **-isomorphism.*

Continuity of *-homomorphisms is, of course, equivalent to their boundedness. We will adopt the following terminology. A *-homomorphism Φ is called *contractive* if $\|\Phi(a)\| \le \|a\|$, for every $a \in \mathfrak{A}$. A *-homomorphism Φ can be continuous without being contractive. A contractive *-isomorphism, whose inverse is also contractive is called an *isometric *-isomorphism.*

If Φ is a *-homomorphism, then $\|\Phi_\#(x)\|_\# \le \|x\|_\#$, for every $x \in \mathfrak{A}_\#$, since each *-homomorphism of C*-algebras is contractive. Analogously, if $\Phi_\#$ is a *-isomorphism, then it is necessarily isometric (Proposition A.6.13).

Finally, we notice that, if Φ is a *-isomorphism and $\{\mathfrak{A}, *, \mathfrak{A}_\#, \#\}$, $\{\mathfrak{B}, *, \mathfrak{B}_\#, \#\}$ have units $e_{\mathfrak{A}}$, $e_{\mathfrak{B}}$, respectively (see Remark 5.2.4), then by (iii), $\Phi(e_{\mathfrak{A}}) = e_{\mathfrak{B}}$.

Taking into account that a continuous invertible linear map Φ from the Banach space $\mathfrak{A}$ onto the Banach space $\mathfrak{B}$ has a continuous inverse, we get

Proposition 5.4.2 *Let Φ be a continuous *-isomorphism from the CQ*-algebra $\{\mathfrak{A}, *, \mathfrak{A}_\#, \#\}$ onto the CQ*-algebra $\{\mathfrak{B}, *, \mathfrak{B}_\#, \#\}$. Then, there exist $\delta, \gamma > 0$, such that*

$$\delta\|a\| \le \|\Phi(a)\| \le \gamma\|a\|, \quad \forall\, a \in \mathfrak{A}.$$

On the other hand, we have the following

Proposition 5.4.3 *Let $\{\mathfrak{A}, *, \mathfrak{A}_\#, \#\}$ and $\{\mathfrak{B}, *, \mathfrak{B}_\#, \#\}$ be CQ*-algebras. Let $\Phi_\#$ be a *-isomorphism of $\mathfrak{A}_\#$ onto $\mathfrak{B}_\#$, for which there exist $0 < \delta, \gamma$, such that*

$$\delta\|x\| \le \|\Phi_\#(x)\| \le \gamma\|x\|, \ \forall\, x \in \mathfrak{A}_\#. \tag{5.4.16}$$

*Then, $\Phi_\#$ can be continuously extended to a continuous *-isomorphism Φ of $\mathfrak{A}$ onto $\mathfrak{B}$, which is contractive, if $\gamma \le 1$.*

Moreover, if $\|\Phi_\#(x)\| = \|x\|$, for all $x \in \mathfrak{A}_\#$, the extension is isometric.

Proof For $a \in \mathfrak{A}$ there exists a sequence $\{x_n\} \subset \mathfrak{A}_\#$ converging to a. Then, one defines

$$\Phi(a) := \|\cdot\| - \lim_{n\to\infty} \Phi_\#(x_n).$$

It is easily seen that $\Phi(a)$ is well-defined and satisfies the conditions of Definition 5.4.1. By (5.4.16) it follows that Φ is a bijection and so it is a continuous *-isomorphism. By the same relation $\Phi_\#$ is $\|\cdot\|$-homeomorphism, so the same also holds for Φ. □

Proposition 5.4.4 *Let Φ be a *-isomorphism from the CQ*-algebra $(\mathfrak{A}, *, \mathfrak{A}_\#, \#)$ onto the CQ*-algebra $\{\mathfrak{B}, *, \mathfrak{B}_\#, \#\}$. Then, it is possible to define a new norm $\|\cdot\|_\Phi$ on $\mathfrak{B}$, such that $\{\mathfrak{B}, *, \mathfrak{B}_\#, \#\}$ is still a CQ*-algebra and Φ is isometric.*

Proof We define $\|b\|_\Phi := \|\Phi^{-1}(b)\|$, for every $b \in \mathfrak{B}$. It is very easy to prove that

(i) $\mathfrak{B}[\|\cdot\|_\Phi]$ is a Banach space;
(ii) $\|b^*\|_\Phi = \|b\|_\Phi, \ \forall\, b \in \mathfrak{B}$;
(iii) $\mathfrak{B}_\#$ is $\|\cdot\|_\Phi$-dense in $\mathfrak{B}$.

Let us now define a new norm on $\mathfrak{B}_\#$:

$$\|x\|_\#^\Phi := \sup_{\|b\|_\Phi \le 1} \|xb\|_\Phi, \quad x \in \mathfrak{B}_\#, \ b \in \mathfrak{B}.$$

We prove that $\|x\|_\# = \|x\|_\#^\Phi$, for all $x \in \mathfrak{B}_\#$. Since Φ^{-1} is necessarily an isometry between $\mathfrak{A}_\#$ and $\mathfrak{B}_\#$, for $b \in \mathfrak{B}$ and $x \in \mathfrak{B}_\#$ we have

$$\|xb\|_\Phi = \|\Phi^{-1}(xb)\| \le \|\Phi^{-1}(x)\|_\# \|\Phi^{-1}(b)\| \le \|x\|_\# \|b\|_\Phi.$$

Therefore, $\|x\|_\#^\Phi \le \|x\|_\#$, for all $x \in \mathfrak{B}_\#$. This inequality and the inverse mapping theorem imply that the two norms are equivalent. To show that they are exactly equal we can proceed as in the proof of Proposition 5.3.1, after checking that $\|\cdot\|_\#^\Phi$ makes of $\mathfrak{B}_\#$ a normed algebra; i.e., we need first the inequality

$$\|xy\|_\#^\Phi \le \|x\|_\#^\Phi \|y\|_\#^\Phi, \quad \forall\, x, y \in \mathfrak{B}_\#.$$

But this can be easily derived from the definition of $\|\cdot\|_\#^\Phi$ itself. The equality $\|x^\#\|_\#^\Phi = \|x\|_\#^\Phi$, for all $x \in \mathfrak{B}_\#$, is easy to be proved. □

From the previous proof one can also deduce the following

Proposition 5.4.5 *Let $\{\mathfrak{A}, *, \mathfrak{A}_\#, \#\}$ be a CQ*-algebra and $\mathfrak{B}$ a left $\mathfrak{B}_\#$-module, where $\mathfrak{B}_\#$ is a C*-algebra with involution # and norm $\|\cdot\|_\#$. Moreover, let $\mathfrak{B}_\# \subset \mathfrak{B}$ and Φ be an injective linear map from $\mathfrak{A}$ into $\mathfrak{B}$ with the properties:*

(i) $\Phi(a^*) = \Phi(a)^*$, $\forall\, a \in \mathfrak{A}$;
(ii) $\Phi_\# := \Phi\lceil_{\mathfrak{A}_\#}$ *maps $\mathfrak{A}_\#$ into $\mathfrak{B}_\#$ and is a *-homomorphism of C*-algebras;*
(iii) $\Phi(xa) = \Phi(x)\Phi(a)$, $\forall\, a \in \mathfrak{A}$, $x \in \mathfrak{A}_\#$.

Let us define a norm $\|\cdot\|_\Phi$ on $\Phi(\mathfrak{A})$ by

$$\|b\|_\Phi := \|\Phi^{-1}(b)\|, \quad b \in \Phi(\mathfrak{A}).$$

Then, $\{\Phi(\mathfrak{A})[\|\cdot\|_\Phi], *, \mathfrak{B}_\#, \#\}$ is a CQ-algebra.*

Now we can answer the following *question*: given the CQ*-algebra $\{\mathfrak{A}, *, \mathfrak{A}_\#, \#\}$, is it possible to endow $\mathfrak{A}$ with a non equivalent norm, which makes of it a CQ*-algebra on the same C*-algebra $\mathfrak{A}_\#$? The following explicit construction shows that the answer is positive. In particular, we shall give a general strategy to build up different proper CQ*-algebras over the same C*-algebra and, after that, we shall give an explicit example.

Our starting point is a proper CQ*-algebra $(\mathfrak{A}[\|\cdot\|], \mathfrak{A}_0)$. The C*-norm of $\mathfrak{A}_0$ is denoted again by $\|\cdot\|_0$.

Proposition 5.4.6 *Let T be an unbounded, invertible linear map from $\mathfrak{A}$ onto $\mathfrak{A}$, such that $T\mathfrak{A}_0 = \mathfrak{A}_0$, $T(x^*) = (Tx)^*$ and $\|T(xy)\| \leq \|Tx\|\|y\|_0$, for all x, y in $\mathfrak{A}_0$. Then, defining $\|a\|' := \|Ta\|$, $a \in \mathfrak{A}$, $(\mathfrak{A}[\|\cdot\|'], \mathfrak{A}_0)$ is a proper CQ*-algebra with norm $\|\cdot\|'$ non equivalent to $\|\cdot\|$.*

Proof We begin with proving that $\mathfrak{A}[\|\cdot\|']$ is a Banach space. Let $\{a_n\}$ be a $\|\cdot\|'$-Cauchy sequence in $\mathfrak{A}$. This implies that $\{Ta_n\}$ is a $\|\cdot\|$-Cauchy sequence in $\mathfrak{A}$, therefore it converges to an element $b \in \mathfrak{A}$. Set $a = T^{-1}b$. Then,

$$\|a_n - a\|' = \|Ta_n - TT^{-1}b\| = \|Ta_n - b\| \to 0,$$

which proves the statement.

Now we show that $\mathfrak{A}_0$ is dense in $\mathfrak{A}[\|\cdot\|']$. Indeed, let $a \in \mathfrak{A}$ and put $b = T^{-1}a$. Let $\{x_n\}$ be a sequence of elements of $\mathfrak{A}_0$, such that $\|x_n - a\| \to 0$. Then, we have

$$\|T^{-1}x_n - b\|' = \|TT^{-1}x_n - Tb\| = \|TT^{-1}x_n - TT^{-1}a\| = \|x_n - a\| \to 0.$$

By the assumption it follows that $\{T^{-1}x_n\} \subset \mathfrak{A}_0$. Hence $\mathfrak{A}_0$ is dense in $\mathfrak{A}[\|\cdot\|']$. The assumptions on T allow to construct a proper CQ*-algebra with the help of Proposition 5.1.3 by completing $\mathfrak{A}_0$ with respect to $\|\cdot\|'$. The completeness of $\mathfrak{A}[\|\cdot\|']$ and the density of $\mathfrak{A}_0$ in it, proved above, imply that the outcome of this construction is $(\mathfrak{A}[\|\cdot\|'], \mathfrak{A}_0)$, which therefore is a proper CQ*-algebra. □

It is worth noticing that, in this construction, the two Banach spaces $\mathfrak{A}[\|\cdot\|]$ and $\mathfrak{A}[\|\cdot\|']$ coincide, while the two norms, $\|\cdot\|'$ and $\|\cdot\|$ differ, for all those elements belonging to $\mathfrak{A}$, but not to $\mathfrak{A}_0$, and for this reason they are not equivalent.

An explicit example of an operator T, as above, can be constructed by means of the Hamel basis of the Banach spaces $\mathfrak{A}$ and $\mathfrak{A}_0$.

Indeed, let $e_{\alpha\{\alpha\in J\}}$ be a Hamel basis for $\mathfrak{A}$, which contains a Hamel basis $\{e_\alpha\}_{\alpha\in I}$, for $\mathfrak{A}_0$. We can always choose e_α, such that $e_\alpha = e_\alpha^*$. In order to simplify things, we suppose that the set J is a subset of the positive reals with no upper bound. We define T through its action on the basis vectors e_α, i.e.,

$$Te_\alpha := e_\alpha, \quad \text{if } \alpha \in I \text{ and } Te_\alpha := \alpha e_\alpha, \quad \text{if } \alpha \in J\backslash I.$$

With this definition, it is clear that T is unbounded and invertible. Moreover, since T is the identity map on $\mathfrak{A}_0$, it is also evident that $T(x^*) = (Tx)^*$, for all $x \in \mathfrak{A}_0$, as well as that the inequality $\|T(xy)\| \le \|Tx\|\|y\|_0$ holds, for all x, y in $\mathfrak{A}_0$. However, it is also clear that the norm $\|\cdot\|'$ is not equivalent to $\|\cdot\|$, as one can check considering the values of these norms on the basis vectors. In this way, we have constructed an operator T satisfying all the properties required in Proposition 5.4.6.

In the rest of this section we discuss briefly the problem of *-semisimplicity of CQ*-algebras in relation with *-isomorphisms. Note that when we say that *a CQ*-algebra* $\{\mathfrak{A}, *, \mathfrak{A}_\#, \#\}$ *is *-semisimple*, we mean that the Banach quasi *-algebra $(\mathfrak{A}[\|\cdot\|], \mathfrak{A}_0)$, with $\mathfrak{A}_0 \equiv \mathfrak{A}_\# \cap \mathfrak{A}_\flat$, is *-semisimple (for the notation and the fact that $(\mathfrak{A}[\|\cdot\|], \mathfrak{A}_0)$ is a Banach quasi *-algebra, see Definition 5.2.2 and the discussion after it). To begin with, let $\{\mathfrak{A}, *, \mathfrak{A}_\#, \#\}$ and $\{\mathfrak{B}, *, \mathfrak{B}_\#, \#\}$ be CQ*-algebras and $\Phi : \mathfrak{A} \to \mathfrak{B}$ a *-homomorphism of $\{\mathfrak{A}, *, \mathfrak{A}_\#, \#\}$ into $\{\mathfrak{B}, *, \mathfrak{B}_\#, \#\}$. If $\psi \in \mathcal{S}_{\mathfrak{B}_0}(\mathfrak{B})$ and Φ is contractive, then the sesquilinear form $\psi \circ \Phi$ defined by $(\psi \circ \Phi)(a, b) = \psi(\Phi(a), \Phi(b))$, $a, b \in \mathfrak{A}$ is an element of $\mathcal{S}_{\mathfrak{A}_0}(\mathfrak{A})$. Suppose that $\{\mathfrak{B}, *, \mathfrak{B}_\#, \#\}$ is *-semisimple and that $\varphi(a, a) = 0$, for every $\varphi \in \mathcal{S}_{\mathfrak{A}_0}(\mathfrak{A})$. Then in particular, for every $\psi \in \mathcal{S}_{\mathfrak{B}_0}(\mathfrak{B})$, $(\psi \circ \Phi)(a, a) = 0$. Hence $\Phi(a) = 0$. Thus, if Φ is injective, $a = 0$ (see also Definition 3.1.17 and Corollary 3.1.24). If Φ is bounded but not contractive, we can replace it with $\Psi := \Phi/\|\Phi\|$, which is obviously contractive. We have then proved the following

Proposition 5.4.7 *Let* $\{\mathfrak{A}, *, \mathfrak{A}_\#, \#\}$ *and* $\{\mathfrak{B}, *, \mathfrak{B}_\#, \#\}$ *be CQ*-algebras and* $\Phi : \mathfrak{A} \to \mathfrak{B}$ *an injective *-homomorphism of* $\{\mathfrak{A}, *, \mathfrak{A}_\#, \#\}$ *into* $\{\mathfrak{B}, *, \mathfrak{B}_\#, \#\}$. *If* $\{\mathfrak{B}, *, \mathfrak{B}_\#, \#\}$ *is *-semisimple, then* $\{\mathfrak{A}, *, \mathfrak{A}_\#, \#\}$ *is *-semisimple. In particular, if* Φ *is a *-isomorphism,* $\{\mathfrak{A}, *, \mathfrak{A}_\#, \#\}$ *is *-semisimple, if and only if,* $\{\mathfrak{B}, *, \mathfrak{B}_\#, \#\}$ *is *-semisimple.*

If Φ is an isometric *-isomorphism then the following equalities are easily proved:

$$\mathcal{S}_{\mathfrak{A}_0}(\mathfrak{A}) = \{\psi \circ \Phi : \psi \in \mathcal{S}_{\mathfrak{B}_0}(\mathfrak{B})\}, \qquad \mathcal{S}_{\mathfrak{B}_0}(\mathfrak{B}) = \{\varphi \circ \Phi^{-1} : \varphi \in \mathcal{S}_{\mathfrak{A}_0}(\mathfrak{A})\}.$$

Corollary 5.4.8 *Let* $\{\mathfrak{A}, *, \mathfrak{A}_\#, \#\}$ *and* $\{\mathfrak{B}, *, \mathfrak{B}_\#, \#\}$ *be CQ*-algebras with* $\mathfrak{B} \subset \mathfrak{A}$ *and* $\mathfrak{B}_\# \subset \mathfrak{A}_\#$. *If* $\mathfrak{B}$ *is continuously embedded in* $\mathfrak{A}$ *and* $\mathfrak{A}$ *is *-semisimple, then* $\mathfrak{B}$ *is also *-semisimple.*

Proof It follows immediately from Proposition 5.4.7. □

Remark 5.4.9 Let $\{\mathfrak{A}, *, \mathfrak{A}_{\#}, \#\}$ and $\{\mathfrak{B}, *, \mathfrak{B}_{\#}, \#\}$ be CQ*-algebras with units $e_{\mathfrak{A}}$, $e_{\mathfrak{B}}$, respectively. We can also introduce a more general notion than that of *-homomorphism. A **-bimorphism* of $\{\mathfrak{A}, *, \mathfrak{A}_{\#}, \#\}$ into $\{\mathfrak{B}, *, \mathfrak{B}_{\#}, \#\}$ is a *pair* $(\pi, \pi_{\#})$ of linear maps $\pi : \mathfrak{A} \to \mathfrak{B}$ and $\pi_{\#} : \mathfrak{A}_{\#} \to \mathfrak{B}_{\#}$, such that

(i) $\pi_{\#}$ is a homomorphism of algebras with $\pi_{\#}(x^{\#}) = \pi_{\#}(x)^{\#}$, $x \in \mathfrak{A}_{\#}$; i.e., $\pi_{\#}$ is a #-homomorpism.
(ii) $\pi(a^*) = \pi(a)^*$, $\quad \forall\, a \in \mathfrak{A}$;
(iii) $\pi(xa) = \pi_{\#}(x)\pi(a)$, $\quad \forall\, a \in \mathfrak{A},\ x \in \mathfrak{A}_{\#}$.

In general, the restriction of π to $\mathfrak{A}_{\#}$ is different from $\pi_{\#}$. For this reason, *-homomorphisms and *-bimorphisms are different objects. Of course any *-homomorphism defines, in trivial way, a *-bimorphism, but the converse is not true, in general. If $(\pi, \pi_{\#})$ is a *-bimorphism, then $\pi(a) = \pi(e_{\mathfrak{A}})\pi_{\#}(a)$, for each $a \in \mathfrak{A}_{\#}$; but, in general, $\pi(e_{\mathfrak{A}})$ is different from $e_{\mathfrak{B}}$. Obviously, if $\pi(e_{\mathfrak{A}}) = e_{\mathfrak{B}}$, then π is a *-homomorphism.

If $(\pi, \pi_{\#})$ is a *-bimorphism of $\{\mathfrak{A}, *, \mathfrak{A}_{\#}, \#\}$ into $(\mathfrak{B}, *, \mathfrak{B}_{\#}, \#)$, then

$$\pi(ax) = \pi(a)\pi_{\flat}(x), \qquad \forall\, a \in \mathfrak{A},\ x \in \mathfrak{A}_{\flat},$$

where $\pi_{\flat}(x) = \pi_{\#}(x^*)^*$. Moreover, $\pi_{\flat}$ is a homomorphism of $\mathfrak{A}_{\flat}$ into $\mathfrak{B}_{\flat}$ preserving the involution $\flat$ of $\mathfrak{A}_{\flat}$.

5.5 CQ*-Algebras and Left Hilbert Algebras

In this section we shall consider again CQ*-algebras constructed from left Hilbert algebras and we shall examine their link with pseudo (strict) CQ*-algebras of Sect. 5.3.2. The starting point is a Hilbertian quasi *-algebra $(\mathfrak{A}[\|\cdot\|], \mathfrak{A}_0)$, as defined in Remark 3.1.9.

Definition 5.5.1 A Hilbertian quasi *-algebra $(\mathfrak{A}[\|\cdot\|], \mathfrak{A}_0)$ is said to be an *HCQ*-algebra* if $\mathfrak{A}$ is endowed with another involution #, such that $L_x^* = L_{x^{\#}}$ and $\|x\| \leq \|L_x\|$, for each $x \in \mathfrak{A}_0$. A Hilbertian quasi *-algebra will be simply denoted by $(\mathfrak{A}[\|\cdot\|], \#)$.

HCQ*-algebras are closely related to left Hilbert algebras (see Example 5.2.6 for notations, basic definitions and facts).

Proposition 5.5.2 *Suppose that $(\mathfrak{A}[\|\cdot\|], \#)$ is an HCQ*-algebra with involution operator $J_{\mathfrak{A}}$; i.e., $J_{\mathfrak{A}}a = a^*$, $a \in \mathfrak{A}$. Then, the following statements hold:*

(i) *$(\mathfrak{A}[\|\cdot\|], \#)$ is a strict CQ*-algebra under the C*-norm $\|x\|_{\#} = \|L_x\|$, $x \in \mathfrak{A}_0$;*
(ii) *$\mathfrak{A}_0$ is a left Hilbert algebra in the Hilbert space $\mathcal{H} \equiv \mathfrak{A}[\|\cdot\|]$, whose full left Hilbert algebra $\mathfrak{A}_0''$ has a unit u.*

Proof

(i) The proof is mostly trivial. We only prove that, in this case, condition (a.4) is satisfied, (see discussion before Definition 5.3.6). Indeed, if $\{x_n\} \subset \mathfrak{A}_0$ is a sequence such that $\|x_n\| \to 0$ and $x_n \to x$ in $\mathfrak{A}_\#[\|\cdot\|_\#]$, then $L_{x_n} \to L_x$, with respect to the operator norm. Continuity of multiplication in $\mathfrak{A}_0[\|\cdot\|_0]$ easily implies that $L_x = 0$; hence, $\|x\|_\# = 0$ and $x = 0$.

(ii) We first show that $\mathfrak{A}_0$ is a left Hilbert algebra in $\mathcal{H}$ with involution $\#$. Since the C*-algebra $\mathfrak{A}_\#$ has an approximate unit, say $\{u_\alpha\}$, $\mathfrak{A}_0$ is dense in the C*-algebra $\mathfrak{A}_\#$ and $\|x\| \leq \|x\|_\#$, for each $x \in \mathfrak{A}_0$, then it follows that $\mathfrak{A}_0^2$ is total in $\mathfrak{A}_0[\|\cdot\|_0]$. The assumption, $L_x^* = L_{x^\#}$, for all $x \in \mathfrak{A}_0$, implies that

$$\langle xy|z\rangle = \langle y|x^\# z\rangle, \quad \forall\, x, y, z \in \mathfrak{A}_0,$$

where $\langle\cdot|\cdot\rangle$ is the inner product defined by the Hilbertian norm $\|\cdot\|$. Moreover, for every $x \in \mathfrak{A}_0$, the operator L_x is bounded. Take any sequence $\{x_n\}$ in $\mathfrak{A}_0$, such that $\lim_{n\to\infty} \|x_n\| = 0$ and $\lim_{n\to\infty} \|{x_n}^\# - y\| = 0$, $y \in \mathfrak{A}_0$. Then, it follows that

$$\langle y|x_1x_2^\flat\rangle = \lim_{n\to\infty} \langle x_n^\# |x_1x_2^\flat\rangle = \lim_{n\to\infty} \langle x_2x_1^\flat|x_n\rangle = 0, \quad \forall\, x_1, x_2 \in \mathfrak{A}_0,$$

which implies that the involution map $x \in \mathfrak{A}_0 \mapsto x^\# \in \mathfrak{A}_0$ is closable. Thus, $\mathfrak{A}_0$ is a left Hilbert algebra in $\mathcal{H}$ under the involution $\#$. Following the notations of Example 5.2.6, we denote by S the closure of the involution $\#$ of $\mathfrak{A}_0$, and by J, Δ, the modular conjugation and the modular operator of $\mathfrak{A}_0$, respectively. We next show that the full left Hilbert algebra $\mathfrak{A}_0''$ has a unit u. For any $\varepsilon > 0$ and for any finite subsets $\{x_1, \ldots, x_m\}$ and $\{y_1, \ldots, y_m\}$ of $\mathfrak{A}_0$, we define the set

$$\begin{aligned} &K\big(\varepsilon : \{x_1, \ldots, x_m\},\ \{y_1, \ldots, y_m\}\big) \\ &\quad := \big\{a \in \mathcal{H} : \|a\| \leq 1,\ \ |\langle ax_k - x_k|y_k\rangle| \leq \varepsilon \\ &\qquad\qquad \text{and } |\langle x_k a - x_k|y_k\rangle| \leq \varepsilon,\ k = 1, \ldots, m\big\}. \end{aligned}$$

Since the C*-algebra $\mathfrak{A}_\#$ has an approximate unit and $\|x\| \leq \|x\|_\#$, for each $x \in \mathfrak{A}_\#$, it follows that the set $K\big(\varepsilon : \{x_1, \ldots, x_m\}, \{y_1, \ldots, y_m\}\big)$ is nonempty. Let now $\mathcal{K}$ be the family of all such subsets, where $\varepsilon > 0$ and $\{x_1, \ldots, x_m\}, \{y_1, \ldots, y_m\}$ are finite subsets of $\mathfrak{A}_0$. Then, $\mathcal{K}$ is a family of non-empty weakly closed subsets of the weakly compact set $\mathcal{H}_1 \equiv \{a \in \mathcal{H} : \|a\| \leq 1\}$. Hence, the intersection of all the sets in $\mathcal{K}$ is non empty. Therefore, an element u of this intersection is such that u is a *quasi-unit* of the topological quasi *-algebra $(\mathfrak{A}[\|\cdot\|], \mathfrak{A}_0)$, that is $u \in \mathfrak{A}[\|\cdot\|]$ and $ux = xu = x$, for every $x \in \mathfrak{A}_0$. Since

$$\langle Sx|u\rangle = \langle x^\#|u\rangle = \langle u|L_x u\rangle = \langle u|x\rangle, \quad \forall\, x \in \mathfrak{A}_0,$$

it follows that $u \in D(S^*)$ and $S^*u = u$. Thus, $u \in \mathfrak{A}_0'$ and therefore

$$\langle S^* y|u\rangle = \langle L_{S^*y}u|u\rangle = \langle u|R_y u\rangle = \langle u|y\rangle, \quad \forall\, y \in \mathfrak{A}_0'.$$

Consequently, we have $u \in \mathfrak{A}_0''$ and $Su = u$. This completes the proof. □

By Proposition 5.5.2, the situation of HCQ*-algebras is represented by the following scheme

$$\begin{array}{ccccc} & \subset \mathfrak{A}_\# \subset \mathfrak{A}_0'' = L(\mathfrak{A}_0)''u \subset & D(S) & \subset & \\ \mathfrak{A}_0 \ \updownarrow J_{\mathfrak{A}} & \updownarrow J & \updownarrow J & & \mathfrak{A}[\|\cdot\|]. \\ & \subset \mathfrak{A}_\flat \subset \mathfrak{A}_0' = L(\mathfrak{A}_0)'u & \subset D(S^*) & \subset & \end{array}$$

We now look for conditions under which $J_{\mathfrak{A}} = J$.

Lemma 5.5.3 *Let* $(\mathfrak{A}[\|\cdot\|], \#)$ *be an HCQ*-algebra. Then, the following statements are equivalent:*

(i) $J_{\mathfrak{A}} = J$;
(ii) $\langle x^\#|x^*\rangle \geq 0$, *for each* $x \in \mathfrak{A}_0$.

Proof (i) ⇒ (ii) This follows from

$$\langle x^\#|x^*\rangle = \langle J\Delta^{\frac{1}{2}}x|Jx\rangle = \langle x|\Delta^{\frac{1}{2}}x\rangle \geq 0, \quad \forall\, x \in \mathfrak{A}_0.$$

(ii) ⇒ (i) By the assumption (ii) we have $S = J_{\mathfrak{A}}(J_{\mathfrak{A}}J\Delta^{\frac{1}{2}})$ and $J_{\mathfrak{A}}J\Delta^{\frac{1}{2}} \geq 0$. The uniqueness of the polar decomposition of S implies $J = J_{\mathfrak{A}}$. □

If anyone of the two equivalent statements of Lemma 5.5.3 holds, we say that the HCQ*-algebra $(\mathfrak{A}[\|\cdot\|], \#)$ is *standard*.

Let $(\mathfrak{A}[\|\cdot\|], \#)$, $(\mathfrak{B}[\|\cdot\|], \#)$ be two HCQ*-algebras, with $\mathfrak{B} = \tilde{\mathfrak{B}}_0[\|\cdot\|]$ (completion of $\mathfrak{B}_0[\|\cdot\|]$), for some left Hilbert algebra $\mathfrak{B}_0$. In general, if $\mathfrak{A}[\|\cdot\|] = \mathfrak{B}[\|\cdot\|]$ as Hilbert spaces, $(\mathfrak{A}[\|\cdot\|], \#)$ and $(\mathfrak{B}[\|\cdot\|], \#)$ need not coincide as HCQ*-algebras. For this reason we introduce the following notion:

Definition 5.5.4 An HCQ*-algebra $(\mathfrak{A}[\|\cdot\|], \#_{\mathfrak{A}})$ is said to be an *extension* of an HCQ*-algebra $(\mathfrak{B}[\|\cdot\|], \#_{\mathfrak{B}})$ (with $\mathfrak{B} = \tilde{\mathfrak{B}}_0[\|\cdot\|]$, for some left Hilbert algebra $\mathfrak{B}_0$) if $\mathfrak{B}_0$ is a dense *-subalgebra of $\mathfrak{A}_0$ and the involutions $\#_{\mathfrak{A}}, \#_{\mathfrak{B}}$ coincide on $\mathfrak{B}_0$.

Proposition 5.5.5 *Let* $(\mathfrak{A}[\|\cdot\|], \#)$ *be a standard HCQ*-algebra and* $\mathfrak{B}_0 := \mathfrak{A}_{00}$ *the maximal Tomita algebra* (Example 5.2.6) *of the full left Hilbert algebra* $\mathfrak{A}_0''$. *Then,* $(\mathfrak{B}[\|\cdot\|], \#)$ *is a standard HCQ*-algebra and it is an extension of* $(\mathfrak{A}[\|\cdot\|], \#)$. *Further,* $\{\Delta^{it}\}_{t\in\mathbb{R}}$ *is a one-parameter group of *-automorphisms of the Hilbert quasi *-algebra* $\mathfrak{B}[\|\cdot\|]$, *that is* $\Delta^{it}\mathfrak{B}_0 = \mathfrak{B}_0$, $(\Delta^{it}a)^* = \Delta^{it}a^*$, $\Delta^{it}(ax) = (\Delta^{it}a)(\Delta^{it}x)$ *and* $\Delta^{it}(xa) = (\Delta^{it}x)(\Delta^{it}a)$, *for all* $a \in \mathfrak{B}[\|\cdot\|]$, $x \in \mathfrak{B}_0$ *and* $t \in \mathbb{R}$.

Proof It is almost clear that $\mathfrak{B}[\|\cdot\|]$ is a Hilbert quasi *-algebra with the involution $J_{\mathfrak{A}} = J_{\mathfrak{B}}$ and further $(\mathfrak{B}[\|\cdot\|], \#_{\mathfrak{A}})$ is a standard HCQ*-algebra. Since $\{\Delta_{\mathfrak{A}}^{it}\}_{t\in\mathbb{R}}$ is a one-parameter group of *-automorphisms of the Tomita algebra $\mathfrak{B}_0$, it follows that $\{\Delta^{it}\}_{t\in\mathbb{R}}$ is also a one-parameter group of *-automorphisms of the Hilbert quasi *-algebra $\mathfrak{B}[\|\cdot\|]$. □

Finally we consider the question of when a Hilbert space can be regarded as a standard HCQ*-algebra. By Propositions 5.5.2, 5.5.5 and [26, Theorem 13.1] we have the following

Theorem 5.5.6 *Let $\mathcal{H}$ be a Hilbert space. The following statements are equivalent:*

(i) *$\mathcal{H}$ is a standard HCQ*-algebra;*
(ii) *$\mathcal{H}$ contains a left Hilbert algebra with unit as dense subspace;*
(iii) *There exists a von Neumann algebra on $\mathcal{H}$ with a cyclic and separating vector.*

Remark 5.5.7 The implication (iii) ⇒ (i) shows that the class of standard HCQ*-algebras is rather rich. Moreover, it gives conditions for the existence of a von Neumann algebra on $\mathcal{H}$ with a cyclic and separating vector in the case when $\mathcal{H}$ is a nonseparable space. In fact, in the separable case, every von Neumann algebra $\mathfrak{M}$ is σ-finite; i.e., any collection of mutually orthogonal projections have at most countable cardinality. This is equivalent to the fact that $\mathfrak{M}$ is isomorphic to a von Neumann algebra possessing a cyclic and separating vector, see [6, Section 2.5.1].

5.5.1 The Structure of Strict CQ*-Algebras

In this subsection we study when a strict CQ*-algebra is embedded in a standard HCQ*-algebra (for the definitions of *-homomorphism, *-isomorphism, etc see beginning of Sect. 5.4.). For that, we need a GNS-like construction for a class of positive sesquilinear forms on strict CQ*-algebras $(\mathfrak{A}[\|\cdot\|], \mathfrak{A}_0, \#)$

We remind the reader that a positive sesquilinear form φ is called *faithful* if $\varphi(a, a) = 0$, $a \in \mathfrak{A}[\|\cdot\|]$, implies $a = 0$.

Theorem 5.5.8 *Let $(\mathfrak{A}[\|\cdot\|], \mathfrak{A}_0, \#)$ be a strict CQ*-algebra with quasi-unit u. Then, the following statements are equivalent:*

(i) *there exists a contractive *-homomorphism* (resp. *-isomorphism) Φ *of the strict CQ*-algebra $(\mathfrak{A}[\|\cdot\|], \mathfrak{A}_0, \#)$ into an HCQ*-algebra $(\mathfrak{B}[\|\cdot\|'], \#')$;*
(ii) *there exists a positive* (resp. faithful) *sesquilinear form φ on $\mathfrak{A} \times \mathfrak{A}$ satisfying the following properties:*

$(ii)_1$ $\varphi(x, y) = \varphi(u, x^{\#}y)$, $\quad \forall\, x, y \in \mathfrak{A}_0$;
$(ii)_2$ $|\varphi(x, y)| \leq \|x\|\|y\|$, $\quad \forall\, x, y \in \mathfrak{A}_0$;
$(ii)_3$ $\varphi(x, y) = \varphi(y^*, x^*)$, $\quad \forall\, x, y \in \mathfrak{A}_0$;

further, $(\mathfrak{B}[\|\cdot\|'], \#')$ *is standard, if and only if,*

(ii)$_4$ $\varphi(x^*, x^\#) \geq 0, \quad \forall\, x \in \mathfrak{A}_0.$

Proof (i) $\Rightarrow$ (ii) We put

$$\varphi(a,b) = \langle \Phi(a) | \Phi(b) \rangle,\ a, b \in \mathfrak{A}$$

where $\langle \cdot | \cdot \rangle$ is the inner product defined by the Hilbertian norm $\|\cdot\|'$ on $\mathfrak{B}$. Then, it is easily shown that φ is a positive sesquilinear form on $\mathfrak{A}[\|\cdot\|] \times \mathfrak{A}[\|\cdot\|]$ satisfying the conditions (ii)$_1$–(ii)$_3$. If $(\mathfrak{B}[\|\cdot\|'], \#')$ is standard, then (ii)$_4$ follows from Lemma 5.5.3.

(ii) $\Rightarrow$ (i) We put $\mathcal{N}_\varphi = \{a \in \mathfrak{A} : \varphi(a,a) = 0\}$. Then, $\mathcal{N}_\varphi$ is a subspace of $\mathfrak{A}[\|\cdot\|]$ and due to the positivity of φ, the quotient space $\lambda_\varphi(\mathfrak{A}) \equiv \mathfrak{A}/\mathcal{N}_\varphi = \{\lambda_\varphi(a) \equiv a + \mathcal{N}_\varphi : a \in \mathfrak{A}\}$ is a pre-Hilbert space with inner product given by

$$\langle \lambda_\varphi(a) | \lambda_\varphi(b) \rangle_\varphi = \varphi(a,b), \quad \forall\, a, b \in \mathfrak{A}.$$

We denote by $\|\cdot\|_\varphi$ the norm defined by the inner product $\langle \cdot | \cdot \rangle_\varphi$ and by $\mathcal{H}_\varphi$ the completion of the pre-Hilbert space $\lambda_\varphi(\mathfrak{A})[\|\cdot\|_\varphi]$. Since $\mathfrak{A}_0$ is $\|\cdot\|$-dense in $\mathfrak{A}$, it follows that

(ii)$'_2$ $|\varphi(a,b)| \leq \|a\|\,\|b\|, \quad \forall\, a, b \in \mathfrak{A};$
(ii)$'_3$ $\varphi(a,b) = \varphi(b^*, a^*), \quad \forall\, a, b \in \mathfrak{A}.$

Now by (ii)$_1$ and the inequality $\|x\| \leq \|x\|_\#$, $\forall\, x \in \mathfrak{A}_0$, we have that

(ii)$'_1$ $\varphi(x, y) = \varphi(u, x^\# y), \quad \forall\, x, y \in \mathfrak{A}_\#.$

By (ii)$'_2$, $\mathfrak{A}_\varphi := \lambda_\varphi(\mathfrak{A})$ is a dense subspace of the Hilbert space $\mathcal{H}_\varphi$ and moreover, it is a *-algebra equipped with a multiplication defined by

$$\lambda_\varphi(x)\lambda_\varphi(y) := L_{\lambda_\varphi(x)}\lambda_\varphi(y) = \lambda_\varphi(xy), \quad \forall\, x, y \in \mathfrak{A}_\varphi$$

and an involution given as follows

$$\lambda_\varphi(x)^* := \lambda_\varphi(x^*), \quad \forall\, x \in \mathfrak{A}_\varphi.$$

From (ii)$'_3$, the involution $\lambda_\varphi(x) \to \lambda_\varphi(x)^*$, $x \in \mathfrak{A}_\varphi$, can be extended to an isometric involution J_φ on $\mathcal{H}_\varphi$. From (ii)$'_1$, the linear functional $x \in \mathfrak{A}_\# \to \varphi(x,u) \in \mathbb{C}$ is positive and so

$$\varphi(y^\#(x^\# x)y, u) \leq \|x\|_\#^2 \varphi(y, y), \quad \forall\, x, y \in \mathfrak{A}_0.$$

Hence, from (ii)$_1$, we obtain

$$\|\lambda_\varphi(x)\lambda_\varphi(y)\|_\varphi^2 = \varphi(xy, xy) = \varphi(y^\# x^\# xy, u) \leq \|x\|_\#^2 \|\lambda_\varphi(y)\|_\varphi^2, \quad \forall\, x, y \in \mathfrak{A}_0,$$

so that $L_{\lambda_\varphi(x)}$ is bounded and $\|L_{\lambda_\varphi(x)}\| \leq \|x\|_\#$, for each $x \in \mathfrak{A}_0$. Thus, $\mathcal{H}_\varphi \equiv \widetilde{\mathfrak{A}_\varphi}[\|\cdot\|_\varphi]$ is a Hilbert quasi *-algebra. Further, the map $\lambda_\varphi(x) \to \lambda_\varphi(x)^\# \equiv \lambda_\varphi(x^\#)$ is an involution of $\mathfrak{A}_\varphi$ and by (ii)$_1$, $L^*_{\lambda_\varphi(x)} = L_{\lambda_\varphi(x)^\#}$, for each $x \in \mathfrak{A}_0$. Hence, $(\widetilde{\mathfrak{A}_\varphi}[\|\cdot\|_\varphi], \#)$ is an HCQ*-algebra.

Now, we put $\Phi(a) := \lambda_\varphi(a)$, $a \in \mathfrak{A}$. Then, it is easily shown that Φ is a *-homomorphism of the strict CQ*-algebra into the HCQ*-algebra $(\widetilde{\mathfrak{A}_\varphi}[\|\cdot\|_\varphi], \#)$, and by (ii)$'_2$, it is contractive. Suppose that φ is faithful. Then, the *-representation of the C*-algebra $\mathfrak{A}_\#$ on $\mathcal{H}_\varphi$ defined by $x \mapsto L_{\lambda_\varphi(x)}$, $x \in \mathfrak{A}_\#$, is faithful, which implies that $\|L_{\lambda_\varphi(x)}\| = \|x\|_\#$, for each $x \in \mathfrak{A}_\#$. Moreover, since $\Phi(\mathfrak{A}_0) = \mathfrak{A}_\varphi$, it follows that $\Phi(\mathfrak{A}_\#) = (\mathfrak{A}_\varphi)_\#$ and $\Phi \lceil \mathfrak{A}_\#$ is a *-isomorphism of the C*-algebra $\mathfrak{A}_\#$ onto the C*-algebra $(\mathfrak{A}_\varphi)_\#$. Consequently, Φ is a *-isomorphism of $(\mathfrak{A}[\|\cdot\|], \mathfrak{A}_0, \#)$ into $(\widetilde{\mathfrak{A}_\varphi}[\|\cdot\|_\varphi], \#)$. By Lemma 5.5.3, the HCQ*-algebra $(\widetilde{\mathfrak{A}_\varphi}[\|\cdot\|_\varphi], \#)$ is standard, if and only if, (ii)$_4$ holds. This completes the proof. □

Now a *question* arises as to whether positive sesquilinear forms as described in (ii) do really exist. The *answer* is certainly positive due to the existence of standard HCQ*-algebras stated in Theorem 5.5.6. Indeed, the inner product $\langle\cdot|\cdot\rangle$ of a left Hilbert algebra satisfies conditions (ii)$_1$–(ii)$_4$.

In conclusion, Theorem 5.5.8 gives an answer to the main question of this subsection, in the following way: *any form φ over a strict CQ*-algebra $(\mathfrak{A}[\|\cdot\|], \mathfrak{A}_0, \#)$ with quasi-unit, can be used to construct an HCQ*-algebra, in which $\mathfrak{A}[\|\cdot\|]$ is contractively embedded.*

5.6 Measurable Operators and CQ*-Algebras

Another interesting situation where CQ*-algebras play an important role is that of noncommutative integration and in particular that of noncommutative L^p spaces constructed on a von Neumann algebra possessing a normal faithful, semifinite trace.

5.6.1 Noncommutative Measure and Integration

Let $\mathfrak{M}$ be a von Neumann algebra on a Hilbert space $\mathcal{H}$ and ϱ a normal faithful semifinite trace defined on $\mathfrak{M}^+$. For each $p \geq 1$, let

$$\mathcal{J}_p := \{X \in \mathfrak{M} : \varrho(|X|^p) < \infty\}.$$

Then, $\mathcal{J}_p$ is a *-ideal of $\mathfrak{M}$. Following [70], we denote with $L^p(\varrho)$ the Banach space completion of $\mathcal{J}_p$ with respect to the norm

$$\|X\|_{p,\varrho} := \varrho(|X|^p)^{1/p}, \quad X \in \mathcal{J}_p.$$

➧ *We shall simply write* $\|X\|_p$ *instead of* $\|X\|_{p,\varrho}$, whenever no ambiguity can arise.

One usually defines $L^\infty(\varrho) := \mathfrak{M}$. Thus, if ϱ is a finite trace, then $L^\infty(\varrho) \subset L^p(\varrho)$, for every $p \geq 1$.

The definition of noncommutative L^p-spaces given here follows Nelson [71]. Operators in this kind of spaces constructed by a von Neumann algebra $\mathfrak{M}$ are, however, measurable in Segal sense [76] and then often they are affiliated with $\mathfrak{M}$ [13]. We give now here an outline of Segal approach.

Let $P \in \mathrm{Proj}(\mathfrak{M})$, the lattice of projections of $\mathfrak{M}$. Two projections $P, Q \in \mathrm{Proj}(\mathfrak{M})$ are called *equivalent* and we write $P \sim Q$, if there is $U \in \mathfrak{M}$, such that $U^*U = P$ and $UU^* = Q$. We say that $P \prec Q$ in the case when P is equivalent to a subprojection of Q.

A projection P of a von Neumann algebra $\mathfrak{M}$ is said to be *finite* if $P \sim Q \leq P$ implies $P = Q$. A projection $P \in \mathfrak{M}$ is said to be *purely infinite* if there is no nonzero finite projection $Q \preceq P$ in $\mathfrak{M}$. A von Neumann algebra $\mathfrak{M}$ is said to be *finite* (resp. *purely infinite*) if the identity operator e is finite (resp. purely infinite).

We say that P is *ϱ-finite* if $P \in \mathcal{J}_1$. Any ϱ-finite projection is finite.

In what follows we shall need the following result (see, [64, Vol. IV, Ex. 6.9.12]) that we state as a lemma.

Lemma 5.6.1 *Let $\mathfrak{M}$ be a von Neumann algebra on a Hilbert space $\mathcal{H}$ and ϱ a normal, faithful, semifinite trace on $\mathfrak{M}_+$. Then, there is an orthogonal family $\{Q_j : j \in J\}$ of nonzero central projections in $\mathfrak{M}$, such that $\bigvee_{j\in J} Q_j = e$ and each Q_j is the sum of an orthogonal family of mutually equivalent finite projections in $\mathfrak{M}$.*

A vector subspace $\mathcal{D}$ of $\mathcal{H}$ is said to be *strongly dense* (resp. *strongly ϱ-dense*) if

- $U'\mathcal{D} \subset \mathcal{D}$, for any unitary operator U' in $\mathfrak{M}'$;
- there exists a sequence $\{P_n\} \in \mathrm{Proj}(\mathfrak{M}) : P_n\mathcal{H} \subset \mathcal{D}$, $P_n^\perp \downarrow 0$ and $P_n^\perp$ is a finite projection (resp. $\varrho(P_n^\perp) < \infty$).

Clearly, every strongly ϱ-dense domain is strongly dense.

Throughout this subsection, when we say that an operator T is affiliated with a von Neumann algebra $\mathfrak{M}$ and write $T \,\eta\, \mathfrak{M}$, we always mean that T is closed, densely defined and $TU \supseteq UT$, for every unitary operator $U \in \mathfrak{M}'$.

An operator $T \,\eta\, \mathfrak{M}$ is called

- *measurable* (with respect to $\mathfrak{M}$) if its domain $D(T)$ is strongly dense;
- *ϱ-measurable* if its domain $D(T)$ is strongly ϱ-dense.

From the definition itself it follows that, if T is ϱ-measurable, then there exists $P \in \mathrm{Proj}(\mathfrak{M})$, such that TP is bounded and $\varrho(P^\perp) < \infty$.

Remark 5.6.2 As shown in [70], every operator $X \in L^p(\varrho)$ is measurable. Moreover, every operator affiliated with a finite von Neumann algebra is measurable, but not necessarily ϱ-measurable (see [76, Cor. 4.1]). In the general case, conditions that guarantee the measurability of affiliated operators are discussed in [27, IX, §2].

If A is a measurable operator and $A \geq 0$, one defines the *integral* of A by

$$\mu(A) = \sup\left\{\varrho(X) : 0 \leq X \leq A,\ X \in \mathcal{J}_1\right\}.$$

Then, the space $L^p(\varrho)$ can also be defined [70] as the space of all measurable operators A, such that $\mu(|A|^p) < \infty$.

The integral of an element $A \in L^p(\varrho)$ can be defined, in obvious way, taking into account that any measurable operator A can be decomposed as $A = B_+ - B_- + iC_+ - iC_-$, where $B = \frac{A+A^*}{2}$, $C = \frac{A-A^*}{2i}$ and $B_+,\ B_-$ (resp. $C_+,\ C_-$) are the positive and negative parts of B (resp. C).

Remark 5.6.3 The following statements will be used later:

(i) Let $T\,\eta\,\mathfrak{M}$ and $Q \in \mathfrak{M}$. If $D(TQ) = \left\{\xi \in \mathcal{H} : Q\xi \in D(T)\right\}$ is dense in $\mathcal{H}$, then $TQ\,\eta\,\mathfrak{M}$.
(ii) If $Q \in \mathrm{Proj}(\mathfrak{M})$, then $Q\mathfrak{M}Q = \left\{QSQ \restriction_{Q\mathcal{H}} : S \in \mathfrak{M}\right\}$ is a von Neumann algebra on the Hilbert space $Q\mathcal{H}$; moreover, $(Q\mathfrak{M}Q)' = Q\mathfrak{M}'Q$. If $T\,\eta\,\mathfrak{M}$, $Q \in \mathfrak{M}$ and $D(TQ) = \left\{\xi \in \mathcal{H} : Q\xi \in D(T)\right\}$ is dense in $\mathcal{H}$, then $QTQ\,\eta\,Q\mathfrak{M}Q$.

For (i) and (ii), see for instance, [14, Chapter 5].

5.6.2 Noncommutative L^p-Spaces as Proper CQ*-Algebras

In this subsection we shall discuss the structure of the noncommutative L^p-spaces as quasi *-algebras.

Proposition 5.6.4 *Let $\mathfrak{M}$ be a von Neumann algebra and ϱ a normal faithful semifinite trace on $\mathfrak{M}_+$. Then, $(L^p(\varrho), L^\infty(\varrho) \cap L^p(\varrho))$ is a Banach quasi *-algebra.*

If ϱ is a finite trace and $\varrho(\mathbb{I}) = 1$, then $(L^p(\varrho), L^\infty(\varrho))$ is a proper CQ-algebra.*

Proof Indeed, it is easily seen that the norms $\|\cdot\|_\infty$ on $L^\infty(\varrho) \cap L^p(\varrho)$ and $\|\cdot\|_p$ on $L^p(\varrho)$ satisfy the conditions of Definition 3.1.1. Moreover, if ϱ is finite, then $L^\infty(\varrho) \subset L^p(\varrho)$ and thus $(L^p(\varrho), L^\infty(\varrho))$ is a proper CQ*-algebra. □

Remark 5.6.5 Of course, the condition $\varrho(\mathbb{I}) = 1$ can be easily removed by normalizing the trace.

We shall now focus our attention on the *question* as to whether for the Banach quasi *-algebra $(L^p(\varrho), L^\infty(\varrho) \cap L^p(\varrho))$, the *family* $\mathcal{S}_{L^\infty(\varrho)}(L^p(\varrho))$, that we are going to describe, *is or is not sufficient*.

Before going forth, we remind that many of the familiar results of the ordinary theory of L^p-spaces hold in the very same form for the noncommutative L^p-spaces. This is the case, for instance, of Hölder's inequality and also of the statement that characterizes the dual of L^p: the form defining the duality is the extension of ϱ (this extension will be denoted with the same symbol) to products of the type XY with $X \in L^p(\varrho)$, $Y \in L^{p'}(\varrho)$ and $p^{-1} + p'^{-1} = 1$. Moreover, one has $(L^p(\varrho))^\star \simeq L^{p'}(\varrho)$.

In order to study $\mathcal{S}_{L^\infty(\varrho)}(L^p(\varrho))$, we introduce, for $p \geq 2$, the following notation

$$\mathcal{B}_+^p := \{X \in L^{\frac{p}{p-2}}(\varrho),\ X \geq 0,\ \|X\|_{p/(p-2)} \leq 1\}$$

where $p/(p-2) = \infty$ if $p = 2$.

For each $W \in \mathcal{B}_+^p$, we consider the right multiplication operator

$$R_W : L^p(\varrho) \to L^{\frac{p}{p-2}}(\varrho) : X \mapsto R_W X := XW, \quad X \in L^p(\varrho).$$

Since $L^\infty(\varrho) \cap L^p(\varrho) = \mathcal{J}_p$, we use, for shortness, the latter notation.

Lemma 5.6.6 *Let $p \geq 2$. The following statements hold:*

(i) *for every $W \in \mathcal{B}_+^p$, the sesquilinear form $\varphi(X, Y) := \varrho(X(R_W Y)^*)$ is an element of $\mathcal{S}_{L^\infty(\varrho)}(L^p(\varrho))$;*

(ii) *if ϱ is finite, then for each $\varphi \in \mathcal{S}_{L^\infty(\varrho)}(L^p(\varrho))$, there exists $W \in \mathcal{B}_+^p$, such that*

$$\varphi(X, Y) = \varrho(X(R_W Y)^*), \quad \forall X, Y \in L^p(\varrho).$$

Proof

(i) We check that the sesquilinear form, $\varphi(X, Y) = \varrho(X(R_W Y)^*)$, $X, Y \in L^p(\varrho)$, satisfies the conditions of Sect. 3.1.2.

For every $X \in L^p(\varrho)$, we have

$$\varphi(X, X) = \varrho(X(R_W X)^*) = \varrho((XW)^* X) = \varrho(W|X|^2) \geq 0.$$

For every $X \in L^p(\varrho)$, $A, B \in \mathcal{J}_p$, we get

$$\varphi(XA, B) = \varrho(XA(BW)^*) = \varrho(WB^*XA) = \varrho(A(X^*BW)^*) = \varphi(A, X^*B).$$

Finally, for every $X, Y \in L_p(\varrho)$,

$$|\varphi(X, Y)| \leq \|X\|_p \|Y\|_p \|W\|_{p/p-2} \leq \|X\|_p \|Y\|_p.$$

(ii) Let $\varphi \in \mathcal{S}_{L^\infty(\varrho)}(L^p(\varrho))$. Let $T : L^p(\varrho) \to L^{p'}(\varrho)$ be the operator which represents φ in the sense of Proposition 3.1.16. The finiteness of ϱ implies that $\mathcal{J}_p = \mathfrak{M}$; thus we can put $W = T(\mathbb{I})$. It is easy to check that $R_W = T$. This concludes the proof. □

Proposition 5.6.7 *If $p \geq 2$, $\mathcal{S}_{L^\infty(\varrho)}\big(L^p(\varrho)\big)$ is sufficient.*

Proof Let $X \in L^p(\varrho)$ be such that $\varphi(X, X) = 0$, for every $\varphi \in \mathcal{S}(L_p(\varrho))$. By the previous lemma, since $|X|^{p-2} \in L^{\frac{p}{p-2}}(\varrho)$, the right multiplication operator R_W with $W = \frac{|X|^{p-2}}{\alpha}$, $\alpha \in \mathbb{R} \setminus \{0\}$, satisfying $\|\frac{|X|^{p-2}}{\alpha}\|_{p/p-2} \leq 1$, represents a sesquilinear form $\varphi \in \mathcal{S}_{L^\infty(\varrho)}(L^p(\varrho))$. By the assumption, $\varphi(X, X) = 0$. We then have

$$\begin{aligned}\varphi(X, X) = \varrho\big(X(R_W X)^*\big) &= \tfrac{1}{\alpha}\varrho\big(X(X|X|^{p-2})^*\big)\\ &= \tfrac{1}{\alpha}\varrho\big((X|X|^{p-2})^* X\big) = \tfrac{1}{\alpha}\varrho(|X|^p) = 0 \Rightarrow X = 0,\end{aligned}$$

by the faithfulness of ϱ. □

5.6.3 CQ*-Algebras over Finite von Neumann Algebras

Let $\mathfrak{M}$ be a von Neumann algebra and $\mathbb{F} = \{\varrho_\alpha;\ \alpha \in \mathcal{I}\}$ be a family of normal, *finite* traces on $\mathfrak{M}$. As usual, we say that the family $\mathbb{F}$ is *sufficient* if, for $X \in \mathfrak{M}$, $X \geq 0$ and $\varrho_\alpha(X) = 0$, for every $\alpha \in \mathcal{I}$, then $X = 0$ (clearly, if $\mathbb{F} = \{\varrho\}$, then $\mathbb{F}$ is sufficient, if and only if, ϱ is faithful). In this case, $\mathfrak{M}$ is a finite von Neumann algebra [25, ch.7]. We assume, in addition, that the following condition (5.6.17) is satisfied:

$$\varrho_\alpha(\mathbb{I}) \leq 1, \quad \forall\, \alpha \in \mathcal{I}, \tag{5.6.17}$$

where $\mathbb{I}$ is the identity of $\mathfrak{M}$. Then, we define

$$\|X\|_{p,\mathcal{I}} = \sup_{\alpha \in \mathcal{I}} \|X\|_{p,\varrho_\alpha} = \sup_{\alpha \in \mathcal{I}} \varrho_\alpha(|X|^p)^{1/p}.$$

Since $\mathbb{F}$ is sufficient, $\|\cdot\|_{p,\mathcal{I}}$ is a norm on $\mathfrak{M}$.

In the sequel we shall need the following lemmas, whose simple proofs will be omitted.

Lemma 5.6.8 *Let $\mathfrak{M}$ be a von Neumann algebra in a Hilbert space $\mathcal{H}$ and $\{P_\alpha\}_{\alpha \in \mathcal{I}}$ a family of projections of $\mathfrak{M}$ with*

$$\bigvee_{\alpha \in \mathcal{I}} P_\alpha = \overline{P}.$$

If $A \in \mathfrak{M}$ and $AP_\alpha = 0$, for every $\alpha \in \mathcal{I}$, then $A\overline{P} = 0$.

Lemma 5.6.9 *Let $\mathbb{F} = \{\varrho_\alpha\}_{\alpha \in \mathcal{I}}$ be a sufficient family of normal, finite traces on the von Neumann algebra $\mathfrak{M}$ and let P_α be the support of ϱ_α. Then, $\bigvee_{\alpha \in \mathcal{I}} P_\alpha = \mathbb{I}$, where $\mathbb{I}$ denotes the identity of $\mathfrak{M}$.*

It is well-known that the support of each ϱ_α enjoys the following properties:

(i) $P_\alpha \in \mathcal{Z}(\mathfrak{M})$, the center of $\mathfrak{M}$, for each $\alpha \in \mathcal{I}$;
(ii) $\varrho_\alpha(X) = \varrho_\alpha(XP_\alpha)$, for each $\alpha \in \mathcal{I}$.

From the two preceding lemmas it follows that, if the P_α's are as in Lemma 5.6.9, then

$$AP_\alpha = 0, \quad \forall\, \alpha \in \mathcal{I} \quad \Rightarrow \quad A = 0.$$

If condition (5.6.17) is fulfilled, then

$$\|X\|_{p,\mathcal{I}} := \sup_{\alpha\in\mathcal{I}} \|XP_\alpha\|_{p,\varrho_\alpha}, \quad \forall\, X \in \mathfrak{M}.$$

Clearly, the sufficiency of the family of traces and condition (5.6.17) imply that $\|\cdot\|_{p,\mathcal{I}}$ is a norm on $\mathfrak{M}$.

Proposition 5.6.10 *Let $\mathfrak{M}(p,\mathcal{I})$ denote the Banach space completion of $\mathfrak{M}$ with respect to the norm $\|\cdot\|_{p,\mathcal{I}}$. Then, $(\mathfrak{M}(p,\mathcal{I})[\|\cdot\|_{p,\mathcal{I}}], \mathfrak{M})$ is a proper CQ*-algebra.*

Proof Indeed, from Definition 5.1.1, $\forall\, X \in \mathfrak{M}$, we have

$$\|X^*\|_{p,\mathcal{I}} = \sup_{\alpha\in\mathcal{I}} \|X^*P_\alpha\|_{p,\varrho_\alpha} = \sup_{\alpha\in\mathcal{I}} \|(XP_\alpha)^*\|_{p,\varrho_\alpha} = \|X\|_{p,\mathcal{I}}. \tag{5.6.18}$$

Furthermore, for every $X, Y \in \mathfrak{M}$,

$$\begin{aligned} \|XY\|_{p,\mathcal{I}} = \sup_{\alpha\in\mathcal{I}} \|XYP_\alpha\|_{p,\varrho_\alpha} &\le \|X\|_{\mathcal{B}(\mathcal{H})} \sup_{\alpha\in\mathcal{I}} \|YP_\alpha\|_{p,\varrho_\alpha} \\ &= \|X\|_{\mathcal{B}(\mathcal{H})}\|Y\|_{p,\mathcal{I}}. \end{aligned} \tag{5.6.19}$$

Finally, condition (5.6.17) implies that

$$\|X\|_{p,\mathcal{I}} \le \|X\|, \quad \forall\, X \in \mathfrak{M}.$$

It follows from (5.6.18) and (5.6.19) that $(\mathfrak{M}(p,\mathcal{I})[\|\cdot\|_{p,\mathcal{I}}], \mathfrak{M})$ is a Banach quasi *-algebra and therefore a proper CQ*-algebra. □

- The next step aims at investigating the Banach space $\mathfrak{M}(p,\mathcal{I})[\|\cdot\|_{p,\mathcal{I}}]$. In particular, *we are interested in the question* as to whether $\mathfrak{M}(p,\mathcal{I})[\|\cdot\|_{p,\mathcal{I}}]$ *can be identified with a space of operators affiliated with* $\mathfrak{M}$. For shortness, whenever no ambiguity can arise, *we write* $\mathfrak{M}_p$ *instead of* $\mathfrak{M}(p,\mathcal{I})$

Let $\mathbb{F} = \{\varrho_\alpha\}_{\alpha\in\mathcal{I}}$ be a sufficient family of normal, finite traces on the von Neumann algebra $\mathfrak{M}$ satisfying condition (5.6.17). The traces ϱ_α are not necessarily faithful. Put $\mathfrak{M}_\alpha := \mathfrak{M}P_\alpha$, where, as before, P_α denotes the support of ϱ_α. Each $\mathfrak{M}_\alpha$ is a von Neumann algebra and ϱ_α is faithful in $\mathfrak{M}P_\alpha$ [27, Proposition V. 2.10].

More precisely,

$$\mathfrak{M}_\alpha := \mathfrak{M}P_\alpha = \{Z = XP_\alpha, \text{ for some } X \in \mathfrak{M}\}.$$

The positive cone $\mathfrak{M}_\alpha^+$ of $\mathfrak{M}_\alpha$ equals the set

$$\{Z = XP_\alpha, \text{ for some } X \in \mathfrak{M}^+\}.$$

For $Z = XP_\alpha \in \mathfrak{M}_\alpha^+$, we put

$$\sigma_\alpha(Z) := \varrho_\alpha(XP_\alpha).$$

The definition of $\sigma_\alpha(Z)$ does not depend on the particular choice of X. Each σ_α is a normal, finite, faithful trace on $\mathfrak{M}_\alpha$. It is then possible to consider the spaces $L^p(\mathfrak{M}_\alpha, \sigma_\alpha)$, $p \geq 1$, in the usual way. Recall that the norm of $L^p(\mathfrak{M}_\alpha, \sigma_\alpha)$ is indicated by $\|\cdot\|_{p,\sigma_\alpha}$. Let now $\{X_k\}$ be a Cauchy sequence in $\mathfrak{M}[\|\cdot\|_{p,\mathcal{I}}]$. For each $\alpha \in \mathcal{I}$, we put $Z_k^{(\alpha)} = X_k P_\alpha$. Then, for each $\alpha \in \mathcal{I}$, $\{Z_k^{(\alpha)}\}$ is a Cauchy sequence in $\mathfrak{M}_\alpha[\|\cdot\|_{p,\sigma_\alpha}]$. Indeed, since $|Z_k^{(\alpha)} - Z_h^{(\alpha)}|^p = |X_k - X_h|^p P_\alpha$, we have

$$\begin{aligned}\|Z_k^{(\alpha)} - Z_h^{(\alpha)}\|_{p,\sigma_\alpha} &= \sigma_\alpha\big(|Z_k^{(\alpha)} - Z_h^{(\alpha)}|^p\big)^{1/p} \\ &= \varrho_\alpha\big(|X_k - X_h|^p P_\alpha\big)^{1/p} \\ &= \varrho_\alpha\big(|X_k - X_h|^p\big)^{1/p} \to 0.\end{aligned}$$

Therefore, for each $\alpha \in \mathcal{I}$, there exists an operator $Z^{(\alpha)} \in L^p(\mathfrak{M}_\alpha, \sigma_\alpha)$, such that

$$Z^{(\alpha)} = \|\cdot\|_{p,\sigma_\alpha} - \lim_{k\to\infty} Z_k^{(\alpha)}.$$

It is now natural to ask the question as to whether there exists an operator X closed, densely defined, affiliated with $\mathfrak{M}$, which reduces to $Z^{(\alpha)}$ on $\mathfrak{M}_\alpha$. To begin with, we assume that the projections $\{P_\alpha\}$ are mutually orthogonal. In this case, setting $\mathcal{H}_\alpha = P_\alpha \mathcal{H}$, we have

$$\mathcal{H} = \bigoplus_{\alpha\in\mathcal{I}} \mathcal{H}_\alpha = \Big\{(f_\alpha)_{\alpha\in I} : f_\alpha \in \mathcal{H}_\alpha, \sum_{\alpha\in I} \|f_\alpha\|^2 < \infty\Big\}.$$

We now set

$$D(X) = \Big\{(f_\alpha)_{\alpha\in I} \in \mathcal{H} : f_\alpha \in D(Z^{(\alpha)}), \sum_{\alpha\in I} \|Z^{(\alpha)} f_\alpha\|^2 < \infty\Big\}$$

and for $f = (f_\alpha)_{\alpha\in I} \in D(X)$ we define $Xf := (Z^{(\alpha)} f_\alpha)$. Then,

1. $D(X)$ is dense in $\mathcal{H}$. Indeed, $D(X)$ contains all $f = (f_\alpha)_{\alpha\in I}$, with $f_\alpha = 0$, for all $\alpha \in \mathcal{I}$, except for a finite subset of indices.
2. X is closed in $\mathcal{H}$. Indeed, let $\{f_n = (f_{n,\alpha})_{\alpha\in I}\}$ be a sequence of elements in $D(X)$, with $f_n \to g = (g_\alpha)_{\alpha\in I} \in \mathcal{H}$ and $Xf_n \to h$. Since,

$$f_n \to g \Leftrightarrow f_{n,\alpha} \to g_\alpha \in \mathcal{H}_\alpha, \quad \forall\, \alpha \in \mathcal{I}$$

and

$$Xf_n \to h \Leftrightarrow (Xf_n)_\alpha \to h_\alpha \in \mathcal{H}_\alpha, \quad \forall\, \alpha \in \mathcal{I},$$

by $(Xf_n)_\alpha = Z^{(\alpha)} f_{n,\alpha}$ and the closedness of each $Z^{(\alpha)}$ in $\mathcal{H}_\alpha$, we obtain

$$g_\alpha \in D(Z^{(\alpha)}) \quad \text{and} \quad h_\alpha = Z^{(\alpha)} g_\alpha.$$

It remains to check that $\sum_{\alpha\in\mathcal{I}} \|Z^{(\alpha)} g_\alpha\|^2 < \infty$, but this is clear, since both $(Z^{(\alpha)} g_\alpha)_{\alpha\in I}$ and $h = (h_\alpha)_{\alpha\in I}$ belong to $\mathcal{H}$.
3. $X \,\eta\, \mathfrak{M}$. Let $Y \in \mathfrak{M}'$. Then, for all $f \in \mathcal{H}$, $Yf = (YP_\alpha f)_{\alpha\in I}$ and $YP_\alpha \in (\mathfrak{M}P_\alpha)' = \mathfrak{M}'P_\alpha$, so that

$$XYf = \big((XY)P_\alpha f\big)_{\alpha\in I} = (YXP_\alpha f)_{\alpha\in I} = YXf.$$

In conclusion, X is a measurable operator.

Thus, we have proved the following

Proposition 5.6.11 *Let $\mathbb{F} = \{\varrho_\alpha\}_{\alpha\in\mathcal{I}}$ be a sufficient family of normal, finite traces on the von Neumann algebra $\mathfrak{M}$. Assume that condition* (5.6.17) *is fulfilled* (i.e., $\varrho_\alpha(\mathbb{I}) \leq 1$, $\forall\, \alpha \in \mathcal{I}$) *and that the ϱ_α's have mutually orthogonal supports. Then, $\mathfrak{M}_p$, $p \geq 1$, consists of measurable operators.*

The analysis of the general case would really be simplified if, from a given sufficient family $\mathbb{F}$ of normal, finite traces, one could extract (or construct) a *sufficient* subfamily $\mathcal{G}$ of traces with mutually orthogonal supports. Apart from quite simple situations (for instance, when $\mathbb{F}$ is finite or countable), we do not know if this is possible or not. There is however a relevant case, where this can be fairly easily done. This occurs when $\mathbb{F}$ is a convex and w*-compact family of traces on $\mathfrak{M}$.

Lemma 5.6.12 *Let $\mathbb{F}$ be a convex w*-compact family of normal, finite traces on a von Neumann algebra $\mathfrak{M}$. Assume that, for each central operator Z, with $0 \leq Z \leq \mathbb{I}$, and each $\eta \in \mathbb{F}$, the functional $\eta_Z(X) := \eta(XZ)$ belongs to $\mathbb{F}$. Let $\mathfrak{E}\mathbb{F}$ be the set of all extreme elements of $\mathbb{F}$. If $\eta_1, \eta_2 \in \mathfrak{E}\mathbb{F}$, $\eta_1 \neq n_2$, and P_1 and P_2 are their respective supports, then P_1 and P_2 are orthogonal.*

Proof Let P_1, P_2 be the supports of η_1 and η_2, respectively. We begin proving that either $P_1 = P_2$ or $P_1 P_2 = 0$. Indeed, assume that $P_1 P_2 \neq 0$. We define

$$\eta_{1,2}(X) := \eta_1(X P_2), \quad \forall\, X \in \mathfrak{M}.$$

When $\eta_{1,2} = 0$, then in particular, $\eta_{1,2}(P_2) = 0$, i.e., $\eta_1(P_2) = 0$ and therefore, by definition of support, $P_2 \leq \mathbb{I} - P_1$. This implies that $P_1 P_2 = 0$, which contradicts the assumption. We now show that the support of $\eta_{1,2}$ is $P_1 P_2$. In fact, let Q be a projection, such that $\eta_{1,2}(Q) = 0$. Then,

$$\eta_1(QP_2) = 0 \Rightarrow QP_2 \leq \mathbb{I} - P_1 \Rightarrow QP_2(\mathbb{I} - P_1) = QP_2 \Rightarrow QP_2 P_1 = 0.$$

Then, the largest Q for which this happens is $\mathbb{I} - P_2 P_1$. We conclude that the support of the trace $\eta_{1,2}$ is $P_1 P_2$. Finally, by definition, one has $\eta_{1,2}(X) = \eta_1(X P_2)$, and since $X P_2 \leq X$,

$$\eta_{1,2}(X) = \eta_1(X P_2) \leq \eta_1(X), \quad \forall\, X \in \mathfrak{M}.$$

Thus, η_1 majorizes $\eta_{1,2}$. But η_1 is extreme in $\mathbb{F}$. Therefore, $\eta_{1,2}$ has the form $\lambda\eta_1$ with $\lambda \in]0, 1]$. This implies that $\eta_{1,2}$ has the same support as η_1; hence, $P_1 P_2 = P_1$, i.e., $P_1 \leq P_2$. Starting from $\eta_{2,1}(X) = \eta_2(X P_1)$, we obtain in a similar way that $P_2 \leq P_1$. Consequently, $P_1 P_2 \neq 0$ implies $P_1 = P_2$. However, two different traces of $\mathfrak{E}\mathbb{F}$ cannot have the same support. Indeed, assume that there exist $\eta_1, \eta_2 \in \mathbb{F}$ having the same support P. Since P is central, we can consider the von Neumann algebra $\mathfrak{M}P$. The restrictions of η_1, η_2 to $\mathfrak{M}P$ are normal, faithful, semifinite traces. By [27, Prop. V.2.31] there exist a central element Z in $\mathfrak{M}P$ with $0 \leq Z \leq P$ (P is here considered as the unit of $\mathfrak{M}P$), such that

$$\eta_1(X) = (\eta_1 + \eta_2)(ZX), \quad \forall\, X \in (\mathfrak{M}P)_+. \tag{5.6.20}$$

Then, Z also belongs to the center of $\mathfrak{M}$, since for every $V \in \mathfrak{M}$,

$$ZV = Z(VP + VP^{\perp}) = ZVP = VZP = VZ.$$

Therefore, the functionals

$$\eta_{1,z}(X) := \eta_1(XZ), \quad \eta_{2,z}(X) := \eta_2(XZ), \quad \forall\, X \in \mathfrak{M},$$

belong to the family $\mathbb{F}$ and are majorized respectively, by the extreme elements η_1, η_2. Then, there exist $\lambda, \mu \in [0, 1]$, such that

$$\eta_1(XZ) = \lambda\eta_1(X), \quad \eta_2(XZ) = \mu\eta_1(X), \quad \forall\, X \in \mathfrak{M}.$$

If $\lambda = 1$, from (5.6.20), we would have $\eta_2(ZX) = 0$, for every $X \in (\mathfrak{M}P)_+$; in particular, $\eta_2(|\ Z\ |^2) = 0$; this implies that $Z = 0$. Thus, $\lambda \neq 1$. Analogously,

$\mu \neq 0$; indeed, if $\mu = 0$, then $\eta_1(X) = \lambda\eta_1(X)$ and so $\lambda = 1$. Therefore, there exist $\lambda, \mu \in (0, 1)$, such that

$$\eta_1(X) = \lambda\eta_1(X) + \mu\eta_2(X), \quad \forall X \in \mathfrak{M}P,$$

which in turn, implies

$$\eta_1(X) = \lambda\eta_1(X) + \mu\eta_2(X), \quad \forall X \in \mathfrak{M}.$$

Hence,

$$(1 - \lambda)\eta_1(X) = \mu\eta_2(X), \quad \forall X \in \mathfrak{M}.$$

From the last equality, dividing by $\max\{1 - \lambda, \mu\}$ one gets that one of the two elements is a convex combination of the other and of 0, which is absurd. In conclusion, different supports of extreme traces of $\mathfrak{F}$ are orthogonal. □

Since, for every $X \in \mathfrak{M}$, $\|X\|_{p,\mathcal{I}}$ remains the same if computed either with respect to $\mathfrak{F}$ or to $\mathfrak{C}\mathfrak{F}$, we can deduce the following

Theorem 5.6.13 *Let $\mathbb{F}$ be a convex and w*-compact sufficient family of normal, finite traces on the von Neumann algebra $\mathfrak{M}$. Assume that $\mathbb{F}$ satisfies condition (5.6.17) and that for each central operator Z, with $0 \leq Z \leq \mathbb{I}$, and each $\eta \in \mathbb{F}$, the functional $\eta_Z(X) := \eta(XZ)$ belongs to $\mathbb{F}$. Then, $\mathfrak{M}_p[\|\cdot\|_{p,\mathcal{I}}]$, consists of measurable operators.*

Families of traces satisfying the assumptions of Theorem 5.6.13 will be constructed in the next subsection.

5.6.4 A First Representation Theorem

Once we have constructed in the previous subsection some CQ*-algebras of operators affiliated to a given von Neumann algebra, *it is natural to pose* the question, *under which conditions, can an abstract CQ*-algebra $(\mathfrak{A}[\|\cdot\|], \mathfrak{A}_0)$ be realized as a CQ*-algebra of operators.*

Let $(\mathfrak{A}[\|\cdot\|], \mathfrak{A}_0)$ be a proper CQ*-algebra with unit e and let

$$\mathcal{T}_{\mathfrak{A}_0}(\mathfrak{A}) \equiv \{\varphi \in \mathcal{S}_{\mathfrak{A}_0}(\mathfrak{A}) : \varphi(a, a) = \varphi(a^*, a^*), \ \forall\, a \in \mathfrak{A}\}.$$

Note that if $\varphi \in \mathcal{T}_{\mathfrak{A}_0}(\mathfrak{A})$, then by polarization, $\varphi(b^*, a^*) = \varphi(a, b)$, $\forall\, a, b \in \mathfrak{A}$. It is easy to prove that the set $\mathcal{T}_{\mathfrak{A}_0}(\mathfrak{A})$ is convex.

For each $\varphi \in \mathcal{T}_{\mathfrak{A}_0}(\mathfrak{A})$, we define a linear functional ω_φ on $\mathfrak{A}_0$ by

$$\omega_\varphi(x) := \varphi(x, e), \quad \forall\, x \in \mathfrak{A}_0.$$

For every $x \in \mathfrak{A}_0$, we have

$$\omega_\varphi(x^*x) = \varphi(x^*x, e) = \varphi(x, x) = \varphi(x^*, x^*) = \omega_\varphi(xx^*) \geq 0.$$

This shows at once that ω_φ is positive and tracial.

We put

$$\mathfrak{M}_{\mathcal{T}}(\mathfrak{A}_0) \equiv \{\omega_\varphi : \varphi \in \mathcal{T}_{\mathfrak{A}_0}(\mathfrak{A})\}.$$

From the convexity of $\mathcal{T}_{\mathfrak{A}_0}(\mathfrak{A})$ it follows easily that $\mathfrak{M}_{\mathcal{T}}(\mathfrak{A}_0)$ is convex too. Then, for the norm of the bounded linear functional ω_φ, recalling that we may suppose that $\|e\| = 1$ (Remark 3.1.4), we have

$$\|\omega_\varphi\|^\star = \omega_\varphi(e) = \varphi(e, e) \leq 1.$$

Therefore,

$$\mathfrak{M}_{\mathcal{T}}(\mathfrak{A}_0) \subseteq \{\omega \in \mathfrak{A}_0^\star : \|\omega\|^\star \leq 1\},$$

where $\mathfrak{A}_0^\star$ denotes the topological dual of $\mathfrak{A}_0[\| \cdot \|_0]$.

Setting

$$f_\varphi(x) := \omega_\varphi(x), \ x \in \mathfrak{A}_0,$$

we obtain

$$f_\varphi \in \{\omega \in \mathfrak{A}_0^\star : \|\omega\|^\star \leq 1\}.$$

By the Banach–Alaoglou theorem, the set $\{\omega \in \mathfrak{A}_0^\star : \|\omega\|^\star \leq 1\}$ is a w*-compact subset of $\mathfrak{A}_0^\star$.

Proposition 5.6.14 *$\mathfrak{M}_{\mathcal{T}}(\mathfrak{A}_0)$ is w*-closed and, therefore, a w*-compact set.*

Proof Let $(\omega_{\varphi_\alpha})$ be a net in $\mathfrak{M}_{\mathcal{T}}(\mathfrak{A}_0)$ w*-converging to a functional $\omega \in \mathfrak{A}_0^\star$. We shall show that $\omega = \omega_\varphi$, for some $\varphi \in \mathcal{T}_{\mathfrak{A}_0}(\mathfrak{A})$. Let us begin defining $\varphi_0(x, y) := \omega(y^*x)$, $x, y \in \mathfrak{A}_0$. By the definition itself, $(\omega_{\varphi_\alpha})(x) \to \omega(x) = \varphi_0(x, e)$. Moreover, for every $x, y \in \mathfrak{A}_0$,

$$\varphi_0(x, y) = \omega(y^*x) = \lim_\alpha \omega_{\varphi_\alpha}(y^*x) = \lim_\alpha \varphi_\alpha(x, y).$$

Therefore,

$$\varphi_0(x, x) = \lim_\alpha \varphi_\alpha(x, x) \geq 0.$$

We also have

$$| \varphi_0(x, y) | = \lim_\alpha | \varphi_\alpha(x, y) | \leq \|x\| \, \|y\|.$$

Hence, φ_0 can be extended by continuity to $\mathfrak{A} \times \mathfrak{A}$. Indeed, let $a, b \in \mathfrak{A}$ such that

$$a = \| \cdot \| - \lim_n x_n, \quad b = \| \cdot \| - \lim_n y_n, \quad \text{with } \{x_n\}, \ \{y_n\} \subset \mathfrak{A}_0.$$

Then,

$$\begin{aligned} | \varphi_0(x_n, y_n) - \varphi_0(x_m, y_m) | &= | \varphi_0(x_n, y_n) - \varphi_0(x_m, y_n) + \varphi_0(x_m, y_n) - \varphi_0(x_m, y_m) | \\ &\leq | \varphi_0(x_n - x_m, y_n) | + | \varphi_0(x_m, y_n - y_m) | \\ &\leq \|x_n - x_m\| \, \|y_n\| + \|x_m\| \, \|y_n - y_m\| \to 0, \end{aligned}$$

since $\{\|x_n\|\}$ and $\{\|y_n\|\}$ are bounded sequences. Hence, we can define

$$\varphi(a, b) := \lim_n \varphi_0(x_n, y_n).$$

Clearly, $\varphi(a, a) \geq 0$, for all $a \in \mathfrak{A}$.

It is then easily checked that finally $\varphi \in \mathcal{T}_{\mathfrak{A}_0}(\mathfrak{A})$. This completes the proof. □

Since $\mathfrak{M}_{\mathcal{T}}(\mathfrak{A}_0)$ is convex and w*-compact, by the Krein–Milmann theorem it follows that it has extreme points and it coincides with the w*-closure of the convex hull of the set $\mathfrak{E}\mathfrak{M}_{\mathcal{T}}(\mathfrak{A}_0)$ of its extreme points.

Let π be the universal *-representation of the C*-algebra $\mathfrak{A}_0$ (Remark A.6.16) and $\pi(\mathfrak{A}_0)''$ the von Neumann algebra generated by $\pi(\mathfrak{A}_0)$.

For every $\varphi \in \mathcal{T}_{\mathfrak{A}_0}(\mathfrak{A})$ and $x \in \mathfrak{A}_0$, we put $\Omega_\varphi(\pi(x)) := \omega_\varphi(x)$.

Then, for each $\varphi \in \mathcal{T}_{\mathfrak{A}_0}(\mathfrak{A})$, Ω_φ is a positive bounded linear functional on the operator algebra $\pi(\mathfrak{A}_0)$. Clearly,

$$\Omega_\varphi(\pi(x)) = \omega_\varphi(x) = \varphi(x, e), \quad \forall \, x \in \mathfrak{A}_0 \ \text{ and}$$

$$| \Omega_\varphi(\pi(x)) | = | \omega_\varphi(x) | = | \varphi(x, e) | \leq \|x\| \leq \|x\|_0 = \|\pi(x)\|, \quad \forall \, x \in \mathfrak{A}_0.$$

Thus, Ω_φ is continuous on $\pi(\mathfrak{A}_0)$ and then according to [15, Proposition 10.1.1], Ω_φ is also weakly continuous and so it extends uniquely to $\pi(\mathfrak{A}_0)''$. Moreover, since Ω_φ is a trace on $\pi(\mathfrak{A}_0)$, the extension $\widetilde{\Omega_\varphi}$ is a trace on $\mathfrak{M} := \pi(\mathfrak{A}_0)''$ too. The norm $\|\widetilde{\Omega_\varphi}\|^\star$ of the linear functional $\widetilde{\Omega_\varphi}$ on $\mathfrak{M}$ equals the norm of the linear functional Ω_φ on $\pi(\mathfrak{A}_0)$. Moreover,

$$\|\widetilde{\Omega_\varphi}\|^\star = \widetilde{\Omega_\varphi}(\pi(e)) = \Omega_\varphi(\pi(e)) = \omega_\varphi(e) \leq 1.$$

The set

$$\mathfrak{N}_{\mathcal{T}}(\mathfrak{A}_0) \equiv \{\widetilde{\Omega}_\varphi : \varphi \in \mathcal{T}_{\mathfrak{A}_0}(\mathfrak{A})\}$$

is convex and w*-compact in $\mathfrak{M}^\star$, as it can be easily seen by considering the map

$$\omega_\varphi \in \mathfrak{M}_{\mathcal{T}}(\mathfrak{A}_0) \to \widetilde{\Omega}_\varphi \in \mathfrak{N}_{\mathcal{T}}(\mathfrak{A}_0),$$

which is linear and injective and taking into account the fact that, if $x_n \to x$ in $\mathfrak{A}_0[\|\cdot\|]$, then $\widetilde{\Omega}_\varphi(\pi(x_n) - \pi(x)) = \omega_\varphi(x_n - x) \to 0$.

Let $\mathfrak{E}\mathfrak{N}_{\mathcal{T}}(\mathfrak{A}_0)$ be the set of the extreme points of $\mathfrak{N}_{\mathcal{T}}(\mathfrak{A}_0)$; then $\mathfrak{N}_{\mathcal{T}}(\mathfrak{A}_0)$ coincides with the w*-closure of the convex hull of $\mathfrak{E}\mathfrak{N}_{\mathcal{T}}(\mathfrak{A}_0)$. The extreme elements of $\mathfrak{N}_{\mathcal{T}}(\mathfrak{A}_0)$ are easily characterized by the following

Proposition 5.6.15 *$\widetilde{\Omega}_\varphi$ is extreme in $\mathfrak{N}_{\mathcal{T}}(\mathfrak{A}_0)$, if and only if, ω_φ is extreme in $\mathfrak{M}_{\mathcal{T}}(\mathfrak{A}_0)$.*

Definition 5.6.16 A Banach quasi *-algebra $(\mathfrak{A}[\|\cdot\|], \mathfrak{A}_0)$ is said to be *strongly regular* if $\mathcal{T}_{\mathfrak{A}_0}(\mathfrak{A})$ is sufficient and

$$\|x\| = \sup_{\varphi \in \mathcal{T}_{\mathfrak{A}_0}(\mathfrak{A})} \varphi(x, x)^{1/2}, \quad \forall\, x \in \mathfrak{A}.$$

Example 5.6.17 If $\mathfrak{M}$ is a von Neumann algebra possessing a sufficient family $\mathbb{F}$ of normal, finite traces, then the proper CQ*-algebra $(\mathfrak{M}_p, \mathfrak{M})$ constructed in the previous subsection is strongly regular. This follows from the definition itself in the completion.

Example 5.6.18 If ϱ is a normal, faithful, finite trace on $\mathfrak{M}$, then $\mathcal{T}_{\mathfrak{J}_p}(L^p(\varrho))$, for $p \geq 2$, is sufficient. To see this, we start defining φ_0 on $\mathfrak{M} \times \mathfrak{M}$ by

$$\varphi_0(X, Y) := \varrho(Y^* X), \quad \forall\, X, Y \in \mathfrak{M}.$$

Then,

$$|\varphi_0(X, Y)| = |\varrho(Y^* X)| \leq \|X\|_p \|Y\|_{p'}, \quad \forall\, X, Y \in \mathfrak{M},$$

where p' is the conjugate of p. Since $p \geq 2$, then $L^p(\varrho)$ is continuously embedded into $L^{p'}(\varrho)$. Thus, there exists $\gamma > 0$, such that $\|Y\|_{p'} \leq \gamma \|Y\|_p$, for every $Y \in \mathfrak{M}$. Let us define

$$\widetilde{\varphi}(X, Y) := \frac{1}{\gamma} \varphi_0(X, Y), \quad \forall\, X, Y \in \mathfrak{M}.$$

Then,

$$|\widetilde{\varphi}(X,Y)| \le \|X\|_p \|Y\|_p, \quad \forall\, X, Y \in \mathfrak{M}.$$

Hence, $\widetilde{\varphi}$ has a unique extension to $L^p(\varrho) \times L^p(\varrho)$, denoted with the same symbol. It is easily seen that $\widetilde{\varphi} \in \mathcal{T}_{\mathcal{J}_p}(L^p(\varrho))$.

Assume that there exists $X \in L^p(\varrho)$, such that $\varphi(X,X) = 0$, for every $\varphi \in \mathcal{T}_{\mathcal{J}_p}(L^p(\varrho))$, then, $\widetilde{\varphi}(X,X) = \|X\|_2^2 = 0$. This clearly implies $X = 0$. The equality $\widetilde{\varphi}(X,X) = \|X\|_2^2$ also shows that $L^2(\varrho)$ is strongly regular.

Let now $(\mathfrak{A}[\|\cdot\|], \mathfrak{A}_0)$ be a proper CQ*-algebra with unit e and $\mathcal{T}_{\mathcal{J}_p}(\mathfrak{A})$ sufficient. Let $\pi : \mathfrak{A}_0 \hookrightarrow \mathcal{B}(\mathcal{H})$ be the universal representation of $\mathfrak{A}_0$. Assume that the C*-algebra $\pi(\mathfrak{A}_0) := \mathfrak{M}$ is a von Neumann algebra. In this case, $\mathfrak{M}_{\mathcal{T}}(\mathfrak{A}_0) = \mathfrak{N}_{\mathcal{T}}(\mathfrak{A}_0)$ and $\mathfrak{N}_{\mathcal{T}}(\mathfrak{A}_0)$ is a family of traces satisfying condition (5.6.17). Therefore, by Proposition 5.6.10, we can construct for $p \ge 1$, the proper CQ*-algebras $(\mathfrak{M}_p[\|\cdot\|_{p,\mathfrak{N}_{\mathcal{T}}(\mathfrak{A}_0)}], \mathfrak{M}[\|\cdot\|])$. Clearly, $\mathfrak{A}_0$ can be identified with $\mathfrak{M}$. It is then natural to pose the *question* if also $\mathfrak{A}$ can be identified with some $\mathfrak{M}_p$. *The next Theorem provides an answer* to this question.

Theorem 5.6.19 *Let $(\mathfrak{A}[\|\cdot\|], \mathfrak{A}_0)$ be a proper CQ*-algebra with unit e, such that $\mathcal{T}_{\mathfrak{A}_0}(\mathfrak{A})$ is sufficient. Then, there exist a von Neumann algebra $\mathfrak{M}$ and a monomorphism*

$$\Phi : a \in \mathfrak{A} \mapsto \Phi(a) := \widetilde{X} \in \mathfrak{M}_2,$$

with the following properties:

(i) *Φ extends the universal *-representation π of $\mathfrak{A}_0$;*
(ii) $\Phi(a^*) = \Phi(a)^*$, $\quad \forall\, a \in \mathfrak{A}$;
(iii) $\Phi(ab) = \Phi(a)\Phi(b)$, $\quad \forall\, a, b \in \mathfrak{A}$, *such that $a \in \mathfrak{A}_0$ or $b \in \mathfrak{A}_0$.*

Then, $\mathfrak{A}$ can be identified with a space of operators affiliated with $\mathfrak{M}$.

If, in addition, $(\mathfrak{A}[\|\cdot\|], \mathfrak{A}_0)$ is strongly regular, then

(iv) *Φ is an isometry of $\mathfrak{A}$ into $\mathfrak{M}_2$;*
(v) *if $\mathfrak{A}_0$ is a W*-algebra, then Φ is an isometric *-isomorphism of $\mathfrak{A}$ onto $\mathfrak{M}_2$.*

Proof Let π be the universal representation of $\mathfrak{A}_0$ and assume first that $\pi(\mathfrak{A}_0) =: \mathfrak{M}$ is a von Neumann algebra. By Proposition 5.6.14, the family of traces $\mathfrak{M}_{\mathcal{T}}(\mathfrak{A}_0)$ is convex and w*-compact. Moreover, for each central positive element $Z \in \mathfrak{M}$ with $0 \le Z \le \mathbb{I}$ and for $\omega \in \mathfrak{M}_{\mathcal{T}}(\mathfrak{A}_0)$, the trace $\omega_Z(X) := \omega(ZX)$ yet belongs to $\mathfrak{M}_{\mathcal{T}}(\mathfrak{A}_0)$. Indeed, starting from the form $\varphi \in \mathcal{T}_{\mathfrak{A}_0}(\mathfrak{A})$, such that $\omega = \omega_\varphi$, one can define the sesquilinear form

$$\varphi_Z(a,b) := \varphi\big(a\pi^{-1}(Z^{1/2}), b\pi^{-1}(Z^{1/2})\big), \quad \forall\, a, b \in \mathfrak{A}.$$

We check that $\varphi_Z \in \mathcal{T}_{\mathfrak{A}_0}(\mathfrak{A})$.

(i) $\varphi_Z(a,a) = \varphi(a\pi^{-1}(Z^{1/2}), a\pi^{-1}(Z^{1/2})) \geq 0, \quad \forall\, a \in \mathfrak{A}$.

(ii) For every $a \in \mathfrak{A}$ and for every $x, y \in \mathfrak{A}_0$,

$$\begin{aligned}\varphi_Z(ax, y) &= \varphi\big(ax\pi^{-1}(Z^{1/2}), y\pi^{-1}(Z^{1/2})\big)\\ &= \varphi\big(x\pi^{-1}(Z^{1/2}), a^* y\pi^{-1}(Z^{1/2})\big)\\ &= \varphi_Z(x, a^* y).\end{aligned}$$

(iii) For every $a, b \in \mathfrak{A}$,

$$\begin{aligned}|\,\varphi_Z(a,b)\,| &= |\,\varphi\big(a\pi^{-1}(Z^{1/2}), b\pi^{-1}(Z^{1/2})\big)|\\ &\leq \|a\pi^{-1}(Z^{1/2})\|\,\|b\pi^{-1}(Z^{1/2})b\|\\ &\leq \|a\|\,\|\pi^{-1}(Z^{1/2})\|_0\,\|b\|\,\|\pi^{-1}(Z^{1/2})\|_0\\ &\leq \|a\|\,\|b\|.\end{aligned}$$

The latter inequality follows from the C*-condition $\|\pi^{-1}(Z^{1/2})\|_0^2 = \|\pi^{-1}(Z)\|_0$ and the fact that $\|\pi^{-1}(Z)\|_0 = \|Z\| \leq 1$.

(iv) For every $a \in \mathfrak{A}$,

$$\begin{aligned}\varphi_Z(a^*, a^*) &= \varphi\big(a^*\pi^{-1}(Z^{1/2}), a^*\pi^{-1}(Z^{1/2})\big)\\ &= \varphi\big(a\pi^{-1}(Z^{1/2}), a\pi^{-1}(Z^{1/2})\big) = \varphi_Z(a,a).\end{aligned}$$

Moreover, for every $A = \pi(x) \in \mathfrak{M} = \pi(\mathfrak{A}_0)$, φ_Z defines the following trace

$$\begin{aligned}\phi_{\varphi_Z}(A) = \varphi_Z(x,e) &= \varphi\big(x\pi^{-1}(Z^{1/2}), \pi^{-1}(Z^{1/2})\big)\\ &= \varphi\big(x\pi^{-1}(Z), e\big) = \varphi\big(\pi^{-1}(AZ), e\big) = \phi_\varphi(AZ).\end{aligned}$$

Then, the family of traces $\mathfrak{N}_{\mathcal{T}}(\mathfrak{A}_0)$ $(= \mathfrak{M}_{\mathcal{T}}(\mathfrak{A}_0))$ satisfies the assumptions of Lemma 5.6.12; therefore, if $\eta_1, \eta_2 \in \mathfrak{E}\mathfrak{N}_{\mathcal{T}}(\mathfrak{A}_0)$, denoting with P_1 and P_2 their respective supports, one has $P_1 P_2 = 0$.

By the sufficiency of $\mathcal{T}_{\mathfrak{A}_0}(\mathfrak{A})$ we get

$$\|X\|_{2,\mathfrak{M}_{\mathcal{T}}(\mathfrak{A}_0)} := \sup_{\varphi \in \mathfrak{M}_{\mathcal{T}}(\mathfrak{A}_0)} \|X\|_{2,\varphi} = \sup_{\varphi \in \mathfrak{E}\mathfrak{M}_{\mathcal{T}}(\mathfrak{A}_0)} \|X\|_{2,\varphi}, \quad \forall\, X \in \pi(\mathfrak{A}_0).$$

By Proposition 5.6.10, the Banach space $\mathfrak{M}_2$, completion of $\mathfrak{M}$ with respect to the norm $\|\cdot\|_{2,\,\mathfrak{N}_{\mathcal{T}}(\mathfrak{A}_0)}$, is a proper CQ*-algebra. Moreover, since the supports of the extreme traces satisfy the assumptions of Theorem 5.6.13, the proper CQ*-algebra $(\mathfrak{M}_2[\|\cdot\|_{2,\mathfrak{N}_{\mathcal{T}}(\mathfrak{A}_0)}], \mathfrak{M}[\|\cdot\|])$, consists of operators affiliated with $\mathfrak{M}$. We shall now define the map Φ.

For every element $a \in \mathfrak{A}$, there exists a sequence $\{x_n\}$ of elements of $\mathfrak{A}_0$ converging to a with respect to $\|\cdot\|$. Put $X_n = \pi(x_n)$, $n \in \mathbb{N}$. Then,

$$\begin{aligned}\|X_n - X_m\|_{2,\mathfrak{N}_{\mathcal{T}}(\mathfrak{A}_0)} &:= \sup_{\varphi\in\mathcal{T}_{\mathfrak{A}_0}(\mathfrak{A})} \|\pi(x_n) - \pi(x_m)\|_{2,\widetilde{\Omega}_\varphi} \\ &= \sup_{\varphi\in\mathcal{T}_{\mathfrak{A}_0}(\mathfrak{A})} \varphi\big((x_n - x_m)^*(x_n - x_m), e\big)^{1/2} \\ &= \sup_{\varphi\in\mathcal{T}_{\mathfrak{A}_0}(\mathfrak{A})} \varphi(x_n - x_m, x_n - x_m)^{1/2} \leq \|x_n - x_m\| \to 0.\end{aligned}$$

Let $\widetilde{X}$ be the $\|\cdot\|_{2,\mathfrak{M}_{\mathcal{T}}(\mathfrak{A}_0)}$-limit of the sequence $\{X_n\}$ in $\mathfrak{M}_2$. We define $\Phi(a) := \widetilde{X}$.

For each $a \in \mathfrak{A}$, we put

$$p_{\mathcal{T}_{\mathfrak{A}_0}(\mathfrak{A})}(a) = \sup_{\varphi\in\mathcal{T}_{\mathfrak{A}_0}(\mathfrak{A})} \varphi(a,a)^{1/2}.$$

Then, due to the sufficiency of $\mathcal{T}_{\mathfrak{A}_0}(\mathfrak{A})$, $p_{\mathcal{T}_{\mathfrak{A}_0}(\mathfrak{A})}$ is a norm on $\mathfrak{A}$ weaker than $\|\cdot\|$. This implies that

$$\|\widetilde{X}\|^2_{2,\mathfrak{N}_{\mathcal{T}}(\mathfrak{A}_0)} = \lim_{n\to\infty} \sup_{\varphi\in\mathcal{T}_{\mathfrak{A}_0}(\mathfrak{A})} \varphi(x_n, x_n) = \lim_{n\to\infty} p_{\mathcal{T}_{\mathfrak{A}_0}(\mathfrak{A})}(x_n)^2 = p_{\mathcal{T}_{\mathfrak{A}_0}(\mathfrak{A})}(a)^2.$$

From this equality it follows easily that the linear map Φ is well defined and injective. The condition (iii) can be easily proved. If $(\mathfrak{A}[\|\cdot\|], \mathfrak{A}_0)$ is strongly regular, then $p_{\mathcal{T}_{\mathfrak{A}_0}(\mathfrak{A})}(a) = \|a\|$, for every $a \in \mathfrak{A}$. Thus, Φ is isometric. Moreover, in this case, Φ is surjective; indeed, if $T \in \mathfrak{M}_2$, then there exists a sequence $\{T_n\}$ of bounded operators in $\pi(\mathfrak{A}_0)$, which converges to T, with respect to the norm $\|\cdot\|_{2,\mathfrak{N}_{\mathcal{T}}(\mathfrak{A}_0)}$. The corresponding sequence, say $\{y_n\}$ in $\mathfrak{A}_0$ with $T_n = \Phi(y_n)$, converges to some b in $\mathfrak{A}$, with respect to the norm of $\mathfrak{A}$ and $\Phi(b) = T$, by definition. Therefore, Φ is an isometric *-isomorphism.

To complete the proof, it is enough to show that the given proper CQ*-algebra $(\mathfrak{A}[\|\cdot\|], \mathfrak{A}_0)$ can be embedded in a proper CQ*-algebra $(\mathfrak{B}[\|\cdot\|_{2,\mathfrak{N}_{\mathcal{T}}(\mathfrak{A}_0)}], \mathfrak{B}_0)$, where $\mathfrak{B}_0$ is a W*-algebra. Of course, we may directly work with $\pi(\mathfrak{A}_0)$, where π is the universal representation of $\mathfrak{A}_0$. The family of traces $\mathfrak{N}_{\mathcal{T}}(\mathfrak{A}_0)$ defined on $\pi(\mathfrak{A}_0)''$ is not necessarily sufficient. Let P_φ, $\varphi \in \mathcal{T}_{\mathfrak{A}_0}(\mathfrak{A})$, denote the support of $\widetilde{\Omega}_\varphi$ and let

$$P := \bigvee_{\varphi\in\mathcal{T}_{\mathfrak{A}_0}(\mathfrak{A})} P_\varphi.$$

Then, $\mathfrak{B}_0 := \pi(\mathfrak{A}_0)''P$ is a von Neumann algebra, that we can complete, with respect to the norm

$$\|X\|_{2,\mathfrak{N}_{\mathcal{T}}(\mathfrak{A}_0)} = \sup_{\varphi\in\mathcal{T}_{\mathfrak{A}_0}(\mathfrak{A})} \widetilde{\Omega}_\varphi(X^*X), \quad X \in \pi(\mathfrak{A}_0)''P.$$

We obtain, in this way, a proper CQ*-algebra $(\mathfrak{B}[\|\cdot\|_{2,\mathfrak{N}_{\mathcal{T}}(\mathfrak{A}_0)}], \mathfrak{B}_0)$ with $\mathfrak{B} \equiv \widetilde{\mathfrak{B}_0}^{\|\cdot\|_{2,\mathfrak{N}_{\mathcal{T}}(\mathfrak{A}_0)}}$ (i.e., $\|\cdot\|_{2,\mathfrak{N}_{\mathcal{T}}(\mathfrak{A}_0)}$-completion of $\mathfrak{B}_0$) and $\mathfrak{B}_0$ a W*-algebra. The faithfullness of π on $\mathfrak{A}_0$ implies that

$$\pi(x)P = \pi(x), \quad \forall x \in \mathfrak{A}_0.$$

It remains to prove that $\mathfrak{A}$ can be identified with a subspace of $\mathfrak{K}$. But this can be shown in the very same way as we did in the first part of the proof, concerning definition of Φ; i.e., for each $a \in \mathfrak{A}$ there exists a sequence $\{x_n\}$ in $\mathfrak{A}_0$, such that $\|a - x_n\| \to 0$, as $n \to \infty$. We now put $X_n = \pi(x_n)$. Then, proceeding as before, we determine the element $\widehat{X} \in \mathfrak{B}$, where

$$\widehat{X} = \|\cdot\|_{2,\mathfrak{N}_{\mathcal{T}}(\mathfrak{A}_0)} - \lim \pi(x_n)P.$$

It is easy to see that the map $a \in \mathfrak{A} \mapsto \widehat{X} \in \mathfrak{B}$ is injective. If $(\mathfrak{A}[\|\cdot\|], \mathfrak{A}_0)$ is strongly regular, but $\pi(\mathfrak{A}_0) \subset \pi(\mathfrak{A}_0)''$, then Φ is an isometry of $\mathfrak{A}$ into $\mathfrak{M}_2$, that it needs not be surjective. □

Chapter 6
Locally Convex Quasi *-Algebras

This chapter is devoted to locally convex quasi *-algebras and locally convex quasi C*-algebras. Both these notions generalize what we have discussed in Chaps. 3 and 5. The advantage is, of course, that the range of applications becomes larger and larger; the drawback is that the theory becomes more involved.

6.1 Representable Functionals on Locally Convex Quasi *-Algebras

Definition 6.1.1 Let $(\mathfrak{A}, \mathfrak{A}_0)$ be a quasi *-algebra and τ a locally convex topology on $\mathfrak{A}$. We say that $(\mathfrak{A}[\tau], \mathfrak{A}_0)$ is *a locally convex quasi *-algebra* if

(i) the map $a \in \mathfrak{A} \to a^* \in \mathfrak{A}$ is continuous;
(ii) for every $x \in \mathfrak{A}_0$, the maps $a \mapsto ax$, $a \mapsto xa$ from $\mathfrak{A}[\tau]$ into $\mathfrak{A}[\tau]$ are continuous;
(iii) $\mathfrak{A}_0$ is τ-dense in $\mathfrak{A}$.

Example 6.1.2 If $\mathcal{L}^\dagger(\mathcal{D}, \mathcal{H})$ is endowed with the strong* topology t_{s^*} (see beginning of Sect. 2.1.3), defined by the family of seminorms

$$p_\xi(X) := \|X\xi\| + \|X^\dagger\xi\|, \quad \xi \in \mathcal{D},\ X \in \mathcal{L}^\dagger(\mathcal{D}, \mathcal{H}),$$

then $(\mathcal{L}^\dagger(\mathcal{D}, \mathcal{H})[\mathsf{t}_{s^*}], \mathcal{L}^\dagger(\mathcal{D})_b)$ is a locally convex quasi *-algebra. Recall that $\mathcal{L}^\dagger(\mathcal{D})_b$ is the bounded part of $\mathcal{L}^\dagger(\mathcal{D})$ (ibid.). Indeed, the continuity of the involution $X \mapsto X^\dagger$, $X \in \mathcal{L}^\dagger(\mathcal{D}, \mathcal{H})$, comes immediately from the definition of p_ξ. Moreover, if $\xi \in \mathcal{D}$, $X \in \mathcal{L}^\dagger(\mathcal{D}, \mathcal{H})$ and $Y \in \mathcal{L}^\dagger(\mathcal{D})_b$, the obvious inequality

$$\|XY\xi\| + \|Y^\dagger X^\dagger\xi\| \le \|XY\xi\| + \|X^\dagger Y\xi\| + \|Y\|(\|X\xi\| + \|X^\dagger\xi\|)$$

M. Fragoulopoulou, C. Trapani, *Locally Convex Quasi *-Algebras and their Representations*, Lecture Notes in Mathematics 2257,
https://doi.org/10.1007/978-3-030-37705-2_6

implies the continuity of $X \mapsto XY$ (and of $X \mapsto YX$, by applying involution). Moreover, $\mathcal{L}^\dagger(\mathcal{D})_b$ is dense in $\mathcal{L}^\dagger(\mathcal{D}, \mathcal{H})[\mathsf{t}_{s^*}]$ by [2, Example 2.5.10].

If $\mathcal{L}^\dagger(\mathcal{D}, \mathcal{H})$ is endowed with the weak topology t_w defined by the set of seminorms

$$p_{\xi,\eta}(X) := |\langle X\xi|\eta\rangle|, \quad \xi, \eta \in \mathcal{D}, \; X \in \mathcal{L}^\dagger(\mathcal{D}, \mathcal{H}),$$

then, again, $(\mathcal{L}^\dagger(\mathcal{D}, \mathcal{H})[\mathsf{t}_w], \mathcal{L}^\dagger(\mathcal{D})_b)$ is a locally convex quasi *-algebra.

Example 6.1.3 As proved in [45, Proposition 6.2], $(\mathcal{L}^\dagger(\mathcal{D}, \mathcal{H})[\mathsf{t}_{s^*}], \mathcal{L}^\dagger(\mathcal{D})_b)$ exhibits a richer structure than that of a locally convex quasi *-algebra. It is, indeed, a *locally convex C*-normed algebra*, whose definition we discuss here (for more details, see [45] and Sect. 7.1).

Let $\mathfrak{A}_0[\|\cdot\|_0]$ be a unital C*-normed algebra and τ a locally convex *-algebra topology on $\mathfrak{A}_0$, with $\{p_\lambda\}_{\lambda\in\Lambda}$ a defining family of *-seminorms. Suppose that the following conditions are satisfied:

(T_1) $\mathfrak{A}_0[\tau]$ is a locally convex $*$-algebra with separately continuous multiplication.
(T_2) $\tau \preceq \|\cdot\|_0$, with τ and $\|\cdot\|_0$ being compatible (in the sense that for any Cauchy net $\{x_\alpha\}$ in $\mathfrak{A}_0[\|\cdot\|_0]$, such that $x_\alpha \to 0$ in τ, $x_\alpha \to 0$ in $\|\cdot\|_0$).

Then, if $\widetilde{\mathfrak{A}_0}[\tau]$ denotes the completion of the C*-normed algebra $\mathfrak{A}_0$ with respect to τ, we conclude from (T_2) and (T_1), respectively that

- $\mathfrak{A}_0[\|\cdot\|_0] \hookrightarrow \widetilde{\mathfrak{A}_0}[\|\cdot\|_0] \hookrightarrow \widetilde{\mathfrak{A}_0}[\tau]$; and
- $\widetilde{\mathfrak{A}_0}[\tau]$ is a locally convex quasi *-algebra over the C*-normed algebra $\mathfrak{A}_0[\|\cdot\|_0]$, but it is not necessarily a locally convex quasi *-algebra over the C*-algebra $\widetilde{\mathfrak{A}_0}[\|\cdot\|_0]$ (completion of $\mathfrak{A}_0[\|\cdot\|_0]$), since $\widetilde{\mathfrak{A}_0}[\|\cdot\|_0]$ is not a locally convex *-algebra under the topology τ.

Furthermore, consider the conditions:

(T_3) For all $\lambda \in \Lambda$, there exists $\lambda' \in \Lambda$, such that $p_\lambda(xy) \leq \|x\|_0 p_{\lambda'}(y)\|$, for all $x, y \in \mathfrak{A}_0$, with $xy = yx$.
(T_4) The set $\mathcal{U}(\widetilde{\mathfrak{A}_0}^+) := \{x \in \widetilde{\mathfrak{A}_0}^+ : \|x\|_0 \leq 1\}$ is τ-closed.
(T_5) $\widetilde{\mathfrak{A}_0}[\tau]^+ \cap \widetilde{\mathfrak{A}_0}[\|\cdot\|_0] = \widetilde{\mathfrak{A}_0}[\|\cdot\|_0]_+$,

where $\widetilde{\mathfrak{A}_0}^+$ denotes the positive cone of the C*-algebra $\widetilde{\mathfrak{A}_0}[\|\cdot\|_0]$ and $\widetilde{\mathfrak{A}_0}[\tau]^+$ the set of all positive elements of the locally convex quasi *-algebra $\widetilde{\mathfrak{A}_0}[\tau]$ (see Definition 7.1.1, in Chap. 7 and discussion at the beginning of Sect. 6.2, from where the respective notation is adopted); i.e., *$\widetilde{\mathfrak{A}_0}[\tau]^+$ is the τ-closure of $\widetilde{\mathfrak{A}_0}^+$*. Note that $\widetilde{\mathfrak{A}_0}[\tau]^+$ is a wedge and not necessarily a positive cone.

Let $\mathfrak{A}_0[\|\cdot\|_0]$ be a unital C*-normed algebra, τ a locally convex topology on $\mathfrak{A}_0$ satisfying the conditions (T_1) – (T_5). Then, a quasi *-subalgebra $\mathfrak{A}$ of the locally convex quasi *-algebra $(\widetilde{\mathfrak{A}_0}[\tau], \mathfrak{A}_0)$, containing $\widetilde{\mathfrak{A}_0}[\|\cdot\|_0]$, is said to be a *locally convex quasi C*-normed algebra over* $\mathfrak{A}_0$ and a *locally convex quasi C*-algebra* if $\mathfrak{A}_0[\|\cdot\|_0]$ is a C*-algebra. The latter case will be discussed in detail in Chap. 7.

➽ If $\pi : \mathfrak{A} \to \mathcal{L}^\dagger(\mathcal{D}_\pi, \mathcal{H}_\pi)$ is a **-representation* of $(\mathfrak{A}, \mathfrak{A}_0)$, which is *continuous* from $\mathfrak{A}[\tau]$ into $\mathcal{L}^\dagger(\mathcal{D}_\pi, \mathcal{H}_\pi)[\mathsf{t}]$, where t is a locally convex topology on $\mathcal{L}^\dagger(\mathcal{D}_\pi, \mathcal{H}_\pi)$, *we write for short*, that π is (τ, t)-*continuous*.

In this regard, the GNS representation (see Sect. 2.4) will play a fundamental role in our analysis.

When $(\mathfrak{A}[\tau], \mathfrak{A}_0)$ is a locally convex quasi *-algebra, taking into account the notation in Definition 2.4.6 (see also beginning of Sect. 3.2), we recall that

$$\mathcal{R}_c(\mathfrak{A}, \mathfrak{A}_0) := \{\omega \in \mathcal{R}(\mathfrak{A}, \mathfrak{A}_0) : \omega \text{ is continuous}\}.$$

We say that a representable linear functional ω on $\mathfrak{A}$ (i.e., $\omega \in \mathcal{R}(\mathfrak{A}, \mathfrak{A}_0)$) is *continuous* if there exists a continuous seminorm p on $\mathfrak{A}$, such that

$$|\omega(a)| \leq p(a), \quad \forall\, a \in \mathfrak{A}. \tag{6.1.1}$$

The notions of closable or closed form, given in Definition 3.1.46 in the Banach case, extend to the present situation too.

Let φ be a positive sesquilinear form defined on $\mathfrak{A}_0 \times \mathfrak{A}_0$. We say that φ is *closable* if for a net $\{x_\delta\}_{\delta \in \Delta}$ in $\mathfrak{A}_0$, one has

$$x_\delta \underset{\tau}{\to} 0 \text{ and } \varphi(x_\delta - x_\gamma, x_\delta - x_\gamma) \to 0 \Rightarrow \varphi(x_\delta, x_\delta) \to 0.$$

Then, $|\varphi(x_\delta, x_\delta)^{1/2} - \varphi(x_\gamma, x_\gamma)^{1/2}| \leq \varphi(x_\delta - x_\gamma, x_\delta - x_\gamma)^{1/2} \to 0$, therefore $\{\varphi(x_\delta, x_\delta)\}_{\delta \in \Delta}$ is a Cauchy net. Thus, if φ is closable, it can be extended to a positive sesquilinear form $\overline{\varphi}$ defined on $D(\overline{\varphi}) \times D(\overline{\varphi})$ by

$$\overline{\varphi}(a, a) := \lim_\delta \varphi(x_\delta, x_\delta),$$

where

$$D(\overline{\varphi}) = \left\{a \in \mathfrak{A} : \exists\, \{x_\delta\} \subset \mathfrak{A}_0 \text{ with } x_\delta \underset{\tau}{\to} a \text{ and } \varphi(x_\delta - x_\gamma, x_\delta - x_\gamma) \to 0\right\}.$$

This definition extends in obvious way to pairs (a, b) with $a, b \in D(\overline{\varphi})$. If ω is a positive linear functional on $\mathfrak{A}_0$, then we can define a positive sesquilinear form φ_ω on $\mathfrak{A}_0 \times \mathfrak{A}_0$ by

$$\varphi_\omega(x, y) := \omega(y^* x), \quad x, y \in \mathfrak{A}_0.$$

Now we prove the following

Proposition 6.1.4 *Let $\omega \in \mathcal{R}_c(\mathfrak{A}, \mathfrak{A}_0)$. Then, φ_ω is closable.*

Proof Let $x_\delta \underset{\tau}{\to} 0$ with $\varphi_\omega(x_\delta - x_\gamma, x_\delta - x_\gamma) \to 0$. Then, $y^* x_\delta \underset{\tau}{\to} 0$, for every $y \in \mathfrak{A}_0$, since the multiplication is continuous (see Definition 6.1.1). The continuity of ω then, implies that $\omega(y^* x_\delta) \to 0$, $y \in \mathfrak{A}_0$. Put

$$N_\omega := \{x \in \mathfrak{A}_0 : \omega(x^* x) = 0\}.$$

Then, $\mathfrak{A}_0/N_\omega$ is a pre-Hilbert space under the well-defined inner product

$$\langle \lambda_\omega(x) | \lambda_\omega(y) \rangle := \omega(y^* x), \quad x, y \in \mathfrak{A}_0,$$

where $\lambda_\omega(z) := z + N_\omega$, $z \in \mathfrak{A}_0$. Let $\mathcal{H}_\omega$ denote the Hilbert space completion of $\mathfrak{A}_0/N_\omega$. The net $\{\lambda_\omega(x_\delta)\}$ is Cauchy, since

$$\|\lambda_\omega(x_\delta) - \lambda_\omega(x_\gamma)\|^2 = \varphi_\omega(x_\delta - x_\gamma, x_\delta - x_\gamma) \to 0.$$

Hence, it converges to some $\xi \in \mathcal{H}_\omega$ and

$$\langle \lambda_\omega(x_\delta) | \lambda_\omega(y) \rangle \to \langle \xi | \lambda_\omega(y) \rangle, \quad \forall\, y \in \mathfrak{A}_0.$$

Moreover,

$$\langle \lambda_\omega(x_\delta) | \lambda_\omega(y) \rangle = \omega(y^* x_\delta) \to 0, \quad \forall\, y \in \mathfrak{A}_0.$$

Thus, $\langle \xi | \lambda_\omega(y) \rangle = 0$, for every $y \in \mathfrak{A}_0$. This implies that $\xi = 0$. Therefore, $\varphi_\omega(x_\delta, x_\delta) \to 0$. □

Note that representability of ω is not used in the proof of Proposition 6.1.4.

Consider now the set

$$\mathfrak{A}_{\mathcal{R}} := \bigcap_{\omega \in \mathcal{R}_c(\mathfrak{A}, \mathfrak{A}_0)} D(\overline{\varphi}_\omega).$$

If $\mathcal{R}_c(\mathfrak{A}, \mathfrak{A}_0) = \{0\}$, we put $\mathfrak{A}_{\mathcal{R}} = \mathfrak{A}$. Note that, if for every $\omega \in \mathcal{R}_c(\mathfrak{A}, \mathfrak{A}_0)$, φ_ω is jointly continuous with respect to τ, we get $\mathfrak{A}_{\mathcal{R}} = \mathfrak{A}$.

Proposition 6.1.5 *$\mathfrak{A}_{\mathcal{R}}$ is a vector subspace of $\mathfrak{A}$ and $\mathfrak{A}_0 \subset \mathfrak{A}_{\mathcal{R}}$. Moreover, if $a \in \mathfrak{A}_{\mathcal{R}}$ and $x \in \mathfrak{A}_0$, then $xa \in \mathfrak{A}_{\mathcal{R}}$. Hence, if $\mathfrak{A}_{\mathcal{R}}$ is *-invariant, then $(\mathfrak{A}_{\mathcal{R}}, \mathfrak{A}_0)$ is a quasi *-algebra.*

Proof We show that $a \in \mathfrak{A}_{\mathcal{R}}$ and $x \in \mathfrak{A}_0$ imply $xa \in \mathfrak{A}_{\mathcal{R}}$. Indeed, if $\omega \in \mathcal{R}_c(\mathfrak{A}, \mathfrak{A}_0)$, then also $\omega_x \in \mathcal{R}_c(\mathfrak{A}, \mathfrak{A}_0)$, by (3.2.19). This implies that $a \in D(\overline{\varphi}_{\omega_x})$ or, equivalently, $xa \in D(\overline{\varphi}_\omega)$. Since ω is arbitrary, the statement is proved. □

6.2 Order Structure

Given a locally convex quasi *-algebra $(\mathfrak{A}[\tau], \mathfrak{A}_0)$, we recall the concept of a positive element of $\mathfrak{A}$ (see e.g., [23, pp. 21, 22]). This notion defines an order on the set $\mathfrak{A}_h$ of all selfadjoint elements of $\mathfrak{A}$. The condition $\mathfrak{A}^+ \cap (-\mathfrak{A}^+) = \{0\}$, on the positive elements $\mathfrak{A}^+$ of $\mathfrak{A}$, implies that every non-zero element in $\mathfrak{A}^+$ gives rise to a non-trivial continuous positive linear functional on $\mathfrak{A}[\tau]$ (Theorem 6.2.2). The preceding condition is characterized in Proposition 6.2.7 and it itself, together with another condition that forces an element of $\mathfrak{A}$ to be positive, show that $(\mathfrak{A}[\tau], \mathfrak{A}_0)$ attains enough (τ, t_w)-continuous *-representations to separate its points (Corollary 6.2.10).

Coming back to the given locally convex quasi *-algebra $(\mathfrak{A}[\tau], \mathfrak{A}_0)$, set (see also discussion after Corollary 3.1.24)

$$\mathfrak{A}_0^+ := \left\{ \sum_{k=1}^{n} x_k^* x_k,\ x_k \in \mathfrak{A}_0,\ n \in \mathbb{N} \right\}.$$

Then, $\mathfrak{A}_0^+$ is a wedge in $\mathfrak{A}_0$ and we call the elements of $\mathfrak{A}_0^+$ *positive elements of* $\mathfrak{A}_0$. We call *positive elements of* $\mathfrak{A}[\tau]$ the elements of $\overline{\mathfrak{A}_0^+}^{\tau}$ and we denote them by $\mathfrak{A}^+$. That is, $\mathfrak{A}^+ := \overline{\mathfrak{A}_0^+}^{\tau}$.

The set $\mathfrak{A}^+$ is a *qm-admissible wedge* (generalization of *m-admissible wedge* given by Schmüdgen [23, p. 22]), in the following sense:

1. $e \in \mathfrak{A}^+$, if $(\mathfrak{A}[\tau], \mathfrak{A}_0)$ has a unit e;
2. $a + b \in \mathfrak{A}^+, \quad \forall\, a, b \in \mathfrak{A}^+$;
3. $\lambda a \in \mathfrak{A}^+, \quad \forall\, a \in \mathfrak{A}^+,\ \lambda \geq 0$;
4. $x^* a x \in \mathfrak{A}^+, \quad \forall\, a \in \mathfrak{A}^+,\ x \in \mathfrak{A}_0$.

Clearly, $\mathfrak{A}^+$ defines an order on the real vector space $\mathfrak{A}_h = \{x \in \mathfrak{A} : x = x^*\}$ by

$$x \leq y \ \Leftrightarrow\ y - x \in \mathfrak{A}^+.$$

For $a \in \mathfrak{A}^+$, we shall often use the notation $a \geq 0$.

The following proposition is straightforward.

Proposition 6.2.1 *If* $a \geq 0$, *then* $\pi(a) \geq 0$, *for every* (τ, t_w)*-continuous* *-*representation of* $(\mathfrak{A}[\tau], \mathfrak{A}_0)$.

The theorem that follows, shows that if the set $\mathfrak{A}^+$ is *proper*, then $(\mathfrak{A}[\tau], \mathfrak{A}_0)$ attains non-trivial continuous positive linear functionals, in the sense of Definition 7.1.3, in Chap. 7.

Theorem 6.2.2 *Assume that* $\mathfrak{A}^+ \cap (-\mathfrak{A}^+) = \{0\}$. *Let* $a \in \mathfrak{A}^+$, $a \neq 0$. *Then, there exists a continuous linear functional* ω *on* $\mathfrak{A}[\tau]$ *with the properties:*

(i) $\omega(b) \geq 0$, $\forall\, b \in \mathfrak{A}^+$;
(ii) $\omega(a) > 0$.

Proof Consider the real vector space $\mathfrak{A}_h$ and $a \in \mathfrak{A}^+ \setminus \{0\}$. The set $\{a\}$ is obviously convex and compact and does not intersect $(-\mathfrak{A}^+)$. Hence by [5, Ch.2, §5, Proposition 4], there exists a closed hyperplane separating these two sets. Let $g(x) = 0$ be the equation of this hyperplane. Then, either $g(a) > 0$ with $g(-\mathfrak{A}^+) < 0$ (in which case we take $\omega = g$), or the contrary (where, in this case, we take $\omega = -g$). □

Theorem 6.2.2 leads to the following (see also discussion before Lemma 3.1.48)

Definition 6.2.3 A linear functional ω on $\mathfrak{A}$ is called *positive* if $\omega(a) \geq 0$, $\forall\, a \in \mathfrak{A}^+$.

The next Proposition 6.2.4, provides conditions under which continuous linear functionals on $(\mathfrak{A}[\tau], \mathfrak{A}_0)$ are positive and hermitian.

Proposition 6.2.4 *Assume that* $\mathfrak{A}_{\mathcal{R}} = \mathfrak{A}$ *and that* $(\mathfrak{A}[\tau], \mathfrak{A}_0)$ *has a unit. Then, every continuous linear functional* ω *on* $\mathfrak{A}$, *such that* $\omega(x^*x) \geq 0$, *for every* $x \in \mathfrak{A}_0$, *is positive and hermitian.*

Proof Since ω is positive on $\mathfrak{A}_0$, it is hermitian on $\mathfrak{A}_0$. Thus, by continuity of ω and continuity of the involution (see Definition 6.1.1) we are done. □

Using representable linear functionals, the following proposition provides invariant positive sesquilinear forms with fixed core, i.e., elements of the set $\mathcal{I}_{\mathfrak{A}_0}(\mathfrak{A})$ (see, Proposition 2.3.2 and Remark 2.3.3).

Proposition 6.2.5 *Assume that* $\mathfrak{A}_{\mathcal{R}} = \mathfrak{A}$. *Then, for every* $\omega \in \mathcal{R}_c(\mathfrak{A}, \mathfrak{A}_0)$, $\overline{\varphi}_\omega \in \mathcal{I}_{\mathfrak{A}_0}(\mathfrak{A})$; *i.e.,* $\overline{\varphi}_\omega$ *is an ips-form on* $\mathfrak{A}$ *with core* $\mathfrak{A}_0$.

Proof Let $N(\overline{\varphi}_\omega) = \{a \in \mathfrak{A} : \overline{\varphi}_\omega(a, a) = 0\}$ and $\mathcal{H}_{\overline{\varphi}_\omega}$ the Hilbert space obtained by completing $\mathfrak{A}/N(\overline{\varphi}_\omega)$ with respect to the (well-defined) inner product

$$\langle \lambda_{\overline{\varphi}_\omega}(a) | \lambda_{\overline{\varphi}_\omega}(b) \rangle := \overline{\varphi}_\omega(a, b), \quad a, b \in \mathfrak{A},$$

where $\lambda_{\overline{\varphi}_\omega}(a) := a + N(\overline{\varphi}_\omega)$. According to the discussion before Remark 2.3.3 and definition of $\overline{\varphi}_\omega$'s, we need to show that $\lambda_{\overline{\varphi}_\omega}(\mathfrak{A}_0)$ is dense in $\mathcal{H}_{\overline{\varphi}_\omega}$. Assume, on the contrary, that there exists $a \in \mathfrak{A}$, such that $\langle \lambda_{\overline{\varphi}_\omega}(x) | \lambda_{\overline{\varphi}_\omega}(a) \rangle = 0$, for every $x \in \mathfrak{A}_0$. Since $a \in \bigcap_{\omega \in \mathcal{R}_c(\mathfrak{A}, \mathfrak{A}_0)} D(\overline{\varphi}_\omega) = \mathfrak{A}$, there exists a net $\{x_\delta\} \subset \mathfrak{A}_0$, such that $x_\delta \xrightarrow{\tau} a$ and $\overline{\varphi}_\omega(x_\delta - a, x_\delta - a) \to 0$. Hence,

$$\overline{\varphi}_\omega(a, a) = \lim_\delta \overline{\varphi}_\omega(x_\delta, x_\delta) = 0.$$

Consequently $\lambda_{\overline{\varphi}_\omega}(a) = 0$. The proof is now complete according to Proposition 2.3.2. □

Definition 6.2.6 A family of positive linear functionals $\mathcal{F}$ on $(\mathfrak{A}[\tau], \mathfrak{A}_0)$ is called *sufficient* if for every $a \in \mathfrak{A}^+$, $a \neq 0$, there exists $\omega \in \mathcal{F}$, such that $\omega(a) > 0$.

Proposition 6.2.7 *Let $(\mathfrak{A}[\tau], \mathfrak{A}_0)$ be a locally convex quasi *-algebra. The following statements are equivalent:*

(i) $\mathfrak{A}^+ \cap (-\mathfrak{A}^+) = \{0\}$;
(ii) $\mathcal{R}_c(\mathfrak{A}, \mathfrak{A}_0)$ *is sufficient.*

Proof (i) ⇒ (ii) This is Theorem 6.2.2.

(ii) ⇒ (i) Let $a \in \mathfrak{A}^+ \cap (-\mathfrak{A}^+)$ and $\omega \in \mathcal{R}_c(\mathfrak{A}, \mathfrak{A}_0)$. Then, ω is a continuous positive linear functional on $\mathfrak{A}$, therefore $\omega(a) \geq 0$ and $\omega(-a) = -\omega(a) \geq 0$. Thus $\omega(a) = 0$. Since ω is arbitrary, we finally get $a = 0$. □

It is clear from Theorem 6.2.2 and Definition 6.2.6 that if $\mathfrak{A}^+ \cap (-\mathfrak{A}^+) = \{0\}$, then *the family of all continuous positive linear functionals on* $(\mathfrak{A}[\tau], \mathfrak{A}_0)$ *is sufficient.*

Proposition 6.2.8 *Let $(\mathfrak{A}[\tau], \mathfrak{A}_0)$ be a locally convex quasi *-algebra with $\mathcal{R}_c(\mathfrak{A}, \mathfrak{A}_0)$ sufficient.*

Assume that the following condition (P) *holds:*

(P) $b \in \mathfrak{A}$ *and* $\omega(x^*bx) \geq 0$, *for all* $\omega \in \mathcal{R}_c(\mathfrak{A}, \mathfrak{A}_0)$ *and* $x \in \mathfrak{A}_0$ *imply* $b \in \mathfrak{A}^+$.

Then, for an element $a \in \mathfrak{A}$, the following statements are equivalent:

(i) $a \in \mathfrak{A}^+$;
(ii) $\omega(a) \geq 0$, *for every* $\omega \in \mathcal{R}_c(\mathfrak{A}, \mathfrak{A}_0)$;
(iii) $\pi(a) \geq 0$, *for every* (τ, t_w)*-continuous *-representation* π *of* $(\mathfrak{A}[\tau], \mathfrak{A}_0)$.

Proof (i) ⇒ (ii) is an easy consequence of the definition of positive elements and the continuity of the elements of $\mathcal{R}_c(\mathfrak{A}, \mathfrak{A}_0)$ with respect to τ.

(ii) ⇒ (iii) Let π be a (τ, t_w)-continuous *-representation of $(\mathfrak{A}[\tau], \mathfrak{A}_0)$. Define $\omega_\xi(a) := \langle \pi(a)\xi|\xi\rangle$, $a \in \mathfrak{A}$, with $\xi \in \mathcal{D}$, $\|\xi\| = 1$. Then, $\omega_\xi \in \mathcal{R}_c(\mathfrak{A}, \mathfrak{A}_0)$, since

$$|\omega_\xi(a)| = |\langle \pi(a)\xi|\xi\rangle| \leq p(a), \ \forall \ a \in \mathfrak{A},$$

for some τ-continuous seminorm p on $\mathfrak{A}$. Thus, if a satisfies (ii), $\langle \pi(a)\xi|\xi\rangle \geq 0$, for every $\xi \in \mathcal{D}$, which proves (iii).

(iii) ⇒ (i) Let $\omega \in \mathcal{R}_c(\mathfrak{A}, \mathfrak{A}_0)$ and let π_ω be the corresponding GNS representation. Then, π_ω is (τ, t_w)-continuous. Indeed, due to the continuity of ω and (ii) of Definition 6.1.1, we get

$$|\langle \pi_\omega(a)\lambda_\omega(x)|\lambda_\omega(y)\rangle| = |\omega(y^*ax)| \leq p(a), \quad \forall \, a \in \mathfrak{A}, \ x, y \in \mathfrak{A}_0,$$

for some τ-continuous seminorm p on $\mathfrak{A}$. Applying (iii) we have $\pi_\omega(a) \geq 0$. This implies that $\omega(x^*ax) \geq 0$, for every $x \in \mathfrak{A}_0$. The statement now follows from the assumption (P). □

Remark 6.2.9

1. If $(\mathfrak{A}[\tau], \mathfrak{A}_0)$ has a unit, then the equivalence of (ii) and (iii) does not depend on (P). In this case, (P) is equivalent to the following

   (P′) If $b \in \mathfrak{A}$ and $\omega(b) \geq 0$, for every $\omega \in \mathcal{R}_c(\mathfrak{A}, \mathfrak{A}_0)$, then $b \in \mathfrak{A}^+$.

   Indeed, since we have unit (P) implies (P′). On the other hand, by (3.2.19), for any $\omega \in \mathcal{R}_c(\mathfrak{A}, \mathfrak{A}_0)$ and $x \in \mathfrak{A}_0$ we have that $\omega_x \in \mathcal{R}_c(\mathfrak{A}, \mathfrak{A}_0)$, where $\omega_x(b) := \omega(x^*bx)$, $b \in \mathfrak{A}$, so that (P′) implies (P).
2. The condition (P) together with $\mathfrak{A}^+ \cap (-\mathfrak{A}^+) = \{0\}$ implies that, for every $0 \neq a \in \mathfrak{A}$, there exists $\omega \in \mathcal{R}_c(\mathfrak{A}, \mathfrak{A}_0)$, such that $\omega(a) \neq 0$. Indeed, if $\omega(a) = 0$, for every $\omega \in \mathcal{R}_c(\mathfrak{A}, \mathfrak{A}_0)$, then (Proposition 6.2.8) $a \in \mathfrak{A}^+$ and $-a \in \mathfrak{A}^+$; hence $a = 0$, a contradiction.

From Remark 6.2.9(2) and the proof of Proposition 6.2.8 we now have the following

Corollary 6.2.10 *Let $(\mathfrak{A}[\tau], \mathfrak{A}_0)$ be a locally convex quasi *-algebra with unit e. Suppose that $\mathcal{R}_c(\mathfrak{A}, \mathfrak{A}_0)$ is sufficient and that condition* (P) *of Proposition* 6.2.8 *is fulfilled. Then, for every $0 \neq a \in \mathfrak{A}$, there is a (τ, t_w)-continuous *-representation π of $(\mathfrak{A}[\tau], \mathfrak{A}_0)$, namely $\pi = \pi_\omega$, $\omega \in \mathcal{R}_c(\mathfrak{A}, \mathfrak{A}_0)$, such that $\pi(a) \neq 0$.*

6.3 Fully Representable Quasi *-Algebras

In Sect. 6.2, given a locally convex quasi *-algebra $(\mathfrak{A}[\tau], \mathfrak{A}_0)$, we have seen that the sufficiency of the set $\mathcal{R}_c(\mathfrak{A}, \mathfrak{A}_0)$ together with the condition $\mathfrak{A}_\mathcal{R} = \mathfrak{A}$ equip the given algebra with important properties (cf., for instance, Theorem 6.2.2, Proposition 6.2.4 and Corollary 6.2.10) that are very close to the properties that C*-algebras enjoy and offer to them their rich structure. All these lead us to the following

Definition 6.3.1 A locally convex quasi *-algebra $(\mathfrak{A}[\tau], \mathfrak{A}_0)$ is called *fully representable* if $\mathcal{R}_c(\mathfrak{A}, \mathfrak{A}_0)$ is sufficient and $\mathfrak{A}_\mathcal{R} = \mathfrak{A}$.

Example 6.3.2 Let $I = [0, 1]$. Consider the Banach quasi *-algebra $(L^p(I), L^\infty(I))$. Then, every $\omega \in \mathcal{R}_c(L^p(I), L^\infty(I))$ has the form

$$\omega(f) = \int_0^1 f(t)v(t)dt, \quad f \in L^p(I),$$

with $v \in L^{p/p-2}(I)$, $v \geq 0$ and $p \geq 2$. One readily checks that ω satisfies the conditions (L1) and (L2) of Definition 2.4.6. It is easily seen that the condition (L3) of the same definition is also fulfilled. Conversely, assume that (L3) is satisfied; i.e., for every $f \in L^p(I)$ there exists $\gamma > 0$, such that

$$|\omega(f^*\alpha)| = \left| \int_0^1 \overline{f(t)}\alpha(t)v(t)dt \right| \leq \gamma \left(\int_0^1 |\alpha(t)|^2 v(t)dt \right)^{1/2}, \quad \forall\, \alpha \in L^\infty(I).$$

This implies that $f \in L^2(I, vdt)$ and $\gamma = \|f\|_{2,v}$. Hence, in order that (L3) be satisfied for every $f \in L^p(I)$, we must have $v \in L^{p/p-2}(I)$. Hence, if $p \geq 1$, ω is representable, if and only if, $v \in L^{p/p-2}(I)$. If $v \in L^{p/p-1}(I) \setminus L^{p/p-2}(I)$, then ω is continuous but not representable.

If $1 \leq p < 2$, the condition $L^p(I) \subset L^2(I, vdt)$ is not satisfied for every non zero v. In this case, there are no continuous representable functionals.

Let us now come back to the case $p \geq 2$. Let $v \in L^{p/p-2}(I)$. We want to determine $\overline{\varphi}_\omega$, where

$$\varphi_\omega(\alpha, \beta) = \int_0^1 \alpha(t)\overline{\beta(t)}v(t)dt, \quad \alpha, \beta \in L^\infty(I).$$

Let $f \in D(\overline{\varphi}_\omega)$. Then, there exists a sequence $\{\alpha_n\} \subset L^\infty(I)$, such that

$$\alpha_n \xrightarrow{p} f \text{ and } \int_0^1 |\alpha_n(t) - \alpha_m(t)|^2 v(t)dt \to 0.$$

Hence, there exists $v_0 \in L^2(I, vdt)$, such that $\alpha_n \to v_0$, in the $L^2(I, vdt)$-norm. This, in turn, implies that $f = v_0$, almost everywhere. Thus, $D(\overline{\varphi}_\omega) = L^p(I) \cap L^2(I, vdt)$. Therefore,

$$L^p(I)_{\mathcal{R}} = \bigcap_{\omega \in \mathcal{R}_c(L^p, L^\infty)} D(\overline{\varphi}_\omega) = L^p(I) \cap \left(\bigcap_{v \in L^{p/p-2}(I)} L^2(I, vdt) \right).$$

But $f \in L^2(I, vdt)$, for every $v \in L^{p/p-2}(I)$, if and only if, $f \in L^p(I)$.

In conclusion, for $p \geq 2$, $L^p(I)_{\mathcal{R}} = L^p(I)$ and $(L^p(I), L^\infty(I))$ is fully representable.

The example above is a particular case of the next Theorem 6.3.3. Since, in the familiar case discussed therein, everything can be easily computed, we have preferred an explicit proof there. It is implicitly proved in this example that $(L^p(I), L^\infty(I))$ has sufficiently many continuous representable linear functionals. Moreover, Theorem 6.3.3 also shows that in the case of Banach quasi *-algebras there is close relationship between full representability and *-semisimplicity.

Theorem 6.3.3 *Let $(\mathfrak{A}[\|\cdot\|], \mathfrak{A}_0)$ be a Banach quasi *-algebra with unit e. The following statements are equivalent:*

(i) $\mathcal{R}_c(\mathfrak{A}, \mathfrak{A}_0)$ *is sufficient;*
(ii) $(\mathfrak{A}, \mathfrak{A}_0)$ *is fully representable.*

If the condition (P) *holds,* (i) *and* (ii) *are equivalent to the following*

(iii) $(\mathfrak{A}[\|\cdot\|], \mathfrak{A}_0)$ *is *-semisimple.*

Proof The equivalence between (i) and (ii) follows from the very definitions and from Proposition 3.2.2.

Suppose now that the condition (P) holds. We shall prove that (iii) is equivalent to (ii) $\sim$ (i).

(ii) $\Rightarrow$ (iii) First, we notice that every $\varphi \in \mathcal{S}_{\mathfrak{A}_0}(\mathfrak{A})$ can be written as $\overline{\varphi}_\omega$, for some $\omega \in \mathcal{R}_c(\mathfrak{A}, \mathfrak{A}_0)$. Indeed, if we put

$$\omega_\varphi(a) := \varphi(a, e), \quad a \in \mathfrak{A},$$

then it is easily seen that ω_φ is continuous and representable, so $\omega_\varphi \in \mathcal{R}_c(\mathfrak{A}, \mathfrak{A}_0)$ and $\overline{\varphi}_{\omega_\varphi} = \varphi$. On the other hand, consider a linear functional $0 \neq \omega \in \mathcal{R}_c(\mathfrak{A}, \mathfrak{A}_0)$ and let $\overline{\varphi}_\omega$ be the sesquilinear form associated to it as in (3.2.16). By Proposition 3.2.2 $D(\overline{\varphi}_\omega) = \mathfrak{A}$, thus $\overline{\varphi}_\omega$ is bounded. If we put $\varphi\prime_\omega = \overline{\varphi}_\omega / \|\overline{\varphi}_\omega\|$, then $\varphi\prime_\omega \in \mathcal{S}_{\mathfrak{A}_0}(\mathfrak{A})$.

Let $a \in \mathfrak{A}$ be such that $\varphi(a, a) = 0$, for every $\varphi \in \mathcal{S}_{\mathfrak{A}_0}(\mathfrak{A})$. For what we have just shown, it is enough to prove that, if $\overline{\varphi}_\omega(a, a) = 0$, for every $\omega \in \mathcal{R}_c(\mathfrak{A}, \mathfrak{A}_0)$, then $a = 0$. We have

$$|\omega(a)| = |\overline{\varphi}_\omega(a, e)| \leq \overline{\varphi}_\omega(e, e)^{1/2}\, \overline{\varphi}_\omega(a, a)^{1/2} = 0.$$

Thus $\omega(a) = 0$, for every $\omega \in \mathcal{R}_c(\mathfrak{A}, \mathfrak{A}_0)$. By condition *(P)*, $a \geq 0$; but since $\mathcal{R}_c(\mathfrak{A}, \mathfrak{A}_0)$ is sufficient, we get $a = 0$.

(iii) $\Rightarrow$ (i) Let $a \in \mathfrak{A}^+$ and suppose that $\omega(a) = 0$, for every $\omega \in \mathcal{R}_c(\mathfrak{A}, \mathfrak{A}_0)$. If $x \in \mathfrak{A}_0$, then the linear functional ω_x defined by $\omega_x(c) := \omega(x^*cx)$, $c \in \mathfrak{A}$, is representable and continuous. Thus, $\omega_x(a) = 0$, for every $x \in \mathfrak{A}_0$. By polarization, one can easily conclude that $\omega(y^*ax) = 0$, for every $x, y \in \mathfrak{A}_0$. This implies that $\overline{\varphi}_\omega(ax, y) = 0$, for every $x, y \in \mathfrak{A}_0$. Now choose $x = e$ and take a sequence $\{y_n\} \subset \mathfrak{A}_0^+$ converging to a. Then $\overline{\varphi}_\omega(a, a) = \lim_{n\to\infty} \overline{\varphi}_\omega(a, y_n) = 0$. As in the proof of (ii) $\Rightarrow$ (iii) we conclude that $\varphi(a, a) = 0$, for every $\varphi \in \mathcal{S}_{\mathfrak{A}_0}(\mathfrak{A})$. Hence $a = 0$. □

Example 6.3.4 The space $\mathcal{S}'(\mathbb{R})$ of tempered distributions may be regarded as a locally convex quasi *-algebra over the *-algebra $\mathcal{S}(\mathbb{R})$. $\mathcal{S}'(\mathbb{R})$ is the dual of $\mathcal{S}(\mathbb{R})$, when the latter is endowed with the locally convex topology $\mathfrak{t}$ defined by the family of seminorms

$$p_{k,r}(f) = \sup_{x\in\mathbb{R}} |x^k D^r f(x)|, \quad f \in \mathcal{S}(\mathbb{R}), \quad k, r \in \mathbb{N}.$$

The (partial) multiplication in $\mathcal{S}'(\mathbb{R})$ is defined by

$$(F \cdot f)(g) = (f \cdot F)(g) = F(fg), \quad F \in \mathcal{S}'(\mathbb{R}), \quad f, g \in \mathcal{S}(\mathbb{R}).$$

The space $\mathcal{S}'(\mathbb{R})$ is endowed with the strong dual topology $\mathfrak{t}'$. Since $\mathcal{S}'(\mathbb{R})[\mathfrak{t}']$ is reflexive, every continuous functional ω on $\mathcal{S}'(\mathbb{R})[\mathfrak{t}']$ has the form $\omega(F) = \omega_f(F) := F(f)$, for some $f \in \mathcal{S}(\mathbb{R})$. Also in this case there are no nontrivial continuous representable functionals on $\mathcal{S}'(\mathbb{R})$. Indeed, (L3) of Definition 2.4.6 is never satisfied by nonzero positive functionals ω_f, $f \geq 0$, since, if for every $F \in \mathcal{S}'(\mathbb{R})$, there exists $\gamma_F > 0$, such that

$$\omega_f(F^* \cdot g) \leq \gamma_F \omega_f(g^* g)^{1/2}, \quad \forall\, g \in \mathcal{S}(\mathbb{R}),$$

then

$$|F^*(gf)| \leq \gamma_F \left(\int_{\mathbb{R}} g^*(x) g(x) f(x) dx \right)^{1/2} = \gamma_F \|g\|_{2,f},$$

where $\|\cdot\|_{2,f}$ denotes the norm of $L^2(\mathbb{R}, f dx)$. This implies that there exists $h \in L^2(\mathbb{R}, f dx)$, such that

$$F^*(gf) = \int_{\mathbb{R}} h(x) \overline{g(x)} f(x) dx, \quad \forall\, g \in \mathcal{S}(\mathbb{R}).$$

Hence, F^* restricted to the linear subspace $\{gf : g \in \mathcal{S}(\mathbb{R})\}$ acts as a function. This is a contradiction if f (and then ω_f) is nonzero.

*Example 6.3.5 (Quasi *-Algebras of Operators)* Let $(\mathcal{L}^\dagger(\mathcal{D}, \mathcal{H}), \mathcal{L}^\dagger(\mathcal{D})_b)$ be the locally convex quasi *-algebra of Example 2.1.3. Let $\xi \in \mathcal{D}$. Then, the positive linear functional

$$\omega_\xi(A) = \langle A\xi | \xi \rangle, \; A \in \mathcal{L}^\dagger(\mathcal{D}, \mathcal{H})$$

is representable. The corresponding sesquilinear form φ_{ω_ξ} on $\mathcal{L}^\dagger(\mathcal{D})_b \times \mathcal{L}^\dagger(\mathcal{D})_b$ is jointly continuous with respect to the topology τ_{s*}, so that $D(\overline{\varphi_{\omega_\xi}}) = \mathcal{L}^\dagger(\mathcal{D}, \mathcal{H})$.

The same is true in the more general case, where

$$\omega(A) = \sum_{i=1}^{n} \langle A\xi_i | \xi_i \rangle, \quad \xi_i \in \mathcal{D}, \; i = 1, \ldots, n.$$

Let us now consider $(\mathcal{L}^\dagger(\mathcal{D},\mathcal{H}), \mathcal{L}^\dagger(\mathcal{D})_b)$, where $\mathcal{L}^\dagger(\mathcal{D},\mathcal{H})$ is endowed with the weak topology t_w. Then, the following statements hold [39, 83]:

(i) Every weakly continuous (or strongly*-continuous) linear functional ω on $\mathcal{L}^\dagger(\mathcal{D},\mathcal{H})$ has the form

$$\omega(A) = \sum_{i=1}^{n} \langle A\xi_i | \eta_i \rangle, \quad \xi_i,\ \eta_i \in \mathcal{D}, i = 1, \ldots, n. \tag{6.3.2}$$

(ii) Every weakly continuous positive linear functional ω on $\mathcal{L}^\dagger(\mathcal{D},\mathcal{H})$ has the form

$$\omega(A) = \sum_{i=1}^{n} \langle A\zeta_i | \zeta_i \rangle, \quad \zeta_i \in \mathcal{D},\ i = 1, \ldots, n. \tag{6.3.3}$$

Both locally convex quasi *-algebras $(\mathcal{L}^\dagger(\mathcal{D},\mathcal{H})[\mathsf{t}_{s^*}], \mathcal{L}^\dagger(\mathcal{D})_b)$ and $(\mathcal{L}^\dagger(\mathcal{D},\mathcal{H})[\mathsf{t}_w], \mathcal{L}^\dagger(\mathcal{D})_b)$ satisfy the equality $\mathcal{L}^\dagger(\mathcal{D},\mathcal{H})_{\mathcal{R}} = \mathcal{L}^\dagger(\mathcal{D},\mathcal{H})$ and therefore both $(\mathcal{L}^\dagger(\mathcal{D},\mathcal{H})[\mathsf{t}_{s^*}], \mathcal{L}^\dagger(\mathcal{D})_b)$ and $(\mathcal{L}^\dagger(\mathcal{D},\mathcal{H})[\mathsf{t}_w], \mathcal{L}^\dagger(\mathcal{D})_b)$ are fully representable.

Indeed, let ω be t_{s^*}-continuous and representable. If $A \in \mathcal{L}^\dagger(\mathcal{D},\mathcal{H})$, there exists a net $\{A_\delta\}$ of elements of $\mathcal{L}^\dagger(\mathcal{D})_b$, such that $A_\delta \xrightarrow[\mathsf{t}_{s^*}]{} A$. Then, using the representation (6.3.3) we have,

$$\begin{aligned}\varphi_\omega(A_\delta - A_\gamma, A_\delta - A_\gamma) &= \omega((A_\delta - A_\gamma)^*(A_\delta - A_\gamma)) \\ &= \sum_{i=1}^{n} \langle (A_\delta - A_\gamma)\zeta_i | (A_\delta - A_\gamma)\zeta_i \rangle \\ &= \sum_{i=1}^{n} \|(A_\delta - A_\gamma)\zeta_i\|^2 \to 0.\end{aligned}$$

This proves that every $A \in \mathcal{L}^\dagger(\mathcal{D},\mathcal{H})$ is in the domain of the closure of φ_ω, with respect to the topology t_{s^*}. The statement for the weak topology follows by observing that if ω is weakly continuous, then it is automatically t_{s^*}-continuous.

Note that in the Examples 6.3.2 and 6.3.5 the condition (P) is satisfied, while for the Example 6.3.4 it is meaningless. We do not know whether (P) always holds, when $\mathcal{R}_c(\mathfrak{A}, \mathfrak{A}_0)$ is sufficient.

6.4 Ordered Bounded Elements

The concept of a bounded element in a locally convex algebra was first introduced by G.R. Allan [32], (1965), for building a spectral theory for this kind of algebras. Similar definitions had been considered earlier by S. Warner [94], in the case of m-convex algebras, as well as by L. Waelbroeck [93] in a specific framework and under the assumption of quasi-completeness and commutativity of the topological algebras involved. Bounded elements in the context of quasi *-algebras have been already considered in Sect. 4.1. For an extension to partial *-algebras, see [39]. K. Schmüdgen has considered Allan's bounded elements in his research on the unbounded operator algebras called O*-algebras and recently (2005), the same author considered bounded elements in a purely algebraic sense (see also [92]) and studied the structure of the set of the introduced bounded elements, in order to use them for proving a "strict Positivstellensatz for the Weyl algebra" [75]. Motivated from this, and having in hands the results of Sect. 6.2, we introduce the concept of "order boundedness" and what is interesting, is that this concept coincides under some conditions with the usual notion of boundedness, which one gets when the *-algebra under consideration admits *-representations (see, for instance, Proposition 6.4.4, Theorem 6.4.5 and Corollary 6.4.8). Furthermore, for suitable fully representable locally convex quasi *-algebras $(\mathfrak{A}[\tau], \mathfrak{A}_0)$, considering the set $\mathfrak{A}_b^{or}$ of all order bounded elements of $\mathfrak{A}[\tau]$, we prove that $\mathfrak{A}_b^{or}$ becomes either a partial C*-algebra or a C*-algebra (Theorem 6.4.16).

Let $(\mathfrak{A}[\tau], \mathfrak{A}_0)$ be an arbitrary locally convex quasi *-algebra. As we have seen in Sect. 6.2, $(\mathfrak{A}[\tau], \mathfrak{A}_0)$ has a natural order related to the topology τ. This order can be used to define *bounded* elements. In what follows, we shall assume that $(\mathfrak{A}, \mathfrak{A}_0)$ has a unit e.

Let $a \in \mathfrak{A}$; put $\Re(a) = \frac{1}{2}(a + a^*)$, $\Im(a) = \frac{1}{2i}(a - a^*)$. Then, $\Re(a), \Im(a) \in \mathfrak{A}_h$ and $a = \Re(a) + i\Im(a)$.

Definition 6.4.1 An element $a \in \mathfrak{A}$ is called *order bounded* if there exists $\gamma \geq 0$, such that

$$\pm\Re(a) \leq \gamma e, \qquad \pm\Im(a) \leq \gamma e.$$

We denote by $\mathfrak{A}_b^{or}$ the set of all order bounded elements of $\mathfrak{A}[\tau]$.

In this regard, we have

Proposition 6.4.2 *The following statements hold:*

(1) $\alpha a + \beta b \in \mathfrak{A}_b^{or}, \quad \forall\, a, b \in \mathfrak{A}_b^{or}, \alpha, \beta \in \mathbb{C}$;
(2) $a \in \mathfrak{A}_b^{or} \quad \Leftrightarrow \quad a^* \in \mathfrak{A}_b^{or}$;
(3) $a \in \mathfrak{A}_b^{or},\ x \in \mathfrak{A}_b^{or} \cap \mathfrak{A}_0 \ \Rightarrow\ xa \in \mathfrak{A}_b^{or}$;
(4) $x \in \mathfrak{A}_b^{or} \cap \mathfrak{A}_0 \ \Leftrightarrow\ xx^* \in \mathfrak{A}_b^{or} \cap \mathfrak{A}_0$.

Hence, $(\mathfrak{A}_b^{or}, \mathfrak{A}_b^{or} \cap \mathfrak{A}_0)$ *is a quasi *-algebra.*

Proof The proof is similar to that of [75, Lemma 2.1]. □

For $a \in (\mathfrak{A}_b^{or})_h$, put

$$\|a\|_b^{or} := \inf\{\gamma > 0 : -\gamma e \leq a \leq \gamma e\}.$$

Then, $\|\cdot\|_b^{or}$ is a seminorm on the real vector space $(\mathfrak{A}_b^{or})_h$.

Lemma 6.4.3 *If* $\mathfrak{A}^+ \cap (-\mathfrak{A}^+) = \{0\}$, $\|\cdot\|_{\mathrm{b}}^{\mathrm{or}}$ *is a norm on* $(\mathfrak{A}_{\mathrm{b}}^{\mathrm{or}})_h$.

Proof Put $E = \{\gamma > 0 : -\gamma e \leq a \leq \gamma e\}$. If $\inf E = 0$, then for every $\varepsilon > 0$, there exists $\gamma_\varepsilon \in E$, such that $\gamma_\varepsilon < \varepsilon$. This implies that $-\varepsilon e \leq a \leq \varepsilon e$. If $\omega \in \mathcal{R}_c(\mathfrak{A}, \mathfrak{A}_0)$, we get $-\varepsilon\,\omega(e) \leq \omega(a) \leq \varepsilon\,\omega(e)$ (we may suppose $\omega(e) > 0$, for every $\omega \in \mathcal{R}_c(\mathfrak{A}, \mathfrak{A}_0)$, since from (L3) of Definition 2.4.6, if $\omega(e) = 0$, then $\omega \equiv 0$). Hence, $\omega(a) = 0$. By the sufficiency of $\mathcal{R}_c(\mathfrak{A}, \mathfrak{A}_0)$ (Definition 6.2.6), it follows that $a = 0$. □

Proposition 6.4.4 *If* $a \in \mathfrak{A}_{\mathrm{b}}^{\mathrm{or}}$, *then* $\pi(a)$ *is a bounded operator, for every* (τ, t_w)*-continuous *-representation* π *of* $(\mathfrak{A}[\tau], \mathfrak{A}_0)$. *Moreover, if* $a = a^*$, *then* $\|\pi(a)\| \leq \|a\|_{\mathrm{b}}^{\mathrm{or}}$.

Proof It follows easily from Proposition 6.2.1 and the very definitions. □

Theorem 6.4.5 *Let* $(\mathfrak{A}[\tau], \mathfrak{A}_0)$ *be fully representable and assume that condition* (P) *holds. Then, for* $a \in \mathfrak{A}$, *the following statements are equivalent:*

(i) *a is order bounded;*
(ii) *there exists* $\gamma_a > 0$, *such that*

$$|\omega(x^*ax)| \leq \gamma_a\,\omega(x^*x), \quad \forall\,\omega \in \mathcal{R}_c(\mathfrak{A}, \mathfrak{A}_0),\ x \in \mathfrak{A}_0;$$

(iii) *there exists* $\gamma_a > 0$, *such that*

$$|\omega(y^*ax)| \leq \gamma_a\,\omega(x^*x)^{1/2}\,\omega(y^*y)^{1/2}, \quad \forall\,\omega \in \mathcal{R}_c(\mathfrak{A}, \mathfrak{A}_0),\ x, y \in \mathfrak{A}_0.$$

Proof It is sufficient to consider the case $a = a^*$. Also, as in the proof of Lemma 6.4.3, we suppose $\omega(e) > 0$, for every $\omega \in \mathcal{R}_c(\mathfrak{A}, \mathfrak{A}_0)$.

(i) ⇒ (ii) If $a = a^*$ is bounded, there exists $\gamma > 0$, such that $-\gamma e \leq a \leq \gamma e$. Hence, from Proposition 6.2.8, for every $\omega \in \mathcal{R}_c(\mathfrak{A}, \mathfrak{A}_0)$, $\omega(\gamma e - a) \geq 0$. It follows that $\omega(x^*(\gamma e - a)x) \geq 0$. Thus, $\omega(x^*ax) \leq \gamma\omega(x^*x)$, for every $x \in \mathfrak{A}_0$. Similarly we can show that $-\gamma\omega(x^*x) \leq \omega(a^*xa)$, for every $x \in \mathfrak{A}_0$.

(ii) ⇒ (i) Assume now that there exists $\gamma_a > 0$ such that

$$|\omega(x^*ax)| \leq \gamma_a\,\omega(x^*x), \quad \forall\,\omega \in \mathcal{R}_c(\mathfrak{A}, \mathfrak{A}_0),\ x \in \mathfrak{A}_0.$$

Define

$$\widetilde{\gamma} := \sup\{|\omega(x^*ax)| : \omega \in \mathcal{R}_c(\mathfrak{A}, \mathfrak{A}_0),\ x \in \mathfrak{A}_0 \text{ with } \omega(x^*x) = 1\}.$$

Then (see Proposition 6.2.8 and Remark 6.2.9(1)), for an arbitrary $\omega' \in \mathcal{R}_c(\mathfrak{A}, \mathfrak{A}_0)$, we get,

$$\omega'(\widetilde{\gamma}e \pm a) = \widetilde{\gamma}\omega'(e) \pm \omega'(a) = \omega'(e)(\widetilde{\gamma} \pm \omega'(u^*au)) \geq 0,$$

where $u = \frac{e}{\omega'(e)^{1/2}}$.

Hence, $\omega'(\widetilde{\gamma}e \pm a) \geq 0$, for every $\omega' \in \mathcal{R}_c(\mathfrak{A}, \mathfrak{A}_0)$. Then, by Remark 6.2.9, $-\widetilde{\gamma}e \leq a \leq \widetilde{\gamma}e$; i.e., a is order bounded.

(i) ⇒ (iii) The GNS representation π_ω is (τ, t_w)-continuous, hence if $a = a^* \in \mathfrak{A}$, by Proposition 6.4.4, $\pi_\omega(a)$ is bounded. Thus,

$$|\omega(y^*ax)| = |\langle \pi_\omega(a)\lambda_\omega(x)|\lambda_\omega(y)\rangle| \leq \|\pi_\omega(a)\| \, \|\lambda_\omega(x)\| \, \|\lambda_\omega(y)\|$$
$$\leq \|a\|_{\mathrm{b}}^{\mathrm{or}} \, \omega(x^*x)^{1/2} \, \omega(y^*y)^{1/2}, \quad \forall \, \omega \in \mathcal{R}_c(\mathfrak{A}, \mathfrak{A}_0), \; x, y \in \mathfrak{A}_0.$$

(iii) ⇒ (ii) is obvious. □

Remark 6.4.6 The proof above shows that for $a = a^*$,

$$\|a\|_{\mathrm{b}}^{\mathrm{or}} \leq \sup\left\{|\omega(x^*ax)| : \; \omega \in \mathcal{R}_c(\mathfrak{A}, \mathfrak{A}_0), \; x \in \mathfrak{A}_0 \text{ with } \omega(x^*x) = 1\right\}.$$

Corollary 6.4.7 *Let $(\mathfrak{A}[\tau], \mathfrak{A}_0)$ be fully representable. If a is order bounded, there exists $\gamma_a > 0$, such that*

$$|\overline{\varphi}_\omega(ax, c)| \leq \gamma_a \, \omega(x^*x)^{1/2} \, \overline{\varphi}_\omega(c, c)^{1/2}, \quad \forall \, \omega \in \mathcal{R}_c(\mathfrak{A}, \mathfrak{A}_0), \; x \in \mathfrak{A}_0, \; c \in \mathfrak{A}.$$

Proof Let $a \in \mathfrak{A}$ be order bounded. Since $\bigcap_{\omega \in \mathcal{R}_c(\mathfrak{A}, \mathfrak{A}_0)} D(\overline{\varphi}_\omega) = \mathfrak{A}$, for every $c \in \mathfrak{A}$, there exists a net $\{z_\delta\} \subset \mathfrak{A}_0$, such that $z_\delta \xrightarrow{\tau} c$ and $\overline{\varphi}_\omega(c - z_\delta, c - z_\delta) \to 0$. Then, by Theorem 6.4.5(iii), we get

$$|\overline{\varphi}_\omega(ax, c)| = \lim_\delta |\overline{\varphi}_\omega(xa, z_\delta)| \leq \gamma_a \, \omega(x^*x)^{1/2} \lim_\delta \overline{\varphi}_\omega(z_\delta, z_\delta)^{1/2}$$
$$= \gamma_a \, \omega(x^*x)^{1/2} \, \overline{\varphi}_\omega(c, c)^{1/2}, \quad \forall \, x \in \mathfrak{A}_0.$$

□

Corollary 6.4.8 *Let $(\mathfrak{A}[\tau], \mathfrak{A}_0)$ be fully representable and $a \in \mathfrak{A}$. Then, a is order bounded, if and only if, there exists $\gamma_a > 0$ such that*

$$\overline{\varphi_\omega}(ax, ax) \leq \gamma_a^2 \omega(x^*x), \quad \forall \, \omega \in \mathcal{R}_c(\mathfrak{A}, \mathfrak{A}_0), \; x \in \mathfrak{A}_0.$$

Proof The necessity follows by putting $c = ax$ in the inequality of Corollary 6.4.7. The sufficiency is clear. □

Let a be order bounded. Define

$$\mathsf{q}(a) := \sup\left\{|\omega(y^*ax)| : \; \omega \in \mathcal{R}_c(\mathfrak{A}, \mathfrak{A}_0), \; x, y \in \mathfrak{A}_0 \text{ with } \omega(x^*x) = \omega(y^*y) = 1\right\}.$$

Then, we have

Lemma 6.4.9 $\mathsf{q}(a) = \|a\|_{\mathrm{b}}^{\mathrm{or}}$, *for every $a = a^* \in \mathfrak{A}_{\mathrm{b}}^{\mathrm{or}}$.*

Proof The inequality $\|a\|_{\mathrm{b}}^{\mathrm{or}} \leq \mathsf{q}(a)$ follows from Remark 6.4.6. Let $\gamma > 0$, such that $-\gamma e \leq a \leq \gamma e$. Then, by the proof of Theorem 6.4.5, we have, $\mathsf{q}(a) \leq \gamma$; whence the statement follows. □

Lemma 6.4.9 shows that q extends $\|\cdot\|_{\mathrm{b}}^{\mathrm{or}}$. For this reason, we will use the symbol $\|\cdot\|_{\mathrm{b}}^{\mathrm{or}}$ for q too. It is easily seen that $\|\cdot\|_{\mathrm{b}}^{\mathrm{or}}$ is a norm on $\mathfrak{A}_{\mathrm{b}}^{\mathrm{or}}$.

An easy consequence of the above statements is now the following

Proposition 6.4.10 *For every* $a \in \mathfrak{A}_{\mathrm{b}}^{\mathrm{or}}$,

$$\|a\|_{\mathrm{b}}^{\mathrm{or}} = \sup\left\{\overline{\varphi_\omega}(ax, ax) : \omega \in \mathcal{R}_c(\mathfrak{A}, \mathfrak{A}_0),\ x \in \mathfrak{A}_0 \ \textit{with}\ \omega(x^* x) = 1\right\}.$$

6.4.1 Partial Multiplication

If $(\mathfrak{A}[\tau], \mathfrak{A}_0)$ is fully representable we can introduce on $\mathfrak{A}$ a partial multiplication, which makes it into a partial *-algebra. This multiplication extends that one introduced before Proposition 3.1.31 for *-semisimple Banach quasi *-algebras. For convenience, we keep the same symbol, □, as there.

Definition 6.4.11 Let $(\mathfrak{A}[\tau], \mathfrak{A}_0)$ be fully representable. The *weak product* $a \Box b$ of two elements $a, b \in \mathfrak{A}$ is well defined if there exists $c \in \mathfrak{A}$ such that

$$\overline{\varphi}_\omega(bx, a^* y) = \overline{\varphi}_\omega(cx, y), \quad \forall\, \omega \in \mathcal{R}_c(\mathfrak{A}, \mathfrak{A}_0),\ x, y \in \mathfrak{A}_0.$$

In this case, we put $a \Box b := c$. We call □ *weak multiplication* on $\mathfrak{A}$.

Since $\mathcal{R}_c(\mathfrak{A}, \mathfrak{A}_0)$ is sufficient, the element c is unique.

Proposition 6.4.12 $(\mathfrak{A}[\tau], \mathfrak{A}_0)$ *endowed with the weak multiplication* □ *is a* partial *-algebra *with* $\mathfrak{A}_0 \subset R\mathfrak{A}$.

Proposition 6.4.13 *Let* a, b *be order bounded elements of* $\mathfrak{A}$*. The following statements hold:*

(i) a^* *is order bounded too, and* $\|a^*\|_{\mathrm{b}}^{\mathrm{or}} = \|a\|_{\mathrm{b}}^{\mathrm{or}}$;
(ii) *If* $a \Box b$ *is well-defined, then* $a \Box b$ *is order bounded and*

$$\|a \Box b\|_{\mathrm{b}}^{\mathrm{or}} \leq \|a\|_{\mathrm{b}}^{\mathrm{or}}\, \|b\|_{\mathrm{b}}^{\mathrm{or}}.$$

Proof

(i) The first part of (i) is given by Proposition 6.4.2(2). The second part follows from the property (L2) (Definition 2.4.6) of an $\omega \in \mathcal{R}_c(\mathfrak{A}, \mathfrak{A}_0)$, from Corollary 6.4.8 and the definition of $\|\cdot\|_{\mathrm{b}}^{\mathrm{or}}$.

(ii) If $a \Box b, a, b \in \mathfrak{A}$, is well-defined, then for every $\omega \in \mathcal{R}_c(\mathfrak{A}, \mathfrak{A}_0)$, Corollary 6.4.7 implies

$$|\overline{\varphi}_\omega((a \Box b)x, y)| = |\overline{\varphi}_\omega(bx, a^* y)| \leq \overline{\varphi}_\omega(bx, bx)^{1/2} \overline{\varphi}_\omega(a^* y, a^* y)^{1/2}$$
$$\leq \|a\|_{\mathrm{b}}^{\mathrm{or}} \|b\|_{\mathrm{b}}^{\mathrm{or}} \overline{\varphi}_\omega(x, x)^{1/2} \overline{\varphi}_\omega(y, y)^{1/2}, \quad \forall\, x, y \in \mathfrak{A}_0.$$

Taking now sup on the left hand side (see Proposition 6.4.10), we get the desired inequality. □

We recall that an *unbounded C*-seminorm p* on a partial *-algebra $\mathfrak{A}$ is a seminorm defined on a partial *-subalgebra $D(p)$ of $\mathfrak{A}$, the domain of p, with the properties:

- $p(ab) \leq p(a)p(b)$, whenever ab is well-defined;
- $p(a^*a) = p(a)^2$, whenever a^*a is well-defined

(see, e.g., [2, 46, 79]).

Proposition 6.4.14 $\|\cdot\|_{\mathrm{b}}^{\mathrm{or}}$ *is an unbounded C*-norm on* $\mathfrak{A}$ *with domain* $\mathfrak{A}_b$.

Proof This can be deduced from [87, Proposition 2.6]. □

It is worth mentioning here that certain unbounded C*-seminorms give rise to "well-behaved" (unbounded) *-representations (for more details, see [2, Chapter 8] and [46, 79]).

Now having $(\mathfrak{A}[\tau], \mathfrak{A}_0)$ to be fully representable, we can endow $\mathfrak{A}$ with the strong and strong* topology, where both are defined in a natural way through the elements of $\mathcal{R}_c(\mathfrak{A}, \mathfrak{A}_0)$. Indeed:

- the *strong* topology τ_s, is defined by the family of seminorms

$$a \in \mathfrak{A} \to \overline{\varphi}_\omega(a, a)^{1/2}, \quad \omega \in \mathcal{R}_c(\mathfrak{A}, \mathfrak{A}_0),$$

- the *strong** topology τ_{s^*}, is respectively defined by the family of seminorms

$$a \in \mathfrak{A} \to \max\left\{\overline{\varphi}_\omega(a, a)^{1/2}, \ \overline{\varphi}_\omega(a^*, a^*)^{1/2}\right\}, \quad \omega \in \mathcal{R}_c(\mathfrak{A}, \mathfrak{A}_0).$$

Definition 6.4.15 Let $\mathfrak{A}$ be a partial *-algebra. We say that $\mathfrak{A}$ is a *partial C*-algebra* if $\mathfrak{A}$ is a Banach space under a norm $\|\cdot\|$ satisfying the following properties:

(i) $\|a^*\| = \|a\|, \quad \forall\, a \in \mathfrak{A}$;
(ii) $\|ab\| \leq \|a\|\,\|b\|$, whenever ab is well-defined;
(iii) $\|a^*a\| = \|a\|^2$, whenever a^*a is well-defined.

The theorem that follows, shows that the quasi *-algebra $(\mathfrak{A}_b, \mathfrak{A}_b \cap \mathfrak{A}_0)$ (see Proposition 6.4.2), under certain conditions achieves a very rich structure.

Theorem 6.4.16 *Let $(\mathfrak{A}[\tau], \mathfrak{A}_0)$ be a fully representable locally convex quasi *-algebra with unit e. Assume that $\mathfrak{A}$ is τ_{s^*}-complete. Then, $\mathfrak{A}_b$ is a partial C*-algebra under the weak multiplication $\Box$ and the norm $\|\cdot\|_{\mathrm{b}}^{\mathrm{or}}$. Assume, in addition, that*

(R) *If $a, b \in \mathfrak{A}$ and $\pi_\omega(a)\Box\pi_\omega(b)$ is well-defined, for every $\omega \in \mathcal{R}_c(\mathfrak{A}, \mathfrak{A}_0)$, then there exists $c \in \mathfrak{A}$, such that $\pi_\omega(a)\Box\pi_\omega(b) = \pi_\omega(c)$, for every $\omega \in \mathcal{R}_c(\mathfrak{A}, \mathfrak{A}_0)$.*

Then, $\mathfrak{A}_b$ is a C-algebra with the weak multiplication $\Box$ and the norm $\|\cdot\|_{\mathrm{b}}^{\mathrm{or}}$.*

Proof Since $\|\cdot\|_{\mathrm{b}}^{\mathrm{or}}$ satisfies (i)–(iii) of Definition 6.4.15 on $\mathfrak{A}_b$ (see e.g., Proposition 6.4.13), we need only to prove the completeness of $\mathfrak{A}_b$.

Let $\{a_n\}$ be a Cauchy sequence with respect to the norm $\|\cdot\|_{\mathrm{b}}^{\mathrm{or}}$. Then, $\{a_n^*\}$ is $\|\cdot\|_{\mathrm{b}}^{\mathrm{or}}$-Cauchy too. Hence, for every $\omega \in \mathcal{R}_c(\mathfrak{A}, \mathfrak{A}_0)$ and $x \in \mathfrak{A}_0$ we have

$$\overline{\varphi}_\omega\big((a_n - a_m)x, (a_n - a_m)x\big) \to 0, \text{ as } n, m \to \infty$$

and

$$\overline{\varphi}_\omega\big((a_n^* - a_m^*)x, (a_n^* - a_m^*)x\big) \to 0, \text{ as } n, m \to \infty.$$

Therefore, $\{a_n\}$ is also Cauchy with respect to τ_{s^*}. Then, since $\mathfrak{A}$ is τ_{s^*} complete, there exists $a \in \mathfrak{A}$ such that $a_n \underset{\tau_{s^*}}{\to} a$. Moreover,

$$\overline{\varphi}_\omega(ax, ax) = \lim_{n\to\infty} \overline{\varphi}_\omega(a_n x, a_n x) \le \limsup_{n\to\infty} \left(\|a_n\|_{\mathrm{b}}^{\mathrm{or}}\right)^2 \overline{\varphi}_\omega(x, x), \ \forall\ x \in \mathfrak{A}_0,$$

with $\limsup_{n\to\infty} \left(\|a_n\|_{\mathrm{b}}^{\mathrm{or}}\right)^2 < \infty$ (by the boundedness of the sequence $\{\|a_n\|_{\mathrm{b}}^{\mathrm{or}}\}$), so by Corollary 6.4.8, we conclude that a is order bounded. Finally, by the Cauchy condition, for every $\varepsilon > 0$, there exists $n_\varepsilon \in \mathbb{N}$, such that $\|a_n - a_m\|_{\mathrm{b}}^{\mathrm{or}} < \varepsilon$, for every $n, m > n_\varepsilon$. This implies that

$$\overline{\varphi}_\omega\big((a_n - a_m)x, (a_n - a_m)x\big) < \varepsilon\, \overline{\varphi}_\omega(x, x), \quad \forall\, \omega \in \mathcal{R}_c(\mathfrak{A}, \mathfrak{A}_0),\ x \in \mathfrak{A}_0.$$

Then, if we fix $n > n_\varepsilon$ and let $m \to \infty$, we obtain

$$\overline{\varphi}_\omega\big((a_n - a)x, (a_n - a)x\big) \le \varepsilon\, \overline{\varphi}_\omega(x, x), \quad \forall\, \omega \in \mathcal{R}_c(\mathfrak{A}, \mathfrak{A}_0),\ x \in \mathfrak{A}_0.$$

This, in turn, implies that $\|a_n - a\|_{\mathrm{b}}^{\mathrm{or}} \le \varepsilon$, for every $n \ge n_\varepsilon$. So completeness of $\mathfrak{A}_b[\|\cdot\|_{\mathrm{b}}^{\mathrm{or}}]$ is proved.

Now, assume that condition (R) holds. By Proposition 6.4.4 it follows that if $a, b \in \mathfrak{A}_b$, then the operators $\pi_\omega(a), \pi_\omega(b)$ are bounded, therefore the operator $\pi_\omega(a)\Box\pi_\omega(b)$ is well-defined, hence (Proposition 6.4.13) bounded. Thus, by (R), there exists $c \in \mathfrak{A}$, such that $\pi_\omega(a)\Box\pi_\omega(b) = \pi_\omega(c)$, for every $\omega \in \mathcal{R}_c(\mathfrak{A}, \mathfrak{A}_0)$.

Furthermore, for every $\omega \in \mathcal{R}_c(\mathfrak{A}, \mathfrak{A}_0)$ and $x, y \in \mathfrak{A}_0$, we have

$$\begin{aligned}\overline{\varphi}_\omega(bx, a^* y) &= \langle \pi_\omega(b)\lambda_\omega(x) | \pi_\varphi(a^*)\lambda_\omega(y)\rangle \\ &= \langle \pi_\omega(a)\Box\pi_\omega(b)\lambda_\omega(x) | \lambda_\omega(y)\rangle \\ &= \langle \pi_\omega(c)\lambda_\omega(x) | \lambda_\omega(y)\rangle \\ &= \overline{\varphi}_\omega(cx, y).\end{aligned}$$

Hence $a\Box b$ is well-defined (Definition 6.4.11). Thus (see also Proposition 6.4.14), $\mathfrak{A}_b$ is a C*-algebra. □

Example 6.4.17 Let us consider again the locally convex quasi *-algebra

$$(\mathcal{L}^\dagger(\mathcal{D}, \mathcal{H}),\ \mathcal{L}^\dagger(\mathcal{D})_b)$$

of example 6.3.5. As proved there, this quasi *-algebra is fully-representable and from (6.3.3) it follows that the topology τ_{s^*} defined before Definition 6.4.15 coincides with the strong* topology t_{s^*} of $\mathcal{L}^\dagger(\mathcal{D}, \mathcal{H})$. One can prove easily that an element $T \in \mathcal{L}^\dagger(\mathcal{D}, \mathcal{H})$ is order bounded, if and only if, $\overline{T} \in \mathcal{B}(\mathcal{H})$. So that,

$$(\mathcal{L}^\dagger(\mathcal{D}, \mathcal{H}))_b = \left\{T \in \mathcal{L}^\dagger(\mathcal{D}, \mathcal{H}) : T \text{ is a bounded operator}\right\}.$$

This is clearly a C*-algebra as expected by Theorem 6.4.16.

Chapter 7
Locally Convex Quasi C*-Algebras and Their Structure

7.1 Locally Convex Quasi C*-Algebras

Throughout this chapter $\mathfrak{A}_0[\|\cdot\|_0]$ denotes a unital C*-algebra and τ a locally convex topology on $\mathfrak{A}_0$. Let $\widetilde{\mathfrak{A}_0}[\tau]$ denote the completion of $\mathfrak{A}_0$ with respect to the topology τ. Under certain conditions on τ, a subspace $\mathfrak{A}$ of $\widetilde{\mathfrak{A}_0}[\tau]$, containing $\mathfrak{A}_0$, will form (together with $\mathfrak{A}_0$) a locally convex quasi *-algebra $(\mathfrak{A}[\tau], \mathfrak{A}_0)$, which is named locally convex quasi C*-algebra. Examples and basic properties of such algebras are presented. So, let $\mathfrak{A}_0[\|\cdot\|_0]$ and τ be as before, with $\{p_\lambda\}_{\lambda\in\Lambda}$ a defining family of seminorms for τ. Suppose that τ satisfies the properties:

(T_1) $\mathfrak{A}_0[\tau]$ is a locally convex *-algebra with separately continuous multiplication.
(T_2) $\tau \preceq \|\cdot\|_0$.

Then, the identity map $\mathfrak{A}_0[\|\cdot\|_0] \to \mathfrak{A}_0[\tau]$ extends to a continuous *-linear map $\mathfrak{A}_0[\|\cdot\|_0] \to \widetilde{\mathfrak{A}_0}[\tau]$ and by (T_2), the C*-algebra $\mathfrak{A}_0[\|\cdot\|_0]$ can be regarded embedded into $\widetilde{\mathfrak{A}_0}[\tau]$. It is easily shown that $\widetilde{\mathfrak{A}_0}[\tau]$ is a locally convex quasi *-algebra over $\mathfrak{A}_0$ (cf. Definition 6.1.1 and [59, Section 3]).

The next Definition 7.1.1 provides concepts of positivity for elements of a quasi *-algebra $\widetilde{\mathfrak{A}_0}[\tau]$. In this regard, see also Example 6.1.3 and beginning of Sect. 6.2.

Definition 7.1.1 An element a of $\widetilde{\mathfrak{A}_0}[\tau]$ is called *positive* (resp. *commutatively positive*) if there is a net (resp. commuting net) $\{x_\alpha\}_{\alpha\in\Delta}$ of the positive cone $\mathfrak{A}_0^+$ of the C*-algebra $\mathfrak{A}_0[\|\cdot\|_0]$, which converges to a with respect to the topology τ.

The set of all positive (resp. commutatively positive) elements of $\widetilde{\mathfrak{A}_0}[\tau]$, we shall denote by $\widetilde{\mathfrak{A}_0}[\tau]^+$ (resp. $\widetilde{\mathfrak{A}_0}[\tau]_c^+$).

We have already used the symbol $\mathfrak{A}_0^+$ for the set of all positive elements of the C*-algebra $\mathfrak{A}_0[\|\cdot\|_0]$. It is worth noticing that the notion of a positive element of $\widetilde{\mathfrak{A}_0}[\tau]$ given here is exactly the same as that introduced in Sect. 6.2, since in the C*-algebra $\mathfrak{A}_0[\|\cdot\|_0]$ the set of positive elements (i.e., elements with spectrum contained

M. Fragoulopoulou, C. Trapani, *Locally Convex Quasi *-Algebras and their Representations*, Lecture Notes in Mathematics 2257,
https://doi.org/10.1007/978-3-030-37705-2_7

in $[0, +\infty)$ coincides with the set of all elements $y \in \mathfrak{A}_0$, such that $y = x^*x$, for some $x \in \mathfrak{A}_0$). Then, as discussed in Sect. 6.2 $\widetilde{\mathfrak{A}}_0[\tau]^+$, is a wedge, but it is not necessarily a positive cone; i.e.,

$$\widetilde{\mathfrak{A}}_0[\tau]^+ \cap (-\widetilde{\mathfrak{A}}_0[\tau]^+) \neq \{0\}.$$

The set $\widetilde{\mathfrak{A}}_0[\tau]_c^+$ is not even, in general, a wedge. But, if $\mathfrak{A}_0$ is commutative, then of course, $\widetilde{\mathfrak{A}}_0[\tau]^+ = \widetilde{\mathfrak{A}}_0[\tau]_c^+$.

- If $(\mathfrak{A}[\tau], \mathfrak{A}_0)$ is a locally convex quasi C*-algebra, in the sense discussed at the beginning of this section, we shall call *positive* (resp., *commutatively positive*) an element of $\mathfrak{A}$, which is positive (resp., commutatively positive) in $\widetilde{\mathfrak{A}}_0[\tau]$ and, for simplifying the notations, *we shall denote by* $\mathfrak{A}^+$ (resp., $\mathfrak{A}_c^+$) *the corresponding set*. Then, we have

$$\mathfrak{A}^+ := \mathfrak{A} \cap \widetilde{\mathfrak{A}}_0[\tau]^+ = \overline{\mathfrak{A}_0^+}^{\tau},$$

where $\overline{\mathfrak{A}_0^+}^{\tau}$ denotes, as usual, the closure of $\mathfrak{A}_0^+$ in $\mathfrak{A}[\tau]$.

Further we employ the following two extra conditions (T_3), (T_4) for the locally convex topology τ on $\mathfrak{A}_0$ and examine the effect on $\widetilde{\mathfrak{A}}_0[\tau]_c^+$.

(T_3) For each $\lambda \in \Lambda$, there exists $\lambda' \in \Lambda$, such that

$$p_\lambda(xy) \leq \|x\|_0 p_{\lambda'}(y), \quad \forall\, x, y \in \mathfrak{A}_0 \text{ with } xy = yx;$$

(T_4) The set $\mathcal{U}(\mathfrak{A}_0^+) := \{x \in \mathfrak{A}_0^+ : \|x\|_0 \leq 1\}$ is τ-closed.

Proposition 7.1.2 *Let* $\mathfrak{A}_0[\|\cdot\|_0]$ *be a unital C*-algebra and* τ *a locally convex topology on* $\mathfrak{A}_0$. *Suppose that* τ *fulfils the conditions* (T_1)–(T_4). *Then,* $\widetilde{\mathfrak{A}}_0[\tau]$ *is a locally convex quasi *-algebra over* $\mathfrak{A}_0$ *with the properties:*

1. *for every* $a \in \widetilde{\mathfrak{A}}_0[\tau]_c^+$, *the element* $e + a$ *is invertible and its inverse* $(e + a)^{-1}$ *belongs to* $\mathcal{U}(\mathfrak{A}_0^+)$.
2. *For a given* $a \in \widetilde{\mathfrak{A}}_0[\tau]_c^+$ *and any* $\varepsilon > 0$, *let*

$$a_\varepsilon = a(e + \varepsilon a)^{-1}.$$

Then, $\{a_\varepsilon\}_{\varepsilon>0}$ *is a commuting net in* $\mathfrak{A}_0^+$, *such that* $a - a_\varepsilon \in \widetilde{\mathfrak{A}}_0[\tau]_c^+$ *and* $a = \tau\text{-}\lim_{\varepsilon \to 0} a_\varepsilon$.

3. $\widetilde{\mathfrak{A}}_0[\tau]_c^+ \cap (-\widetilde{\mathfrak{A}}_0[\tau]_c^+) = \{0\}$.
4. *If* $a \in \widetilde{\mathfrak{A}}_0[\tau]_c^+$ *and* $b \in \mathfrak{A}_0^+$, *such that* $b - a \in \widetilde{\mathfrak{A}}_0[\tau]^+$, *then* $a \in \mathfrak{A}_0^+$.

Proof

1. Let $a \in \widetilde{\mathfrak{A}}_0[\tau]_c^+$. Then, there is a net $\{x_\alpha\}_{\alpha\in\Delta}$ in $\mathfrak{A}_0^+$, such that $x_\alpha x_\beta = x_\beta x_\alpha$, for all $\alpha, \beta \in \Delta$ and $x_\alpha \underset{\tau}{\to} a$. Using properties of the positive elements of a C^*-algebra and the condition (T$_3$), we have that for every $\lambda \in \Lambda$, there is $\lambda' \in \Lambda$ with

$$\begin{aligned} p_\lambda\big((e+x_\alpha)^{-1} - (e+x_\beta)^{-1}\big) &= p_\lambda\big((e+x_\alpha)^{-1}(x_\alpha - x_\beta)(e+x_\beta)^{-1}\big) \\ &\le \|(e+x_\alpha)^{-1}\|_0\, \|(e+x_\beta)^{-1}\|_0\, p_{\lambda'}(x_\alpha - x_\beta) \\ &\le p_{\lambda'}(x_\alpha - x_\beta) \to 0. \end{aligned}$$

So, $\{(e+x_\alpha)^{-1}\}_{\alpha\in\Delta}$ is a Cauchy net in $\mathfrak{A}_0[\tau]$ consisting of elements of $\mathcal{U}(\mathfrak{A}_0^+)$, which by (T$_4$) is τ-closed. Hence,

$$(e+x_\alpha)^{-1} \underset{\tau}{\to} y \in \mathcal{U}(\mathfrak{A}_0^+). \tag{7.1.1}$$

We shall show that $(e+a)^{-1}$ exists and equals y. Indeed: Using again condition (T$_3$), for each $\lambda \in \Lambda$, there is $\lambda' \in \Lambda$ with

$$\begin{aligned} p_\lambda\big(e - (e+a)(e+x_\alpha)^{-1}\big) &= p_\lambda\big((x_\alpha - a)(e+x_\alpha)^{-1}\big) \\ &\le \|(e+x_\alpha)^{-1}\|_0\, p_{\lambda'}(x_\alpha - a) \le p_{\lambda'}(x_\alpha - a) \to 0. \end{aligned}$$

Therefore,

$$(e+a)(e+x_\alpha)^{-1} \underset{\tau}{\to} e. \tag{7.1.2}$$

On the other hand, since

$$x_\beta y = \tau - \lim_\alpha x_\beta (e+x_\alpha)^{-1} = \tau - \lim_\alpha (e+x_\alpha)^{-1} x_\beta = y x_\beta, \quad \forall\, \beta \in \Delta,$$

we have $ay = ya$. Further, we can show that

$$(e+a)(e+x_\alpha)^{-1} \underset{\tau}{\to} (e+a)y. \tag{7.1.3}$$

Indeed, since $x_\alpha \underset{\tau}{\to} a$, for any $\varepsilon > 0$ there exists $\alpha_0 \in \Delta$, such that for all $\alpha \ge \alpha_0$ and all $\lambda \in \Lambda$ one has $p_\lambda(x_\alpha - a) < \varepsilon$. Now, by (T$_3$) we have that for any $\alpha \in \Delta$

$$\begin{aligned} & p_\lambda\big((e+a)(e+x_\alpha)^{-1} - (e+a)y\big) \\ &\le p_\lambda\big((e+a)(e+x_\alpha)^{-1} - (e+x_{\alpha_0})(e+x_\alpha)^{-1}\big) \\ &\quad + p_\lambda\big((e+x_{\alpha_0})(e+x_\alpha)^{-1} - (e+x_{\alpha_0})y\big) + p_\lambda\big((e+x_{\alpha_0})y - (e+a)y\big) \end{aligned}$$

$$\leq p_{\lambda'}(a - x_{\alpha_0}) + \|e + x_{\alpha_0}\|_0\, p'_\lambda\big((e + x_\alpha)^{-1} - y\big) + p_\lambda(x_{\alpha_0} - a)$$
$$< 2\varepsilon + \|e + x_{\alpha_0}\|_0\, p'_\lambda\big((e + x_\alpha)^{-1} - y\big),$$

which by (7.1.1) implies that $\lim_\alpha p_\lambda\big((e+a)(e+x_\alpha)^{-1} - (e+a)y\big) = 0$. Thus, by (7.1.2) and (7.1.3) we have $(e+a)y = e = y(e+a)$. Hence, $(e+a)^{-1}$ exists and belongs to $\mathcal{U}(\mathfrak{A}_0^+)$ (since y does).

2. It is clear from (1) that for every $\varepsilon > 0$ the element $(e + \varepsilon a)^{-1}$ exists in $\widetilde{\mathfrak{A}}_0[\tau]$ and belongs to $\mathcal{U}(\mathfrak{A}_0^+)$. In particular, applying (T_3) we get that for each $\lambda \in \Lambda$, there is $\lambda' \in \Lambda$ with

$$p_\lambda\big(e - (e + \varepsilon a)^{-1}\big) = \varepsilon p_\lambda\big(a(e + \varepsilon a)^{-1}\big) \leq \varepsilon \|(e + \varepsilon a)^{-1}\|_0\, p_{\lambda'}(a) \leq \varepsilon p_{\lambda'}(a),$$

so that

$$\tau\text{–}\lim_{\varepsilon \to 0} (e + \varepsilon a)^{-1} = e. \tag{7.1.4}$$

On the other hand, from the very definitions one has

$$a_\varepsilon = a(e + \varepsilon a)^{-1} = (e + \varepsilon a)^{-1} a = \frac{1}{\varepsilon}\big(e - (e + \varepsilon a)^{-1}\big), \quad \forall\, \varepsilon > 0, \text{ and}$$

$$a - a_\varepsilon = a\big(e - (e + \varepsilon a)^{-1}\big) = \big(e - (e + \varepsilon a)^{-1}\big)a \in \widetilde{\mathfrak{A}}_0[\tau]_c^+. \tag{7.1.5}$$

Now, by the same way as in (7.1.3), we conclude from (7.1.4) and (7.1.5) that $\tau\text{–}\lim_{\varepsilon\to 0} a_\varepsilon = a$.

3. Let $a \in \widetilde{\mathfrak{A}}_0[\tau]_c^+ \cap (-\widetilde{\mathfrak{A}}_0[\tau]_c^+)$. For any $\varepsilon > 0$, we have by (2) that

$$\mathfrak{A}_0^+ \ni a(e + \varepsilon a)^{-1} \underset{\tau}{\to} a \text{ and } \mathfrak{A}_0^+ \ni (-a)(e - \varepsilon a)^{-1} \underset{\tau}{\to} -a.$$

Thus, if

$$x_\varepsilon := a(e + \varepsilon a)^{-1} - (-a)(e - \varepsilon a)^{-1}, \tag{7.1.6}$$

we obtain

$$x_\varepsilon = a\big((e + \varepsilon a)^{-1} + (e - \varepsilon a)^{-1}\big) = a(e + \varepsilon a)^{-1}(e - \varepsilon a + e + \varepsilon a)(e - \varepsilon a)^{-1}$$
$$= 2a(e + \varepsilon a)^{-1}(e - \varepsilon a)^{-1},$$

where by (1) and (2) we conclude that $(e - \varepsilon a)^{-1} \in \mathfrak{A}_0^+$ and $a(e + \varepsilon a)^{-1} \in \mathfrak{A}_0^+$ respectively. Therefore, $x_\varepsilon \in \mathfrak{A}_0^+$ according to the functional calculus in commutative C*-algebras. Similarly, we have that

$$-x_\varepsilon = 2(-a)(e - \varepsilon a)^{-1}(e + \varepsilon a)^{-1} \in \mathfrak{A}_0^+$$

since $(-a)(e-\varepsilon a)^{-1}$ and $(e+\varepsilon a)^{-1}$ belong to $\mathfrak{A}_0^+$. Thus,

$$x_\varepsilon \in \mathfrak{A}_0^+ \cap (-\mathfrak{A}_0^+) = \{0\}$$

and so (see (7.1.6))

$$a(e+\varepsilon a)^{-1} = -a(e-\varepsilon a)^{-1}.$$

Taking τ-limits with $\varepsilon \to 0$, we get $a = -a$, i.e., $a = 0$.

4. By (2) and the assumptions in (4), $b - a$ and $a - a_\varepsilon$ are contained in $\widetilde{\mathfrak{A}}_0[\tau]^+$. Since, $\widetilde{\mathfrak{A}}_0[\tau]^+$ is a wedge, $b - a_\varepsilon = (b-a) + (a - a_\varepsilon) \in \widetilde{\mathfrak{A}}_0[\tau]^+$. Furthermore, by ($T_4$)

$$b - a_\varepsilon \in \widetilde{\mathfrak{A}}_0[\tau]^+ \cap \mathfrak{A}_0 = \mathfrak{A}_0^+, \quad \forall\, \varepsilon > 0.$$

Hence,

$$\|a_\varepsilon\|_0 \le \|b\|_0, \quad \forall\, \varepsilon > 0,$$

so that if $b = 0$, then $a = 0 \in \mathfrak{A}_0^+$, since $a = \tau\text{-}\lim_{\varepsilon\to 0} a_\varepsilon$. If $b \neq 0$, then $\left\{\frac{1}{\|b\|_0} a_\varepsilon : \varepsilon > 0\right\} \subset \mathcal{U}(\mathfrak{A}_0^+)$ and by (T_4), $\mathcal{U}(\mathfrak{A}_0^+)$ is τ-closed; so again we get that $a \in \mathfrak{A}_0^+$. □

The above lead to the following

Definition 7.1.3 Let $\mathfrak{A}_0[\|\cdot\|_0]$ be a unital C*-algebra and τ a locally convex topology on $\mathfrak{A}_0$ satisfying the conditions (T_1)–(T_4). If $\mathfrak{A}$ is a vector subspace of $\widetilde{\mathfrak{A}}_0[\tau]$, stable under involution and containing $\mathfrak{A}_0$, then we say that the pair $(\mathfrak{A}[\tau], \mathfrak{A}_0)$ is a *locally convex quasi C*-algebra*, or that $\mathfrak{A}[\tau]$ is a locally convex quasi C*-algebra over $\mathfrak{A}_0$.

We present now some examples of locally convex quasi C*-algebras. Before we go on we recall the concept of a GB*-algebra.

Definition 7.1.4 (Allan) Let $\mathfrak{A}[\tau]$ be a locally convex *-algebra with unit e and $\mathfrak{B}_{\mathfrak{A}}^*$ the collection of all subsets B of $\mathfrak{A}$, such that B is absolutely convex, closed and bounded and moreover $e \in B$, $B \subseteq B^2$, and $B = B^* := \{a^* : a \in B\}$.

Then, $\mathfrak{A}[\tau]$ is called a *GB*-algebra* if the following conditions are fulfilled:

1. The collection $\mathfrak{B}_{\mathfrak{A}}^*$ has a greatest member (under the partial ordering of $\mathfrak{B}_{\mathfrak{A}}^*$ by inclusion), denoted by B_0;
2. $\mathfrak{A}[\tau]$ is symmetric, in the sense that every element of the form $e + a^*a$ is invertible for every $a \in \mathfrak{A}$ and its inverse, say b, is a bounded element (i.e., there is $\lambda \in \mathbb{C}\backslash\{0\}$, such that the set $\{(\lambda b)^n, n \in \mathbb{N}\}$ is bounded in $\mathfrak{A}[\tau]$);

3. $\mathfrak{A}[\tau]$ is pseudo-complete, in the sense that, for every $B \in \mathfrak{B}^*_{\mathfrak{A}}$, the *-subalgebra $\mathfrak{A}[B]$ of $\mathfrak{A}$ generated by B is a Banach *-algebra under the norm given by the Minkowski functional of B.

All pro C*-algebras (inverse limits of C*-algebras) and all C*-like locally convex *-algebras [63], like for instance, the Arens algebra $L^{\omega}[0, 1]$ [41] are GB*-algebras.

Example 7.1.5 (GB-Algebras)* Let $\mathfrak{A}[\tau]$ be a GB*-algebra over B_0. Then, $\mathfrak{A}_0[\|\cdot\|_0] \equiv \mathfrak{A}[B_0]$ is a C*-algebra under the C*-norm $\|\cdot\|_0 \equiv \|\cdot\|_{B_0}$ given by the Minkowski functional of B_0. Assume that the locally convex topology τ fulfils the condition (T_3). Then, it is easily checked that $(\mathfrak{A}[\tau], \mathfrak{A}_0)$ is a locally convex quasi C*-algebra over $\mathfrak{A}_0$.

Example 7.1.6 (Banach Quasi C-Algebras)* Let $(\mathfrak{A}, \mathfrak{A}_0)$ be a normal Banach quasi *-algebra (Definition 4.1.8). In this case, $\mathfrak{A}_\flat$ is a Banach *-algebra equipped with the multiplication

$$a \bullet b = \overline{L}_a b, \quad \forall\, a, b \in \mathfrak{A}_\flat$$

and the norm $\|a\|_\flat := \max\left\{\|\overline{L}_a\|, \|\overline{R}_a\|\right\}$, $a \in \mathfrak{A}_\flat$ (see Corollary 4.1.9). Furthermore, we have (see discussion after (4.1.2))

$$\overline{\mathcal{U}(\mathfrak{A}_0[\|\cdot\|_\flat])}^{\|\cdot\|} \subset \mathcal{U}(\mathfrak{A}_\flat). \tag{7.1.7}$$

Indeed, take an arbitrary $x \in \overline{\mathcal{U}(\mathfrak{A}_0[\|\cdot\|_\flat])}^{\|\cdot\|}$. Then, there is a sequence $\{x_n\}$ in $\mathcal{U}(\mathfrak{A}_0[\|\cdot\|_\flat])$, such that $\lim_{n\to\infty} \|x_n - x\| = 0$. On the other hand, using (7.1.5), we have that for each $y \in \mathfrak{A}_0$

$$\|xy\| = \lim_n \|x_n y\| \leq \overline{\lim_{n\to\infty}} \|x_n\|_\flat \|y\| \leq \|y\|$$

and similarly $\|yx\| \leq \|y\|$. Hence, $x \in \mathcal{U}(\mathfrak{A}_\flat)$.

We recall that, if $\mathfrak{A}_0 = \mathfrak{A}_\flat$, then the Banach quasi *-algebra $(\mathfrak{A}[\|\cdot\|], \mathfrak{A}_0)$ is said to be *full*.

Banach quasi C*-algebras are related to proper CQ*-algebras in the following way:

1. If $(\mathfrak{A}[\|\cdot\|], \mathfrak{A}_0)$ is a full proper CQ*-algebra, then $\mathfrak{A}[\|\cdot\|]$ is a Banach quasi C*-algebra over the C*-algebra $\mathfrak{A}_0[\|\cdot\|_\flat = \|\cdot\|_0]$.

   This follows by the very definitions (in this respect, see also Example 7.1.6) and (7.1.4), (7.1.5), (7.1.6).
2. Conversely, suppose that $\mathfrak{A}[\|\cdot\|]$ is a Banach quasi C*-algebra over the C*-algebra $\mathfrak{A}_0[\|\cdot\|_0]$. Then, $(\mathfrak{A}[\|\cdot\|], \mathfrak{A}_0)$ is a proper CQ*-algebra, if and only if, $\|x\|_\flat = \|x\|_0$, for all $x \in \mathfrak{A}_0$.

We consider the following realization of this situation. Let I be a compact interval of $\mathbb{R}$. Then, it is shown that the proper CQ*-algebra $(L^p(I), L^\infty(I))$ is a Banach quasi C*-algebra over $L^\infty(I)$, but the proper CQ*-algebra $(L^p(I), C(I))$ is not a Banach quasi C*-algebra over $C(I)$.

A noncommutative example of a proper CQ*-algebra, which is also a Banach quasi C*-algebra, is provided by the CQ*-algebra $(\mathfrak{C}, (S^{-1})')$ introduced in Example 5.3.3. Making use of the weak operator topology of $\mathcal{B}(\mathcal{H})$, one can prove that T_4 also holds on $(S^{-1})'$. The proof will be given in the next section in a more general context. Then, $\mathfrak{C}$ is a locally convex quasi C*-algebra.

Example 7.1.7 In order to describe certain quantum physical models with an infinite number of degrees of freedom, Lassner introduced in [65, 66] some locally convex topologies in a noncommutative C*-algebra that he called *physical topologies*. In this example we are going to discuss how these topologies can be cast in the framework developed in this chapter.

Let $\mathfrak{A}_0[\|\cdot\|_0]$ be a C*-algebra and $\Sigma = \{\pi_\alpha : \alpha \in I\}$ a system of *-representations of $\mathfrak{A}_0$ on a dense subspace $\mathcal{D}_\alpha$ of a Hilbert space $\mathcal{H}_\alpha$, i.e., each π_α is a *-homomorphism of $\mathfrak{A}_0$ into the O*-algebra $\mathcal{L}^\dagger(\mathcal{D}_\alpha)$ (see Sect. 7.2). Since $\mathfrak{A}_0[\|\cdot\|_0]$ is a C*-algebra, each π_α is a bounded *-representation, i.e., $\overline{\pi_\alpha(x)} \in \mathcal{B}(\mathcal{H}_\alpha)$, for every $x \in \mathfrak{A}_0$. The system Σ is supposed to be *faithful*, in the sense that if $x \in \mathfrak{A}_0$, $x \neq 0$, then there exists $\alpha \in \Sigma$, such that $\pi_\alpha(x) \neq 0$. The physical topology τ_Σ is the coarsest locally convex topology on $\mathfrak{A}_0$, such that every $\pi_\alpha \in \Sigma$ is continuous from $\mathfrak{A}[\tau_\Sigma]$ into $\mathcal{L}^\dagger(\mathcal{D}_\alpha)[\tau_u(\mathcal{L}^\dagger(\mathcal{D}_\alpha))]$, where $\tau_u(\mathcal{L}^\dagger(\mathcal{D}_\alpha))$ is the $\mathcal{L}^\dagger(\mathcal{D}_\alpha)$-uniform topology of $\mathcal{L}^\dagger(\mathcal{D}_\alpha)$ (see Sect. 7.2). This topology depends, of course, on the choice of an appropriate system Σ of *-representations of $\mathfrak{A}_0$; these *-representations are, in general, nothing but the *GNS* representations constructed starting from a family ω_α of states which are *relevant* (and they are usually called in this way) for the physical model under consideration. Every physical topology satisfies the conditions T_1, T_2 and T_4, but it does not necessarily satisfy T_3. Here we show that $\widetilde{\mathfrak{A}_0}[\tau_\Sigma]$ is a locally convex quasi C*-algebra over $\mathfrak{A}_0$ for some special choice of the system Σ of *-representations of $\mathfrak{A}_0$. Suppose that

$$\mathcal{D}_\alpha = \mathcal{D}^\infty(M_\alpha) = \bigcap_{n \in \mathbb{N}} \mathcal{D}(M_\alpha^n),$$

where M_α is a selfadjoint unbounded operator. Without loss of generality we may assume that $M_\alpha \geq \mathbb{I}_\alpha$, with $\mathbb{I}_\alpha$ the identity operator in $\mathcal{B}(\mathcal{H}_\alpha)$. Let Σ be a system of *-representations π_α of $\mathfrak{A}_0$ on $\mathcal{D}_\alpha$, such that $\pi_\alpha(x)M_\alpha\xi = M_\alpha\pi_\alpha(x)\xi$, for every $x \in \mathfrak{A}_0$ and for every $\xi \in \mathcal{D}_\alpha$. Then, $\widetilde{\mathfrak{A}_0}[\tau_\Sigma]$ is a locally convex quasi C*-algebra over $\mathfrak{A}_0$. This follows, on the one hand, from the fact that, in this case, the physical topology τ_Σ is defined by the family of seminorms

$$p_\alpha^f(x) := \left\| f(M_\alpha)\overline{\pi_\alpha(x)} \right\|, \ \forall\, x \in \mathfrak{A}_0,$$

where $\pi_\alpha \in \Sigma$ and f is running in the set $\mathcal{F}$ of all positive, bounded and continuous functions on $\mathbb{R}^+$, such that $\sup_{x\in\mathbb{R}^+} x^k f(x) < \infty$, for every $k = 0, 1, 2, \ldots$ [66, Lemma 2.8] and, on the other hand, from the inequality

$$p_\alpha^f(xy) = \|f(M_\alpha)\overline{\pi_\alpha(x)\pi_\alpha(y)}\| \leq \|\overline{\pi_\alpha(x)}\| p_\alpha^f(y), \ \forall\, x, y \in \mathfrak{A}_0.$$

7.2 Locally Convex Quasi C*-Algebras of Operators

Let $\mathcal{D}$ be a dense subspace in a Hilbert space $\mathcal{H}$ and $\mathcal{L}^\dagger(\mathcal{D}, \mathcal{H})$ the partial *-algebra of operators on $\mathcal{D}$ defined in Sect. 2.1.3.

Let now $\mathfrak{M}_0[\|\cdot\|_0]$ be a unital C*-algebra over $\mathcal{H}$ that leaves $\mathcal{D}$ invariant, i.e., $\mathfrak{M}_0\mathcal{D} \subset \mathcal{D}$. Then, the restriction $\mathfrak{M}_0 \upharpoonright \mathcal{D}$ of $\mathfrak{M}_0$ to $\mathcal{D}$ is an O*-algebra on $\mathcal{D}$, therefore an element T of $\mathfrak{M}_0$ is regarded as an element $T \upharpoonright \mathcal{D}$ of $\mathfrak{M}_0 \upharpoonright \mathcal{D}$. Moreover, let

$$\mathfrak{M}_0 \subset \mathfrak{M} \subset \mathcal{L}^\dagger(\mathcal{D}, \mathcal{H}),$$

where $\mathfrak{M}$ is an O*-vector space on $\mathcal{D}$, that is a *-invariant subspace of $\mathcal{L}^\dagger(\mathcal{D}, \mathcal{H})$. Denote by $\mathcal{B}(\mathfrak{M})$ the set of all bounded subsets of $\mathcal{D}[t_{\mathfrak{M}}]$, where $t_{\mathfrak{M}}$ is the graph topology on $\mathfrak{M}$ (see discussion after Definition 2.2.5). Further, denote by $\mathcal{B}_f(\mathcal{D})$ the set of all finite subsets of $\mathcal{D}$. Then, $\mathcal{B}_f(\mathcal{D}) \subset \mathcal{B}(\mathfrak{M})$. A subset $\mathcal{B}$ of $\mathcal{B}(\mathfrak{M})$ is called *admissible* if the following hold:

(i) $\mathcal{B}_f(\mathcal{D}) \subset \mathcal{B}$,
(ii) $\forall\, B_1, B_2 \in \mathcal{B}\ \exists\, B_3 \in \mathcal{B} : B_1 \cup B_2 \subset B_3$,
(iii) $AB \in \mathcal{B}$, $\forall\, A \in \mathfrak{M}_0$ and $\forall\, B \in \mathcal{B}$.

It is clear that $\mathcal{B}_f(\mathcal{D})$ and $\mathcal{B}(\mathfrak{M})$ are admissible. Consider now an arbitrary admissible subset $\mathcal{B}$ of $\mathcal{B}(\mathfrak{M})$. Then, for any $B \in \mathcal{B}$ define the following seminorms on $\mathfrak{M}$:

$$p_B(T) := \sup_{\xi,\eta\in B} |\langle T\xi|\eta\rangle|, \quad T \in \mathfrak{M} \tag{7.2.8}$$

$$p^B(T) := \sup_{\xi\in B} \|T\xi\|, \quad T \in \mathfrak{M} \tag{7.2.9}$$

$$p_\dagger^B(T) := \sup_{\xi\in B} \left\{\|T\xi\| + \|T^\dagger\xi\|\right\}, \quad T \in \mathfrak{M}. \tag{7.2.10}$$

We call the corresponding locally convex topologies on $\mathfrak{M}$ defined by the families (7.2.8), (7.2.9) and (7.2.10) of seminorms, *$\mathcal{B}$-uniform topology*, *strongly $\mathcal{B}$-uniform topology*, respectively *strongly* $\mathcal{B}$-uniform topology on* $\mathfrak{M}$ and denote them by $\mathrm{t}_u(\mathcal{B})$, $\mathrm{t}^u(\mathcal{B})$, respectively $\mathrm{t}_*^u(\mathcal{B})$. In particular, the $\mathcal{B}(\mathfrak{M})$-uniform topology, the strongly $\mathcal{B}(\mathfrak{M})$-uniform topology, respectively the strongly* $\mathcal{B}(\mathfrak{M})$-uniform topology will be simply called *$\mathfrak{M}$-uniform topology*, strongly *$\mathfrak{M}$-uniform topology*, respectively *strongly* $\mathfrak{M}$-uniform topology* and will be denoted by $\mathrm{t}_u(\mathfrak{M})$, $\mathrm{t}^u(\mathfrak{M})$, respectively $\mathsf{t}^u_*(\mathfrak{M})$. In the book of Schmüdgen [23], these topologies are called *bounded topologies* and $\mathsf{t}_u(\mathcal{B})$, $\mathsf{t}^u(\mathcal{B})$ are denoted by $\mathsf{t}_{\mathcal{B}}$, $\mathsf{t}^{\mathcal{B}}$, while $\mathsf{t}_u(\mathfrak{M})$, $\mathsf{t}^u(\mathfrak{M})$ are denoted by $\mathsf{t}_{\mathcal{D}}$, $\mathsf{t}^{\mathcal{D}}$, respectively. The $\mathcal{B}_f(\mathcal{D})$-uniform topology, strongly $\mathcal{B}_f(\mathcal{D})$-uniform topology, resp. strongly* $\mathcal{B}_f(\mathcal{D})$-uniform topology is called *weak topology*, *strong topology*, respectively *strong *-topology on* $\mathfrak{M}$, denoted by t_w, t_s and t_{s^*}, respectively, as we did for the space $\mathcal{L}^\dagger(\mathcal{D}, \mathcal{H})$ in Sect. 2.1.3. All these topologies are related in the following way:

$$\begin{array}{ccccc} \mathsf{t}_w & \preceq & \mathsf{t}_u(\mathcal{B}) & \preceq & \mathsf{t}_u(\mathfrak{M}) \\ \curlywedge\!\mid & & \curlywedge\!\mid & & \curlywedge\!\mid \\ \mathsf{t}_s & \preceq & \mathsf{t}^u(\mathcal{B}) & \preceq & \mathsf{t}^u(\mathfrak{M}) \\ \curlywedge\!\mid & & \curlywedge\!\mid & & \curlywedge\!\mid \\ \mathsf{t}_{s^*} & \preceq & \mathsf{t}^u_*(\mathcal{B}) & \preceq & \mathsf{t}^u_*(\mathfrak{M}). \end{array} \qquad (7.2.11)$$

We investigate now whether $\widetilde{\mathfrak{M}}_0[\mathsf{t}_u(\mathcal{B})]$ and $\widetilde{\mathfrak{M}}_0[\mathsf{t}^u_*(\mathcal{B})]$ are locally convex quasi C*-algebras over $\mathfrak{M}_0$. We must check the properties (T_1)–(T_4) (stated before and after Definition 7.1.1) for the locally convex topologies $\mathsf{t}_u(\mathcal{B})$, $\mathsf{t}^u_*(\mathcal{B})$ and the operator C*-norm $\|\cdot\|_0$ on $\mathfrak{M}_0$.

(T_1) follows easily for both topologies, since $\mathcal{B}$ is admissible and $\mathfrak{M}_0\mathcal{D} \subset \mathcal{D}$.

For (T_2) notice that for all $T \in \mathfrak{M}_0$ and $B \subset \mathcal{B}$ we have

$$p^B_\dagger(T) = \sup_{\xi\in B}\{\|T\xi\| + \|T^\dagger\xi\|\} \le \big(2\sup_{\xi\in B}\|\xi\|\big)\|T\|_0,$$

so by (7.2.11) we conclude that $\mathsf{t}_u(\mathcal{B}) \preceq \mathsf{t}^u_*(\mathcal{B}) \preceq \|\cdot\|_0$.

Concerning $\mathsf{t}^*_u(\mathcal{B})$, the property (T_3) follows easily from the very definitions. Now, notice the following: for any $T, S \in \mathfrak{M}_0$ with $TS = ST$ and $S^* = S$, one concludes that

$$p_B(TS) \le \|T\|_0 \sup_{\xi\in B}\langle |S|\xi|\xi\rangle, \quad \forall\, B \in \mathcal{B}, \qquad (7.2.12)$$

where $|S| := (S^2)^{1/2}$. Then, it follows that for any $T, S \in \mathfrak{M}_0$ with $TS = ST$ and $S \ge 0$, one has

$$p_B(TS) \le \|T\|_0 \sup_{\xi\in B}\langle S\xi|\xi\rangle, \quad \forall\, B \in \mathcal{B}.$$

We prove (7.2.12). From the polar decomposition of S, there is a unique partial isometry V from $\mathcal{H}$ to $\mathcal{H}$, such that

$$S = V|S| = |S|V, \quad \mathrm{Ker}(V) = \mathrm{Ker}(S) \text{ and } VS = |S|.$$

By continuous functional calculus it follows that: T commutes with both $|S|$ and $|S|^{1/2}$, but also $V|S|^{1/2} = |S|^{1/2}V$. Thus,

$$\begin{aligned}
p_B(TS) &= \sup_{\xi,\eta\in B} \left|\langle TS\xi|\eta\rangle\right| = \sup_{\xi,\eta\in B} \left|\langle V|S|T\xi|\eta\rangle\right| \\
&= \sup_{\xi,\eta\in B} \left|\langle T|S|^{1/2}\xi\,\big|\,|S|^{1/2}V\eta\rangle\right| \le \sup_{\xi,\eta\in B} \|T\|_0 \big\||S|^{1/2}\xi\big\|\big\||S|^{1/2}\eta\big\| \\
&\le \frac{1}{2}\|T\|_0 \sup_{\xi,\eta\in B} \left\{\big\||S|^{1/2}\xi\big\|^2 + \big\||S|^{1/2}\eta\big\|^2\right\} \\
&\le \|T\|_0 \sup_{\xi\in B} \langle |S|\xi|\xi\rangle, \quad \forall\, B \in \mathcal{B}.
\end{aligned}$$

But, we can not say whether (T_3) holds for $\mathsf{t}_u(\mathcal{B})$. In the case when $\mathfrak{M}_0[\|\cdot\|_0]$ is a von Neumann algebra we have the following:

- If $\mathfrak{M}_0$ is commutative, then (T_3) holds for the topology t_w.
- If $\mathfrak{M}$ is a commutative O*-algebra on $\mathcal{D}$ in $\mathcal{H}$, containing $\mathfrak{M}_0$, then (T_3) holds for the topology $\mathsf{t}_u(\mathfrak{M})$.

Indeed, suppose that $\mathfrak{M}$ is commutative with $\mathfrak{M}_0 \subset \mathfrak{M}$. For each $B \in \mathcal{B}(\mathfrak{M})$ consider the set

$$B' := \bigcup \left\{ VB : V \text{ partial isometry in } \mathfrak{M}_0 \right\}.$$

Commutativity of $\mathfrak{M}$ implies that $B' \in \mathcal{B}(\mathfrak{M})$. Moreover, $B \subset B'$. Let now $T, S \in \mathfrak{M}_0$. Let $S = V|S|$ be the polar decomposition of S. Since $\mathfrak{M}_0$ is a von Neumann algebra, we have $V \in \mathfrak{M}_0$, which implies that

$$\begin{aligned}
p_B(TS) &= \sup_{\xi,\eta\in B} \left|\langle TS\xi|\eta\rangle\right| = \sup_{\xi,\eta\in B} \left|\langle VT|S|^{1/2}\xi\,\big|\,|S|^{1/2}\eta\rangle\right| \\
&\le \|VT\|_0 \sup_{\xi,\eta\in B} \big\||S|^{1/2}\xi\big\|\,\big\||S|^{1/2}\eta\big\| \\
&= \|T\|_0 \sup_{\xi\in B} \langle |S|\xi|\xi\rangle \\
&= \|T\|_0 \sup_{\xi\in B} \langle S\xi|V^*\xi\rangle \\
&\le \|T\|_0 \sup_{\xi,\eta\in B'} \left|\langle S\xi|\eta\rangle\right| = \|T\|_0\, p_{B'}(S).
\end{aligned}$$

Hence, (T_3) holds for $\mathsf{t}_u(\mathfrak{M})$.

(T_4) This property holds for all topologies in (7.2.11). It suffices to prove (T_4) for the topology t_w. So, let $T \in \overline{\mathcal{U}(\mathfrak{M}_0)}^{\mathsf{t}_w}$ be arbitrary. Then, there is a net $\{T_\alpha\}$ in $\mathcal{U}(\mathfrak{M}_0)$ with $T_\alpha \underset{\mathsf{t}_w}{\longrightarrow} T$. Notice that the sesquilinear form defined on $\mathcal{D} \times \mathcal{D}$ by

$$\mathcal{D} \times \mathcal{D} \ni (\xi, \eta) \mapsto \lim_\alpha \langle T_\alpha\xi|\eta\rangle \in \mathbb{C},$$

is bounded. Hence, T can be regarded as a bounded linear operator on $\mathcal{H}$, such that

$$\|T\|_0 = 1 \text{ and } \langle T\xi|\eta\rangle = \lim_\alpha \langle T_\alpha \xi|\eta\rangle, \quad \forall\, \xi, \eta \in \mathcal{D}.$$

Since $\mathcal{D}$ is dense in $\mathcal{H}$, an easy computation shows that

$$\langle Tx|y\rangle = \lim_\alpha \langle T_\alpha x|y\rangle, \quad \forall\, x, y \in \mathcal{H}. \tag{7.2.13}$$

This proves that $T \in \mathfrak{M}_0 \cap \mathcal{U}(\mathcal{B}(\mathcal{H})) = \mathcal{U}(\mathfrak{M}_0)$, which means that $\mathcal{U}(\mathfrak{M}_0)$ is t_w-closed. A consequence of (7.2.13) is now that $\mathcal{U}(\mathfrak{M}_0^+)$ is weakly closed, so that (T_4) holds for the topology t_w on $\mathfrak{M}_0$. From (7.2.11), (T_4) also holds for the topologies $\mathsf{t}_u(\mathcal{B})$ and $\mathsf{t}_*^u(\mathcal{B})$.

From the preceding discussion we conclude the following

Proposition 7.2.1 *Let $\mathcal{B}$ be an admissible subset of $\mathcal{B}(\mathfrak{M})$. Then, both $(\widetilde{\mathfrak{M}_0}[\mathsf{t}_*^u(\mathcal{B})], \mathfrak{M}_0)$ and $(\widetilde{\mathfrak{M}_0}[\mathsf{t}_{s^*}], \mathfrak{M}_0)$ are locally convex quasi C*-algebras. If $\mathfrak{M}_0$ is a von Neumann algebra and there is a commutative O*-algebra $\mathfrak{M}$ on $\mathcal{D}$ in $\mathcal{H}$, containing $\mathfrak{M}_0$, then $(\widetilde{\mathfrak{M}_0}[\mathsf{t}_w], \mathfrak{M}_0)$ and $(\widetilde{\mathfrak{M}_0}[\mathsf{t}_u(\mathfrak{M})], \mathfrak{M}_0)$ are commutative locally convex quasi C*-algebras.*

Remark 7.2.2

1. In general, we do not know whether $\widetilde{\mathfrak{M}_0}[\mathsf{t}_u(\mathcal{B})]$ and $\widetilde{\mathfrak{M}_0}[\mathsf{t}_w]$ are locally convex quasi C*-algebras.
2. The locally convex quasi C*-algebra $\widetilde{\mathfrak{M}_0}[\mathsf{t}_{s^*}]$ over $\mathfrak{M}_0$, equals to the completion $\widetilde{\mathfrak{M}_0''}[\mathsf{t}_{s^*}]$ of the von Neumann algebra $\mathfrak{M}_0''$ with respect to the topology t_{s^*}, but $\widetilde{\mathfrak{M}_0''}[\mathsf{t}_{s^*}]$ is not necessarily a locally convex quasi C*-algebra over $\mathfrak{M}_0''$, since in general, $\mathfrak{M}_0''\mathcal{D} \not\subset \mathcal{D}$. In the case when $\mathfrak{M}_0''\mathcal{D} \subset \mathcal{D}$, one has the equality

$$\widetilde{\mathfrak{M}_0''}[\mathsf{t}_{s^*}] = \widetilde{\mathfrak{M}_0}[\mathsf{t}_{s^*}],$$

set-theoretically; but, the corresponding locally convex quasi C*-algebras over $\mathfrak{M}_0$ do not coincide. In particular, one has that

$$\widetilde{\mathfrak{M}_0}[\mathsf{t}_{s^*}]_c^+ \subsetneq \widetilde{\mathfrak{M}_0''}[\mathsf{t}_{s^*}]_c^+.$$

We present now some properties of the locally convex quasi C*-algebra $(\widetilde{\mathfrak{M}_0}[\mathsf{t}_{s^*}], \mathfrak{M}_0)$.

Proposition 7.2.3 *Let $A \in \widetilde{\mathfrak{M}_0}[\mathsf{t}_{s^*}]^+$. Consider the following statements:*

(i) *$A \in \widetilde{\mathfrak{M}_0}[\mathsf{t}_{s^*}]_c^+$;*
(ii) *$(\mathbb{I} + A)^{-1}$ exists and belongs to $\mathcal{U}(\mathfrak{M}_0^+)$;*
(iii) *the closure $\overline{A}$ of A is a positive selfadjoint operator.*

Then, one has that (i) $\Rightarrow$ (ii) $\Rightarrow$ (iii).

Proof (i) $\Rightarrow$ (ii) follows from Proposition 7.1.2(1).

(ii) $\Rightarrow$ (iii) Since $(\mathbb{I}+A)^{-1}$ is a bounded selfadjoint operator and $(\mathbb{I}+A)^{-1}\mathcal{D} \subset \mathcal{D}$, it follows that

$$\big\langle(\mathbb{I}+A)^{-1}(\mathbb{I}+A^*)\xi\,|\,\eta\big\rangle = \big\langle\mathbb{I}+A^*)\xi\,|\,(\mathbb{I}+A)^{-1}\eta\big\rangle = \langle\xi\,|\,\eta\rangle,$$

for all $\xi \in \mathcal{D}(A^*)$ and $\eta \in \mathcal{D}$, which implies

$$\begin{aligned}\big\langle A^*\xi\,|\,\zeta\big\rangle &= \big\langle(\mathbb{I}+A^*)\xi\,|\,(\mathbb{I}+A)^{-1}(\mathbb{I}+A^*)\zeta\big\rangle - \langle\xi\,|\,\zeta\rangle\\ &= \big\langle\xi\,|\,(\mathbb{I}+A^*)\zeta\big\rangle - \langle\xi\,|\,\zeta\rangle = \big\langle\xi\,|\,A^*\zeta\big\rangle, \quad \forall\, \xi,\zeta \in \mathcal{D}(A^*).\end{aligned}$$

Hence, $\xi \in \mathcal{D}(\overline{A})$ and $\overline{A}\xi = A^*\xi$. It is now easily seen that $\overline{A}$ is a positive selfadjoint operator. □

Corollary 7.2.4 *Suppose that* $A \in \widetilde{\mathfrak{M}_0''}[\mathrm{t}_{s^*}]$ *and* $\mathfrak{M}_0''\mathcal{D} \subset \mathcal{D}$. *Then, the following statements are equivalent:*

(i) $A \in \widetilde{\mathfrak{M}_0''}[\mathrm{t}_{s^*}]_c^+$;
(ii) $(\mathbb{I}+A)^{-1} \in \mathcal{U}\big((\mathfrak{M}_0'')^+\big)$;
(iii) $\overline{A}$ *is a positive selfadjoint operator.*

Proof From Proposition 7.2.3 we have that (i) $\Rightarrow$ (ii) $\Rightarrow$ (iii).

(iii) $\Rightarrow$ (i) follows easily by considering the spectral decomposition of $\overline{A}$. □

It is natural now to ask whether there exists an extended C*-algebra (abbreviated to EC*-algebra; see right after for the definition) $\mathfrak{M}$ on $\mathcal{D}$, such that

$$\mathfrak{M}_0 \subset \mathfrak{M} \subset \widetilde{\mathfrak{M}_0}[\mathrm{t}_{s^*}].$$

If $\mathfrak{M}$ is a closed O*-algebra on $\mathcal{D}$ in $\mathcal{H}$, let $\mathfrak{M}_b := \{T \in \mathfrak{M} : \overline{T} \in \mathcal{B}(\mathcal{H})\}$ be the bounded part of $\mathfrak{M}$, where $\mathcal{B}(\mathcal{H})$ is the C*-algebra of all bounded linear operators on $\mathcal{H}$. Then, when $\overline{\mathfrak{M}_b} \equiv \{\overline{T} : T \in \mathfrak{M}_b\}$ is a C*-algebra on $\mathcal{H}$ and $(\mathbb{I}+T^*T)^{-1} \in \mathfrak{M}_b$, for each $T \in \mathfrak{M}$, $\mathfrak{M}$ is said to be an *EC*-algebra* on $\mathcal{D}$.

For more details on EC*-algebras, see [10, Section 4.3]. The next Proposition 7.2.5 gives a characterization of certain EC*-algebras on $\mathcal{D}$, through the set of commutatively positive elements of $(\widetilde{\mathfrak{M}_0}[\mathrm{t}_{s^*}], \mathfrak{M}_0)$.

Proposition 7.2.5 *Let* $\mathfrak{M}$ *be a closed O*-algebra on* $\mathcal{D}$, *such that* $\mathfrak{M}_0 \subset \mathfrak{M} \subset \widetilde{\mathfrak{M}_0}[\mathrm{t}_{s^*}]$ *and* $\mathfrak{M}_b = \mathfrak{M}_0$. *Then,* $\mathfrak{M}$ *is an EC*-algebra on* $\mathcal{D}$, *if and only if,* $\mathfrak{M}^+ \subset \widetilde{\mathfrak{M}_0}[\mathrm{t}_{s^*}]_c^+$.

Proof Suppose that $\mathfrak{M}$ is an EC*-algebra on $\mathcal{D}$ and let $A \in \mathfrak{M}^+$ be arbitrary. Then, since $\mathfrak{M}_b = \mathfrak{M}_0$, $\overline{A}$ is a bounded positive selfdjoint operator with $(\mathbb{I}+\overline{A})^{-1} \in \mathcal{U}(\mathfrak{M}_0^+)$. But, $(\widetilde{\mathfrak{M}_0}[\mathrm{t}_{s^*}], \mathfrak{M}_0)$ is a locally convex quasi C*-algebra (Proposition 7.2.1), therefore $\mathcal{U}(\mathfrak{M}_0^+)$ is t_{s^*}-closed. Note that for each $n \in \mathbb{N}$, the elements $A_n := \overline{A}(\mathbb{I}+\frac{1}{n}\overline{A})^{-1}$ belonging to $\mathfrak{M}_0^+$, are commuting and $A_n \underset{\mathrm{t}_{s^*}}{\longrightarrow} A$; so Definition 7.1.1 implies that $A \in \widetilde{\mathfrak{M}_0}[\mathrm{t}_{s^*}]_c^+$.

Conversely, suppose that $\mathfrak{M}^+ \subset \widetilde{\mathfrak{M}_0}[t_{s^*}]_c^+$. So, $A \in \mathfrak{M}$ implies $A^\dagger A \in \widetilde{\mathfrak{M}_0}[t_{s^*}]_c^+$, therefore $(\mathbb{I} + A^\dagger A)^{-1} \in \mathcal{U}(\mathfrak{M}_0^+)$ from Proposition 7.1.2(1). Now, since $\mathfrak{M}_b = \mathfrak{M}_0$ we finally get that $\mathfrak{M}$ is an EC*-algebra on $\mathcal{D}$. □

7.3 Structure of Commutative Locally Convex Quasi C*-Algebras

Throughout this section $(\mathfrak{A}[\tau], \mathfrak{A}_0)$ is a commutative locally convex quasi C*-algebra, where the C*-algebra $\mathfrak{A}_0$ is supposed to have a unit element. If the multiplication of $\mathfrak{A}_0$ with respect to the topology τ is jointly continuous, then $\mathfrak{A}[\tau]$ is a commutative GB*-algebra [59, Theorem 2.1], and so $\mathfrak{A}[\tau]$ is isomorphic to a *-algebra of $\mathbb{C}^*$-valued continuous functions on a compact space, which take the value ∞ on at most a nowhere dense subset [33, Theorem 3.9]; $\mathbb{C}^*$ denotes the extended complex plane in its usual topology as the one-point compactification of $\mathbb{C}$. The purpose of this section is to consider a generalization of the above result in the case when the multiplication of $\mathfrak{A}_0[\tau]$ is not jointly continuous. As a^*a is not necessarily defined for $a \in \mathfrak{A}[\tau]$, it is impossible to extend any nonzero multiplicative linear functional φ on $\mathfrak{A}_0$ to $\mathfrak{A}[\tau]$, like in the case of [32, Proposition 6.8]. Here we show that φ is extendable to a $\mathbb{C}^*$-valued partial multiplicative linear functional φ' on $\mathfrak{A}^+ = \overline{\mathfrak{A}_0^+}^{\tau}$ (see beginning of Sect. 6.2 and discussion before Proposition 7.1.2) and that $\mathfrak{A}^+$ is isomorphic to a wedge of $\mathbb{C}^*$-valued positive functions on a compact space, which take the value ∞ on at most a nowhere dense subset. This result will be applied in Sect. 7.4 for studying a functional calculus for positive elements. Consider now the set

$$\mathfrak{M}(\mathfrak{A}_0, \mathfrak{A}^+) := \{ax + y : a \in \mathfrak{A}^+,\ x, y \in \mathfrak{A}_0\}$$

and denote by $\mathfrak{M}(\mathfrak{A}_0)$ the *Gelfand space* of $\mathfrak{A}_0$ (see also beginning of Sect. A.6.1), i.e., the set of all nonzero multiplicative linear functionals on $\mathfrak{A}_0$, endowed with the weak*-topology $\sigma(\mathfrak{M}(\mathfrak{A}_0), \mathfrak{A}_0)$. Now, let $a \in \mathfrak{A}^+$ and $x, y \in \mathfrak{A}_0$. Suppose x is *hermitian* (i.e., $x^* = x$). Then, by continuous functional calculus, x is uniquely decomposed in the following way:

$$x = x_+ - x_-,\quad x_+, x_- \in \mathfrak{A}_0^+,\quad x_+x_- = 0$$

$$|x| \equiv (x^*x)^{1/2} = x_+ + x_- \in \mathfrak{A}_0^+.$$

Hence, $a|x|$, ax_+, $ax_- \in \mathfrak{A}^+$, and by (1) and (2) of Proposition 7.1.2, $(e+a|x|)^{-1}$, $a|x|(e + a|x|)^{-1} \in \mathfrak{A}_0^+$. Furthermore, since

$$a|x|(e + a|x|)^{-1} - ax_+(e + a|x|)^{-1} = ax_-(e + a|x|)^{-1} \in \mathfrak{A}^+,$$

Proposition 7.1.2(4) implies that $ax_+(e + a|x|)^{-1} \in \mathfrak{A}_0^+$. Similarly, $ax_-(e + a|x|)^{-1} \in \mathfrak{A}_0^+$. Hence, we have that the element

$$(ax + y)(e + a|x|)^{-1} = ax_+(e + a|x|)^{-1} - ax_-(e + a|x|)^{-1} + y(e + a|x|)^{-1}$$

belongs to $\mathfrak{A}_0$. Since a general element x of $\mathfrak{A}_0$ is a linear combination of two hermitian elements of $\mathfrak{A}_0$, we finally obtain that

$$(ax + y)(e + a|x|)^{-1} \in \mathfrak{A}_0, \quad \forall\, a \in \mathfrak{A}^+ \text{ and } x, y \in \mathfrak{A}_0.$$

Indeed, let x be arbitrary in $\mathfrak{A}_0$. Then, $x = x_1 + ix_2$, with x_1 and x_2 hermitian. An easy computation shows that

$$|x| \leq |x_1| + |x_2|,\ |x_j| \leq |x| \text{ and } (e + a|x_j|)(e + a|x|)^{-1} \leq e,\ j = 1, 2.$$

The latter together with Proposition 7.1.2(4) gives $(e + a|x_j|)(e + a|x|)^{-1} \in \mathfrak{A}_0^+$; moreover, from the above $(ax_j + y)(e + a|x_j|)^{-1} \in \mathfrak{A}_0$. Thus, for $j = 1, 2$, we obtain

$$(ax_j + y)(e + a|x|)^{-1} = \big((ax_j + y)(e + a|x_j|)^{-1}\big)\big((e + a|x_j|)(e + a|x|)^{-1}\big) \in \mathfrak{A}_0,$$

which implies

$$(ax + y)(e + a|x|)^{-1} = (ax_1 + y)(e + a|x|)^{-1} + iax_2(e + a|x|)^{-1} \in \mathfrak{A}_0.$$

Hence, the elements $\varphi((e + a|x|)^{-1})$ and $\varphi\big((ax + y)(e + a|x|)^{-1}\big)$ are complex numbers, for each $\varphi \in \mathfrak{M}(\mathfrak{A}_0)$, so that we can consider the correspondence

$$\varphi' : \mathfrak{M}(\mathfrak{A}_0, \mathfrak{A}^+) \longrightarrow \mathbb{C}^* \equiv \mathbb{C} \cup \{\infty\}, \text{ with}$$

$$ax + y \mapsto \varphi'(ax + y) = \begin{cases} \dfrac{\varphi\big((ax+y)(e+a|x|)^{-1}\big)}{\varphi\big((e+a|x|)^{-1}\big)}, & \text{if } \varphi\big((e + a|x|)^{-1}\big) \neq 0 \\ \infty, & \text{if } \varphi\big((e + a|x|)^{-1}\big) = 0. \end{cases}$$

Then, we have

Lemma 7.3.1 *The following statements hold:*

1. *For every $\varphi \in \mathfrak{M}(\mathfrak{A}_0)$ the correspondence φ', given above, is well-defined.*
2. *Let $a \in \mathfrak{A}^+$ and $x \in \mathfrak{A}_0$. Then, $(e + a)^{-1}$ exists in $\mathfrak{A}_0$* (from Proposition 7.1.2(1)) *and we have:*

   (i) *$\varphi\big((e + a|x|)^{-1}\big) = 0$ implies $\varphi\big((e + a)^{-1}\big) = 0$, $\varphi \in \mathfrak{M}(\mathfrak{A}_0)$;*

   (ii) *$\varphi\big((e + a)^{-1}\big) = 0$ and $\varphi(x) \neq 0$ imply $\varphi\big((e + a|x|)^{-1}\big) = 0$, $\varphi \in \mathfrak{M}(\mathfrak{A}_0)$.*

Proof

1. Let $a, b \in \mathfrak{A}^+$ and $x, y, z, w \in \mathfrak{A}_0$, such that $ax + y = bz + w$. Then, for every $\varphi \in \mathfrak{M}(\mathfrak{A}_0)$ one has that

$$\varphi\big((e + a|x|)^{-1}\big) = 0 \Leftrightarrow \varphi\big((e + b|z|)^{-1}\big) = 0. \tag{7.3.14}$$

Indeed, we first show (7.3.14) in case x and z are hermitian. Since $ax + y = bz + w$, we have

$$(e + a|x|) - 2ax_- + y = (e + b|z|) - 2bz_- + w.$$

We multiply the last equality by $(e + a|x|)^{-1}(e + b|z|)^{-1}$ and obtain

$$\begin{aligned}&(e + b|z|)^{-1} - 2ax_-(e + a|x|)^{-1}(e + b|z|)^{-1} + y(e + a|x|)^{-1}(e + b|z|)^{-1}\\ &= (e + a|x|)^{-1} - 2bz_-(e + b|z|)^{-1}(e + a|x|)^{-1}\\ &+ w(e + a|x|)^{-1}(e + b|z|)^{-1}.\end{aligned}$$

This implies that for every $\varphi \in \mathfrak{M}(\mathfrak{A}_0)$

$$\varphi\big((e + a|x|)^{-1}\big) = 0 \Leftrightarrow \varphi\big((e + b|z|)^{-1}\big) = 0. \tag{7.3.15}$$

We next prove (7.3.14) in the case when x and z are arbitrary elements of $\mathfrak{A}_0$. Then, the elements x, y, z and w are decomposed into

$$x = x_1 + ix_2, \quad y = y_1 + iy_2, \quad z = z_1 + iz_2, \quad w = w_1 + iw_2,$$

where x_j, y_j, z_j, w_j $(j = 1, 2)$ are hermitian elements in $\mathfrak{A}_0$ that satisfy the equations:

$$ax_1 + y_1 = bz_1 + w_1, \quad ax_2 + y_2 = bz_2 + w_2. \tag{7.3.16}$$

We show now that

$$\varphi\big((e + a|x|)^{-1}\big) = 0 \Leftrightarrow \text{ either } \varphi\big((e + a|x_1|)^{-1}\big) = 0 \text{ or } \varphi\big((e + a|x_2|)^{-1}\big) = 0. \tag{7.3.17}$$

Suppose that $\varphi\big((e + a|x_1|)^{-1}\big) \neq 0$ and $\varphi\big((e + a|x_2|)^{-1}\big) \neq 0$. Then, inserting the quantity $\big(e+a(|x_1|+|x_2|)\big)^{-1}\big)(e+a|x_1|+a|x_2|)$ in front of the second term of the left hand-side equality in (7.3.18) and doing calculations by repeating this pattern, you obtain the right hand-side of the equality that follows:

$$\begin{aligned}&\big(e + a(|x_1| + |x_2|)\big)^{-1} - (e + a|x_1|)^{-1}(e + a|x_2|)^{-1}\\ &= \big(e + a(|x_1| + |x_2|)\big)^{-1}\big(a|x_1|(e + a|x_1|)^{-1}\big)\big(a|x_2|(e + a|x_2|)^{-1}\big) \in \mathfrak{A}_0^+,\end{aligned} \tag{7.3.18}$$

whence

$$\varphi\big((e+a(|x_1|+|x_2|))^{-1}\big) \geq \varphi\big((e+a|x_1|)^{-1}(e+a|x_2|)^{-1}\big) > 0.$$

Furthermore, since $|x| \leq |x_1| + |x_2|$, we have

$$0 < \varphi\big((e+a(|x_1|+|x_2|))^{-1}\big) \leq \varphi\big((e+a|x|)^{-1}\big).$$

Therefore, $\varphi\big((e+a|x|)^{-1}\big) \neq 0$. Conversely, suppose that $\varphi\big((e+a|x_1|)^{-1}\big) = 0$ or $\varphi\big((e+a|x_2|)^{-1}\big) = 0$. Then, since $(e+a|x_j|)^{-1} \geq (e+a|x|)^{-1}$, $j = 1, 2$, we have that $\varphi\big((e+a|x|)^{-1}\big) = 0$.

Now from (7.3.15), (7.3.16) and (7.3.17) we get

$$\begin{aligned}\varphi\big((e+a|x|)^{-1}\big) = 0 &\Leftrightarrow \varphi\big((e+a|x_1|)^{-1}\big) = 0 \text{ or } \varphi\big((e+a|x_2|)^{-1}\big) = 0\\ &\Leftrightarrow \varphi\big((e+b|z_1|)^{-1}\big) = 0 \text{ or } \varphi\big((e+b|z_2|)^{-1}\big) = 0\\ &\Leftrightarrow \varphi\big((e+b|z|)^{-1}\big) = 0.\end{aligned}$$

Thus, (7.3.14) has been shown. Furthermore, by assumption $ax + y = bz + w$, consequently

$$\varphi'(ax+y) = \infty \Leftrightarrow \varphi'(bz+w) = \infty.$$

On the other hand, from (7.3.14) it follows that

$$\varphi'(ax+y) < \infty \Leftrightarrow \varphi'(bz+w) < \infty.$$

In this case,

$$\varphi'(ax+y) = \frac{\varphi\big((ax+y)(e+a|x|)^{-1}(e+b|z|)^{-1}\big)}{\varphi\big((e+a|x|)^{-1}\big)\varphi\big((e+b|z|)^{-1}\big)} = \varphi'(bz+w)$$

and this completes the proof of (1).

2. (i) Suppose $\varphi\big((e+a|x|)^{-1}\big) = 0$, $\varphi \in \mathfrak{M}(\mathfrak{A}_0)$, Then,

$$\begin{aligned}(e+a)^{-1} &= (e+a|x|)^{-1}(e+a|x|)(e+a)^{-1}\\ &= (e+a|x|)^{-1}\big((e+a)^{-1} + a|x|(e+a)^{-1}\big)\\ &= (e+a|x|)^{-1}\big((e+a)^{-1} + |x| - |x|(e+a)^{-1}\big)\\ &= (e+a|x|)^{-1}\big((e-|x|)(e+a)^{-1} + |x|\big),\end{aligned}$$

where $(e-|x|)(e+a)^{-1}+|x| \in \mathfrak{A}_0$. So applying φ we have $\varphi((e+a)^{-1}) = 0$.

(ii) Suppose that $\varphi\big((e+a)^{-1}\big) = 0$ and $\varphi(x) \neq 0$, $\varphi \in \mathfrak{M}(\mathfrak{A}_0)$. Then, we apply φ to the final result of the preceding calculation in (i) and we take

$$\varphi\big((e+a|x|)^{-1}\big)\varphi(|x|) = 0.$$

Since $\varphi(x) \neq 0$, if and only if, $\varphi(|x|) \neq 0$, clearly we have $\varphi\big((e+a|x|)^{-1}\big) = 0$. □

Proposition 7.3.2 *For $\varphi \in \mathfrak{M}(\mathfrak{A}_0)$, the well defined map φ' has the following properties:*

(1) *$\varphi' \supset \varphi$ (i.e., φ' is an extension of φ);*
(2) *$\varphi'(ax+y) = \varphi'(a)\varphi(x) + \varphi(y)$ and $\varphi'(ax) = \varphi'(a)\varphi(x)$, whenever $a \in \mathfrak{A}^+$ and $x, y \in \mathfrak{A}_0$, such that $\varphi'(a)\varphi(x) \neq \infty \cdot 0$;*
(3) *$\varphi'(a+b) = \varphi'(a) + \varphi'(b)$, for all $a, b \in \mathfrak{A}^+$;*
(4) *$\varphi'(\lambda a) = \lambda\varphi'(a)$, for all $\lambda \in \mathbb{C}$ and $a \in \mathfrak{A}^+$, where $0 \cdot \infty = 0$.*

Proof

1. is trivial.
2. Suppose that $\varphi'(a)\varphi(x) \neq \infty \cdot 0$, $\varphi \in \mathfrak{M}(\mathfrak{A}_0)$. Then, from the definition of φ' and Lemma 7.3.1(2), we have the following implications (considering separately the cases where $\varphi'(a)$ is infinite or not):

•
$$\begin{array}{ccc} \varphi'(ax+y) = \infty & \Leftrightarrow & \varphi'(a) = \infty \\ \Updownarrow & & \Updownarrow \\ \varphi\big((e+a|x|)^{-1}\big) = 0 & & \varphi\big((e+a)^{-1}\big) = 0 \\ & & \Downarrow \\ & & \varphi'(a)\varphi(x) + \varphi(y) = \infty. \end{array}$$

•
$$\begin{array}{ccc} \varphi'(ax+y) < \infty & \Leftrightarrow & \varphi'(a) < \infty \\ \Updownarrow & & \Updownarrow \\ \varphi\big((e+a|x|)^{-1}\big) \neq 0 & & \varphi\big((e+a)^{-1}\big) \neq 0. \end{array}$$

So, in this case we also obtain

$$\begin{aligned} \varphi'(ax+y) &= \frac{\varphi\big(ax(e+a|x|)^{-1}\big)}{\varphi\big((e+a|x|)^{-1}\big)} + \varphi(y) \\ &= \frac{\varphi\big(a(e+a)^{-1}\big)\varphi(x)}{\varphi\big((e+a)^{-1}\big)} + \varphi(y) = \varphi'(a)\varphi(x) + \varphi(y), \end{aligned}$$

and this completes the proof of (2).

(3) Observe that for any $a, b \in \mathfrak{A}^+$, one has

$$\begin{aligned} (e+a)^{-1}(e+b)^{-1} = (e+a+b)^{-1}\big(&(e+a)^{-1}(e+b)^{-1} \\ &+ a(e+a)^{-1}(e+b)^{-1} + (e+a)^{-1}b(e+b)^{-1}\big), \end{aligned}$$

where $(e+a)^{-1}(e+b)^{-1}+a(e+a)^{-1}(e+b)^{-1}+(e+a)^{-1}b(e+b)^{-1} \in \mathfrak{A}_0$ (see Proposition 7.1.2). Thus, applying any $\varphi \in \mathfrak{M}(\mathfrak{A}_0)$ to the last equality we conclude that

$$\varphi\big((e+a+b)^{-1}\big)=0 \text{ implies either } \varphi\big((e+a)^{-1}\big)=0 \text{ or } \varphi((e+b)^{-1})=0. \tag{7.3.19}$$

Conversely, observe that

$$(e+a)^{-1}=(e+a+b)^{-1}+b(e+a+b)^{-1}(e+a)^{-1},$$

where $b(e+a+b)^{-1} \in \mathfrak{A}_0^+$ by Proposition 7.1.2(4), since

$$(a+b)(e+a+b)^{-1}-a(e+a+b)^{-1}=b(e+a+b)^{-1} \in \mathfrak{A}^+ \text{ with } (a+b)(e+a+b)^{-1} \in \mathfrak{A}_0^+.$$

So, taking also into account an analogous equality for $(e+b)^{-1}$, as well as (7.3.19) we have that

$$\varphi\big((e+a+b)^{-1}\big)=0 \quad \Leftrightarrow \quad \text{either } \varphi\big((e+a)^{-1}\big)=0 \text{ or } \varphi\big((e+b)^{-1}\big)=0,$$

for all $\varphi \in \mathfrak{M}(\mathfrak{A}_0)$. Using now the preceding equivalence, clearly we conclude that:

- $\varphi'(a+b)=\infty \Leftrightarrow$ either $\varphi'(a)=\infty$ or $\varphi'(b)=\infty$; thus,
$$\varphi'(a+b)=\varphi'(a)+\varphi'(b)=\infty; \text{ or}$$
- $\varphi'(a+b)<\infty \Leftrightarrow \varphi'(a)<\infty$ and $\varphi'(b)<\infty$.
In this case,

$$\begin{aligned}
&\varphi'(a+b)\\
&=\frac{\varphi\big(a(e+a)^{-1}(e+b)^{-1}(e+a+b)^{-1}+b(e+a)^{-1}(e+b)^{-1}(e+a+b)^{-1}\big)}{\varphi\big((e+a)^{-1}\big)\varphi\big((e+b)^{-1}\big)\varphi\big((e+a+b)^{-1}\big)}\\
&=\frac{\varphi\big(a(e+a)^{-1}\big)}{\varphi\big((e+a)^{-1}\big)}+\frac{\varphi\big(b(e+b)^{-1}\big)}{\varphi\big((e+b)^{-1}\big)}\\
&=\varphi'(a)+\varphi'(b).
\end{aligned}$$

(4) It follows from (2) by replacing x with λe, $\lambda \in \mathbb{C}$, and y with 0. □

Remark 7.3.3 In order to have all the values of φ' fully determined, we need to define the following:

- $\varphi'(a)\varphi(x)$, $\varphi'(ax) + \varphi'(bx)$ and $\varphi'(a)\varphi(x_1) + \varphi'(a)\varphi(x_2)$, for any $a, b \in \mathfrak{A}^+$ and $x_1, x_2 \in \mathfrak{A}_0$.

From Proposition 7.3.2 we conclude that:

(i) $\varphi'(a)\varphi(x) = \varphi'(ax)$, for any $a \in \mathfrak{A}^+$ and $x \in \mathfrak{A}_0$ with $\varphi'(a)\varphi(x) \neq \infty \cdot 0$.
(ii) $\varphi'(ax) + \varphi'(bx) = \varphi'\big((a+b)x\big)$, for any $a,\ b \in \mathfrak{A}^+$ and $x \in \mathfrak{A}_0$ with either $\varphi'(a)\varphi(x) \neq \infty \cdot 0$ or $\varphi'(b)\varphi(x) \neq \infty \cdot 0$.
(iii) $\varphi'(a)\varphi(x_1 + x_2) = \varphi'\big(a(x_1 + x_2)\big)$, for any $a \in \mathfrak{A}^+$ and $x_1,\ x_2 \in \mathfrak{A}_0$ with $\varphi'\big(a(x_1 + x_2)\big) \neq \infty \cdot 0$.

Furthermore, the definition of φ' and Proposition 7.3.2 imply that:

1. When $\varphi'(a) = \infty$ and $\varphi(x) = 0$, the value $\varphi'(ax)$ of φ' depends upon a and x. For instance,
   - $x = 0 \Rightarrow \varphi'(ax) = \varphi'(0) = \varphi(0) = 0$;
   - $x = (e+a)^{-1} \Rightarrow \varphi'\big(a(e+a)^{-1}\big) = \varphi\big(a(e+a)^{-1}\big) = \varphi\big(e-(e+a)^{-1}\big) = 1$.
2. For $a, b \in \mathfrak{A}^+$ and $x \in \mathfrak{A}_0$, such that either $\varphi'(a)\varphi(x) = \infty \cdot 0$ or $\varphi'(b)\varphi(x) = \infty \cdot 0$, the value $\varphi'((a+b)x)$ clearly depends upon a, b and x.
3. For $a \in \mathfrak{A}^+$ and $x_1, x_2 \in \mathfrak{A}_0$, such that either $\varphi'(a)\varphi(x_1) = \infty \cdot 0$ or $\varphi'(a)\varphi(x_2) = \infty \cdot 0$, then again the value $\varphi'\big(a(x_1 + x_2)\big)$ depends upon a, x_1 and x_2.

Conclusion We define the requested values of φ' by (i), (ii) and (iii), for any $a,\ b \in \mathfrak{A}^+$ and $x_1, x_2 \in \mathfrak{A}_0$.

Remark 7.3.4 We do not know whether φ' is defined or not on the linear span of $\mathfrak{M}(\mathfrak{A}_0, \mathfrak{A}^+)$.

Now, for any $a \in \mathfrak{A}^+$ and $x,\ y \in \mathfrak{A}_0$, we fix the notation:

$$\widehat{ax+y}(\varphi) \equiv \varphi'(ax+y), \quad \varphi \in \mathfrak{M}(\mathfrak{A}_0).$$

Then, we have the following

Proposition 7.3.5 *With a, x, y, φ as before, $\widehat{ax+y}$ is a $\mathbb{C}^*$-valued continuous function on the compact Hausdorff space $\mathfrak{M}(\mathfrak{A}_0)$, which takes the value ∞ on at most a nowhere dense subset of $\mathfrak{M}(\mathfrak{A}_0)$.*

Proof We shall show that the set

$$N_{\widehat{ax+y}} \equiv \{\varphi \in \mathfrak{M}(\mathfrak{A}_0) : \widehat{ax+y}(\varphi) = \infty\},$$

is a nowhere dense closed subset of $\mathfrak{M}(\mathfrak{A}_0)$. Notice that

$$N_{\widehat{ax+y}} = \{\varphi \in \mathfrak{M}(\mathfrak{A}_0) : \varphi\big((e+a|x)^{-1}\big) = 0\}, \tag{7.3.20}$$

from which it follows that $N_{\widehat{ax+y}}$ is closed. Now, suppose that

$$\exists\ \mathcal{U} \text{ non-empty open subset of } \mathfrak{M}(\mathfrak{A}_0) \text{ with } \mathcal{U} \subset N_{\widehat{ax+y}}.$$

From the commutative Gelfand–Naimark theorem, $\mathfrak{A}_0 = \mathcal{C}(\mathfrak{M}(\mathfrak{A}_0))$, with respect to an isometric *-isomorphism. Thus, using Urysohn's lemma for $\mathfrak{M}(\mathfrak{A}_0)$ we get that

$$\exists\ b \in \mathfrak{A}_0 : \|b\|_0 = 1 \text{ and } \widehat{b}(\varphi) = \varphi(b) = 0, \quad \forall\, \varphi \notin \mathcal{U}.$$

But this together with (7.3.20) and the fact that $\mathcal{U} \subset N_{\widehat{ax+y}}$, implies

$$\varphi\big(b(e + a|x|)^{-1}\big) = 0, \quad \forall\, \varphi \in \mathfrak{M}(\mathfrak{A}_0).$$

The afore-mentioned identification $\mathfrak{A}_0 = \mathcal{C}(\mathfrak{M}(\mathfrak{A}_0))$ gives now $b(e + a|x|)^{-1} = 0$, which clearly yields $b = 0$, a contradiction to $\|b\|_0 = 1$. Hence, $N_{\widehat{ax+y}}$ is a nowhere dense closed subset of $\mathfrak{M}(\mathfrak{A}_0)$.

Next we show that $\widehat{ax + y}$ is continuous on $\mathfrak{M}(\mathfrak{A}_0)$. Put

$$z \equiv (e + a|x|)^{-1} \text{ and } w \equiv ax(e + a|x|)^{-1}.$$

Take an arbitrary $\varphi_0 \in \mathfrak{M}(\mathfrak{A}_0)$ and consider the cases:

- $\widehat{ax + y}(\varphi_0) \neq \infty$, i.e., $\widehat{z}(\varphi_0) \neq 0$.

From the continuity of $\widehat{z}$ there is a neighborhood $\mathcal{U}_{\varphi_0}$ of φ_0 with $\widehat{z}(\varphi) \neq 0$, for all $\varphi \in \mathcal{U}_{\varphi_0}$. Thus, we get

$$\widehat{ax + y}(\varphi) = \frac{\widehat{w}(\varphi)}{\widehat{z}(\varphi)} + \widehat{y}(\varphi), \quad \forall\, \varphi \in \mathcal{U}_{\varphi_0},$$

where all functions $\widehat{w}, \widehat{z}, \widehat{y}$ are continuous at φ_0, so that the same is true for $\widehat{ax + y}$.

- $\widehat{ax + y}(\varphi_0) = \infty$, i.e., $\widehat{z}(\varphi_0) = 0$.

Take an arbitrary net $\{\varphi_\alpha\}$ in $\mathfrak{M}(\mathfrak{A}_0)$, such that $\varphi_\alpha \to \varphi_0$, with respect to the weak*-topology $\sigma(\mathfrak{M}(\mathfrak{A}_0), \mathfrak{A}_0)$. Then,

$$\widehat{z}(\varphi_\alpha) \to \widehat{z}(\varphi_0) = 0,$$

where $\widehat{z}(\varphi_\alpha) \neq 0$, since $N_{\widehat{ax+y}}$ is a nowhere dense subset of $\mathfrak{M}(\mathfrak{A}_0)$. Since

$$|\widehat{ax}(\varphi_\alpha)| = \frac{\varphi_\alpha\big((ax(e + a|x|)^{-1})(ax(e + a|x|)^{-1})\big)^{1/2}}{\varphi_\alpha\big((e + a|x|)^{-1}\big)}$$

$$= \frac{\varphi_\alpha\big((a(e+a|x|)^{-1})(x^*xa(e+a|x|)^{-1})\big)^{1/2}}{\varphi_\alpha\big((e+a|x|)^{-1}\big)}$$

$$= \frac{\varphi_\alpha\big((a|x|(e+a|x|)^{-1})^2\big)^{1/2}}{\varphi_\alpha\big((e+a|x|)^{-1}\big)}$$

$$= \frac{\varphi_\alpha\big(a|x|(e+a|x|)^{-1}\big)}{\varphi_\alpha\big((e+a|x|)^{-1}\big)}$$

$$= \frac{\varphi_\alpha\big(e-(e+a|x|)^{-1}\big)}{\varphi_\alpha\big((e+a|x|)^{-1}\big)}$$

$$= \frac{1}{\widehat{z}(\varphi_\alpha)} - 1,$$

it follows that $\lim_\alpha \widehat{ax}(\varphi_\alpha) = \infty$, which implies

$$\lim_\alpha \widehat{ax+y}(\varphi_\alpha) = \infty = \widehat{ax+y}(\varphi_0).$$

This completes the proof. □

All the above lead to the following

Definition 7.3.6 Let W be a completely regular topological space and $\mathcal{F}(W)^+$ the set of all $\mathbb{C}^*$-valued positive continuous functions on W, which take the value ∞ on at most a nowhere dense subset W_0 of W. Then, $\mathcal{F}(W)^+$ is said to be a *wedge* on W, if for any $f, g \in \mathcal{F}(W)^+$ and $\lambda \geq 0$, the functions $f+g$ and λf defined pointwise on W_0, on which f and g are both finite, are extendible to $\mathbb{C}^*$-valued positive continuous functions on W that also belong to $\mathcal{F}(W)^+$. We keep the same symbols $f+g$ and λf for the respective extensions.

Consider now the set

$$\mathcal{F}(W) \equiv \{fg_0 + h_0 : f \in \mathcal{F}(W)^+,\ g_0, h_0 \in \mathcal{C}(W)\},$$

where $\mathcal{C}(W)$ is the *-algebra of all continuous $\mathbb{C}$-valued functions on W. Then, the set $\mathcal{F}(W)$ fulfils the following conditions:

- $(f_1+f_2)g_0 = f_1g_0 + f_2g_0$,
- $(\lambda f)g_0 = \lambda(fg_0)$,
- $f(g_0+h_0) = fg_0 + fh_0$,

for all $f, f_1, f_2 \in \mathcal{F}(W)^+$, $g_0, h_0 \in \mathcal{C}(W)$ and $\lambda \geq 0$.

Definition 7.3.7 We call $\mathcal{F}(W)$ *the set of $\mathbb{C}^*$-valued positive continuous functions on W generated by the wedge $\mathcal{F}(W)^+$ and the *-algebra $\mathcal{C}(W)$.*

In this regard (see also Remark 7.3.3), we have the following

Theorem 7.3.8 *Let* $\mathcal{F}(\mathfrak{M}(\mathfrak{A}_0))^+ \equiv \{\widehat{a} : a \in \mathfrak{A}^+\}$. *Then,*

1. $\mathcal{F}(\mathfrak{M}(\mathfrak{A}_0))^+$ *is a wedge on* $\mathfrak{M}(\mathfrak{A}_0)$.
2. *The map* $\Phi : \mathfrak{M}(\mathfrak{A}_0, \mathfrak{A}^+) \to \mathcal{F}(\mathfrak{M}(\mathfrak{A}_0)) : ax + y \mapsto \widehat{ax + y}$, *is a bijection satisfying the properties:*
   (i) $\Phi(\mathfrak{A}^+) = \mathcal{F}(\mathfrak{M}(\mathfrak{A}_0))^+$, *with* $\Phi(a + b) = \Phi(a) + \Phi(b)$ *and* $\Phi(\lambda a) = \lambda\Phi(a)$, *for all* $a, b \in \mathfrak{A}^+$ *and* $\lambda \geq 0$.
   (ii) $\Phi(\mathfrak{A}_0) = \mathcal{C}(\mathfrak{M}(\mathfrak{A}_0))$, Φ *being an isometric *-isomorphism from* $\mathfrak{A}_0$ *onto* $\mathcal{C}(\mathfrak{M}(\mathfrak{A}_0))$.
   (iii) $\Phi(ax) = \Phi(a)\Phi(x)$, *for all* $a \in \mathfrak{A}^+$ *and* $x \in \mathfrak{A}_0$. $\Phi\big((a + b)x\big) = \big(\Phi(a) + \Phi(b)\big)\Phi(x)$, *for all* $a, b \in \mathfrak{A}^+$ *and* $x \in \mathfrak{A}_0$. $\Phi(\lambda ax) = \lambda\Phi(a)\Phi(x)$, *for all* $a \in \mathfrak{A}^+$, $x \in \mathfrak{A}_0$ *and* $\lambda \geq 0$. $\Phi\big(a(x_1 + x_2)\big) = \Phi(a)\big(\Phi(x_1) + \Phi(x_2)\big)$, *for all* $a \in \mathfrak{A}^+$ *and* $x_1, x_2 \in \mathfrak{A}_0$.

Proof The statements (1), (2)(i) and (2)(ii) follow from Propositions 7.3.2 and 7.3.5. We show the statement (2)(iii). Let $a \in \mathfrak{A}^+$ and $x \in \mathfrak{A}_0$. From Proposition 7.3.5, $\widehat{a}$ and $\widehat{ax}$ are $\mathbb{C}^*$-valued continuous functions on $\mathfrak{M}(\mathfrak{A}_0)$ that take the value ∞ on at most a nowhere dense subset of $\mathfrak{M}(\mathfrak{A}_0)$. Hence, the set

$$\mathcal{K} \equiv \{\varphi \in \mathfrak{M}(\mathfrak{A}_0) : \widehat{a}(\varphi) < \infty \text{ and } \widehat{ax}(\varphi) < \infty\}$$

is dense in $\mathfrak{M}(\mathfrak{A}_0)$ and

$$\widehat{ax}(\varphi) = \widehat{a}(\varphi)\widehat{x}(\varphi), \quad \forall\, \varphi \in \mathcal{K},$$

therefore by the continuity of $\widehat{a}$ and $\widehat{ax}$ we conclude that $\widehat{ax} = \widehat{a}\widehat{x}$, from which it follows that $\Phi(ax) = \Phi(a)\Phi(x)$. The rest of the properties in (2)(iii) are similarly proved. □

7.4 Functional Calculus for Positive Elements

Throughout this section $(\mathfrak{A}[\tau], \mathfrak{A}_0)$ is a commutative locally convex quasi C*-algebra with unit $e \in \mathfrak{A}_0$. Here we shall consider a functional calculus for the positive elements of $\mathfrak{A}[\tau]$, resulting, for instance, to consideration of the nth-root of an element $a \in \mathfrak{A}^+$ (see Corollary 7.4.7). For this purpose, we first need to extend the multiplication of $\mathfrak{A}[\tau]$.

Definition 7.4.1 Let $a, b \in \mathfrak{A}^+$; a is called *left-multiplier* of b, and we write $a \in L(b)$, if there exist nets $\{x_\alpha\}$, $\{y_\beta\}$ in $\mathfrak{A}_0^+$, such that $x_\alpha \underset{\tau}{\to} a$, $y_\beta \underset{\tau}{\to} b$ and $x_\alpha y_\beta \underset{\tau}{\to} c$ (in the sense that the double indexed net $\{x_\alpha y_\beta\}$ converges to c). The product of a, b denoted by ab is given as follows

$$ab := c = \tau - \lim_{\alpha,\beta} x_\alpha y_\beta.$$

Lemma 7.4.2 *The product ab is well-defined, in the sense that it is independent of the selection of the nets* $\{x_\alpha\}$, $\{y_\beta\}$.

Proof Let $\{x_\alpha\}$, $\{y_\beta\}$ be two nets in $\mathfrak{A}_0^+$, such that

$$x_\alpha \underset{\tau}{\to} a, \ y_\beta \underset{\tau}{\to} b \ \text{and} \ x_\alpha y_\beta \underset{\tau}{\to} c.$$

Then, (see also Proposition 7.1.2)

$$\begin{aligned}
&(e+x_\alpha)^{-1}x_\alpha y_\beta(e+y_\beta)^{-1}(e+c)^{-1} - (e+a)^{-1}c(e+c)^{-1}(e+b)^{-1}\\
&\quad = \big((e+x_\alpha)^{-1}x_\alpha y_\beta(e+y_\beta)^{-1}(e+c)^{-1} - (e+x_\alpha)^{-1}c(e+c)^{-1}(e+y_\beta)^{-1}\big)\\
&\qquad + \big((e+x_\alpha)^{-1}c(e+c)^{-1}(e+y_\beta)^{-1} - (e+a)^{-1}c(e+c)^{-1}(e+y_\beta)^{-1}\big)\\
&\qquad + \big((e+a)^{-1}c(e+c)^{-1}(e+y_\beta)^{-1} - (e+a)^{-1}c(e+c)^{-1}(e+b)^{-1}\big).
\end{aligned}$$

As we have seen in the proof of Proposition 7.1.2, (1) $(e+x_\alpha)^{-1} \underset{\tau}{\to} a$, so taking τ-limits in the preceding equality, we conclude that

$$(e+x_\alpha)^{-1}x_\alpha y_\beta(e+y_\beta)^{-1}(e+c)^{-1} \underset{\tau}{\to} (e+a)^{-1}c(e+c)^{-1}(e+b)^{-1}.$$

On the other hand,

$$\begin{aligned}
&(e+x_\alpha)^{-1}x_\alpha y_\beta(e+y_\beta)^{-1}(e+c)^{-1} - \big((e+a)^{-1}a\big)\big(b(e+b)^{-1}\big)(e+c)^{-1}\\
&\quad = \big((e+x_\alpha)^{-1}x_\alpha - (e+a)^{-1}a\big)y_\beta(e+y_\beta)^{-1}(e+c)^{-1}\\
&\qquad + (e+a)^{-1}a\big(y_\beta(e+y_\beta)^{-1} - b(e+b)^{-1}\big)(e+c)^{-1},
\end{aligned}$$

from which, as before, we take that

$$(e+x_\alpha)^{-1}x_\alpha y_\beta(e+y_\beta)^{-1}(e+c)^{-1} \underset{\tau}{\to} \big((e+a)^{-1}a\big)\big(b(e+b)^{-1}\big)(e+c)^{-1}.$$

Hence, we finally obtain

$$(e+a)^{-1}c(e+b)^{-1} = \big((e+a)^{-1}a\big)\big(b(e+b)^{-1}\big). \tag{7.4.21}$$

Suppose now that two other nets $\{x'_\lambda\}$, $\{y'_\mu\}$ exist in $\mathfrak{A}_0^+$, such that

$$x'_\lambda \underset{\tau}{\to} a, \ y'_\mu \underset{\tau}{\to} b \ \text{and} \ x'_\lambda y'_\mu \underset{\tau}{\to} c'.$$

Working exactly as before we are led to the equality

$$(e+a)^{-1}c'(e+b)^{-1} = \big((e+a)^{-1}a\big)\big(b(e+b)^{-1}\big),$$

which together with (7.4.21) gives

$$(e+a)^{-1}c(e+b)^{-1} = (e+a)^{-1}c'(e+b)^{-1} \Leftrightarrow c = c'.$$

□

We may now set the following

Definition 7.4.3 Let $a, b \in \mathfrak{A}^+$ with $a \in L(b)$ and $x, y \in \mathfrak{A}_0$. The product of the elements ax, by is defined as follows:

$$(ax)(by) := (ab)xy.$$

Further, we consider the spectrum of an element $a \in \mathfrak{A}^+$.

Definition 7.4.4 Let $a \in \mathfrak{A}^+$. The *spectrum* of a denoted by $\sigma_{\mathfrak{A}_0}(a)$, is that subset of $\mathbb{C}^*$, defined in the following way:

- Let $\lambda \in \mathbb{C}$. Then, $\lambda \in \sigma_{\mathfrak{A}_0}(a) \Leftrightarrow \lambda e - a$ has no inverse in $\mathfrak{A}_0$;
- $\infty \in \sigma_{\mathfrak{A}_0}(a) \Leftrightarrow a \notin \mathfrak{A}_0$.

Lemma 7.4.5 *Let $a \in \mathfrak{A}^+$. Then,*

$$\sigma_{\mathfrak{A}_0}(a) = \big\{\widehat{a}(\varphi) : \varphi \in \mathfrak{M}(\mathfrak{A}_0)\big\} \subset \mathbb{R}^+ \cup \big\{\infty\big\}.$$

In particular, $\sigma_{\mathfrak{A}_0}(a)$ is a locally compact subset of $\mathbb{C}^*$.

Proof Let $\lambda \in \mathbb{C}$. Then, (see also Theorem 7.3.8)

$$\lambda \notin \sigma_{\mathfrak{A}_0}(a) \Leftrightarrow (\lambda e - a)^{-1} \in \mathfrak{A}_0 \Leftrightarrow \lambda \neq \widehat{a}(\varphi), \ \forall \varphi \in \mathfrak{M}(\mathfrak{A}_0).$$

Let now $\lambda = \infty$. Then,

$$\lambda \in \sigma_{\mathfrak{A}_0}(a) \Leftrightarrow a \notin \mathfrak{A}_0 \Leftrightarrow \widehat{a} \notin C\big(\mathfrak{M}(\mathfrak{A}_0)\big) \Leftrightarrow \exists\, \varphi_0 \in \mathfrak{M}(\mathfrak{A}_0) \ : \ \widehat{a}(\varphi_0) = \infty.$$

The rest is clear. □

If $a \in \mathfrak{A}^+$, denote by $C_b\big(\sigma_{\mathfrak{A}_0}(a)\big)$, the C*-algebra of all bounded continuous functions on $\sigma_{\mathfrak{A}_0}(a)$. For $n \in \mathbb{N}$ and $f \in C\big(\sigma_{\mathfrak{A}_0}(a)\big)$, define the function

$$g_n(\lambda) := \frac{f(\lambda)}{(1+\lambda)^n}, \quad \lambda \in \sigma_{\mathfrak{A}_0}(a). \tag{7.4.22}$$

In this regard, set

$$C_n\big(\sigma_{\mathfrak{A}_0}(a)\big) := \{f \in C\big(\sigma_{\mathfrak{A}_0}(a) \cap \mathbb{R}\big) : g_n \in C_b\big(\sigma_{\mathfrak{A}_0}(a)\big)\}. \tag{7.4.23}$$

Then,

$$C_b\big(\sigma_{\mathfrak{A}_0}(a)\big) \subset C_1\big(\sigma_{\mathfrak{A}_0}(a)\big) \subset C_2\big(\sigma_{\mathfrak{A}_0}(a)\big) \subset \cdots .$$

Now, the promised functional calculus for positive elements in $\mathfrak{A}[\tau]$ is given by the following

Theorem 7.4.6 *Let $a \in \mathfrak{A}^+$. Suppose that the element a^n is well-defined for some $n \in \mathbb{N}$. Then, there is a unique *-isomorphism $f \mapsto f(a)$ from $\bigcup_{k=1}^{n} C_k(\sigma_{\mathfrak{A}_0}(a))$ into $\mathfrak{A}[\tau]$, in such a way that:*

(i) *If $u_0 \in \bigcup_{k=1}^{n} C_k(\sigma_{\mathfrak{A}_0}(a))$, with $u_0(\lambda) = 1$, for each $\lambda \in \sigma_{\mathfrak{A}_0}(a)$, then $u_0(a) = e \in \mathfrak{A}_0 \hookrightarrow \mathfrak{A}[\tau]$.*

(ii) *If $u_1 \in \bigcup_{k=1}^{n} C_k(\sigma_{\mathfrak{A}_0}(a))$, with $u_1(\lambda) = \lambda$, for each $\lambda \in \sigma_{\mathfrak{A}_0}(a)$, then $u_1(a) = a \in \mathfrak{A}[\tau]$.*

(iii) *$\widehat{f(a)}(\varphi) = f(\widehat{a}(\varphi))$, for any $f \in \bigcup_{k=1}^{n} C_k(\sigma_{\mathfrak{A}_0}(a))$ and $\varphi \in \mathfrak{M}(\mathfrak{A}_0)$;*

(iv) *$(f_1 + f_2)(a) = f_1(a) + f_2(a)$, for any $f_1, f_2 \in \bigcup_{k=1}^{n} C_k(\sigma_{\mathfrak{A}_0}(a))$, $(\lambda f)(a) = \lambda f(a)$, for any $f \in \bigcup_{k=1}^{n} C_k(\sigma_{\mathfrak{A}_0}(a))$ and $\lambda \in \mathbb{C}$, $(f_1 f_2)(a) = f_1(a) f_2(a)$, for any $f_j \in C_{k_j}(\sigma_{\mathfrak{A}_0}(a))$, $j = 1, 2$, with $k_1 + k_2 \leq n$.*

(v) *Restricted to $C_b(\sigma_{\mathfrak{A}_0}(a))$ the map $f \mapsto f(a)$ is an isometric *-isomorphism of the C*-algebra $C_b(\sigma_{\mathfrak{A}_0}(a))$ onto the closed *-subalgebra of the C*-algebra $\mathfrak{A}_0$ generated by e and $(e + a)^{-1}$.*

Proof Let $f \in \bigcup_{k=1}^{n} C_k(\sigma_{\mathfrak{A}_0}(a))$. Then, $f \in C_k(\sigma_{\mathfrak{A}_0}(a))$, for some k with $1 \leq k \leq n$, and $g_k \in C_b(\sigma_{\mathfrak{A}_0}(a))$ with $g_k(\lambda) := \frac{f(\lambda)}{(1+\lambda)^k}$, $\lambda \in \sigma_{\mathfrak{A}_0}(a)$. From Lemma 7.4.5 we have that $g_k \circ \widehat{a} \in C(\mathfrak{M}(\mathfrak{A}_0))$, therefore (Gelfand–Naimark theorem) there is a unique element $g_k(a) \in \mathfrak{A}_0$, such that

$$\widehat{g_k(a)}(\varphi) = g_k(\widehat{a}(\varphi)), \quad \forall\, \varphi \in \mathfrak{M}(\mathfrak{A}_0). \tag{7.4.24}$$

Now let

$$f(a) := g_k(a)(e + a)^k \in \mathfrak{A}[\tau]. \tag{7.4.25}$$

We shall show that $f(a)$ does not depend on k, $1 \leq k \leq n$. Indeed, let $f \in C_j(\sigma_{\mathfrak{A}_0}(a))$ with:

- $j \leq k$; then, for each $\lambda \in \sigma_{\mathfrak{A}_0}(a)$,

$$g_k(\lambda) = \frac{f(\lambda)}{(1+\lambda)^k} = \frac{f(\lambda)}{(1+\lambda)^j} \frac{1}{(1+\lambda)^{k-j}} = g_j(\lambda) \frac{1}{(1+\lambda)^{k-j}}.$$

Hence, $g_k(a) = g_j(a)(e+a)^{-(k-j)} \in \mathfrak{A}_0$ and

$$g_k(a)(e+a)^k = g_j(a)(e+a)^j; \tag{7.4.26}$$

- $j > k$; in this case too, one takes (7.4.26) in a similar way. So, the element $f(a) \in \mathfrak{A}[\tau]$ is well-defined by (7.4.25). Now, it is easily seen that the map

$$f \mapsto f(a) \text{ from } \bigcup_{k=1}^{n} C_k\big(\sigma_{\mathfrak{A}_0}(a)\big) \text{ into } \mathfrak{A}[\tau]$$

is a *-isomorphism with the properties (i), (ii), (iii).

(iv) Consider the functions $f_1 \in C_{k_1}\big(\sigma_{\mathfrak{A}_0}(a)\big)$, $f_2 \in C_{k_2}\big(\sigma_{\mathfrak{A}_0}(a)\big)$ with $k_1+k_2 \leq n$. Then, (see (7.4.23) and discussion before (7.4.24)) $g_{k_i} \in C_b\big(\sigma_{\mathfrak{A}_0}(a)\big)$ with $g_{k_i}(a)$ unique in $\mathfrak{A}_0$, $i = 1, 2$. Define the function $f(\lambda) := f_1(\lambda)f_2(\lambda)$, $\lambda \in \sigma_{\mathfrak{A}_0}(a)$. Then, $f \in C_{k_1+k_2}\big(\sigma_{\mathfrak{A}_0}(a)\big)$ and

$$g_{k_1+k_2}(\lambda) = \frac{f(\lambda)}{(1+\lambda)^{k_1+k_2}} = g_{k_1}(\lambda)g_{k_2}(\lambda), \quad \lambda \in \sigma_{\mathfrak{A}_0}(a),$$

that is $g_{k_1+k_2} \in C_b\big(\sigma_{\mathfrak{A}_0}(a)\big)$. Hence, $g_{k_1+k_2}(a) = g_{k_1}(a)g_{k_2}(a) \in \mathfrak{A}_0$. Moreover (see also Definition 7.4.3 and (7.4.25))

$$\begin{aligned}(f_1f_2)(a) = f(a) &= g_{k_1+k_2}(a)(e+a)^{k_1+k_2}\\ &= \big(g_{k_1}(a)(e+a)^{k_1}\big)\big(g_{k_2}(a)(e+a)^{k_2}\big)\\ &= f_1(a)f_2(a).\end{aligned}$$

The first two equalities in (iv) are similarly shown.

(v) Arguing as in (7.4.24) and taking into account Lemma 7.4.5, we easily reach at the conclusion. □

Corollary 7.4.7 *Let $a \in \mathfrak{A}^+$ and $n \in \mathbb{N}$. Then, there is unique $b \in \mathfrak{A}^+$, such that $a = b^n$. The element b is called nth-root of a and is denoted by $a^{\frac{1}{n}}$. If, in particular, $n = 2$, the element $a^{\frac{1}{2}}$ is called square-root of a.*

Proof Consider the functions $f_1(\lambda) := \lambda^{\frac{1}{n}}$ and $f_2(\lambda) := \lambda^{1-\frac{1}{n}}$, $\lambda \geq 0$, which clearly belong to $C_1\big(\sigma_{\mathfrak{A}_0}(a)\big)$. Then, (see (7.4.22), (7.4.23)) $g_1, g_2 \in C_b\big(\sigma_{\mathfrak{A}_0}(a)\big)$ with $g_1(\lambda) = f_1(\lambda)(1+\lambda)^{-1}$, $g_2(\lambda) = f_2(\lambda)(1+\lambda)^{-1}$, $\lambda \geq 0$. Theorem 7.4.6 gives that the elements $f_1(a)$, $f_2(a)$ are uniquely defined in $\mathfrak{A}[\tau]$ with

$$f_1(a) = g_1(a)(e+a), \quad f_2(a) = g_2(a)(e+a),$$

where $g_i(a) \in \mathfrak{A}_0^+$, $i = 1, 2$ (see, e.g., (7.4.24)). Moreover (see also Proposition 7.1.2, (1) and (2)), for each $\varepsilon > 0$

$$\mathfrak{A}_0^+ \ni g_1(a)(e+a)(e+\varepsilon a)^{-1} \xrightarrow[\tau]{\varepsilon\to 0} f_1(a), \text{ resp.}$$

$$\mathfrak{A}_0^+ \ni g_2(a)(e+a)(e+\varepsilon a)^{-1} \xrightarrow[\tau]{\varepsilon\to 0} f_2(a).$$

On the other hand, since $(f_1 f_2)(\lambda) = \lambda$, from Theorem 7.4.6(ii) we have that $(f_1 f_2)(a) = a$, therefore (see also Proposition 7.1.2(2))

$$\big(g_1(a)(e+a)(e+\varepsilon a)^{-1}\big)\big(g_2(a)(e+a)(e+\varepsilon a)^{-1}\big) = a(e+\varepsilon a)^{-1} \xrightarrow[\tau]{\varepsilon\to 0} a.$$

So, from Definition 7.4.1, we conclude that

$$f_1(a) \in L\big(f_2(a)\big) \text{ and } a = f_1(a) f_2(a).$$

Now, since $f_2(a) \in \mathfrak{A}^+$, we repeat the previous procedure with $f_2(a)$ in the place of a, so that continuing in this way we finally obtain

$$a = f_1(a) f_1(a) \cdots f_1(a) \quad (n \text{ times}).$$

The proof is completed by taking $b = f_1(a)$. □

7.5 Structure of Noncommutative Locally Convex Quasi C*-Algebras

In this section we consider a noncommutative locally convex quasi C*-algebra $(\mathfrak{A}[\tau], \mathfrak{A}_0)$, with unit $e \in \mathfrak{A}_0$ and we investigate the following: (a) Conditions under which such an algebra is continuously embedded in a locally convex quasi C*-algebra of operators (Theorems 7.5.2, 7.5.4); (b) a functional calculus for the commutatively positive elements in $\mathfrak{A}[\tau]$ (Theorem 7.5.7).

Lemma 7.5.1 *Let π be a *-representation of $(\mathfrak{A}[\tau], \mathfrak{A}_0)$ with domain $\mathcal{D}(\pi)$, dense in $\mathcal{H}_\pi$. Let also $\mathcal{B}$ be an admissible subset of $\mathcal{B}(\pi(\mathfrak{A}))$. The following hold:*

1. *if π is (τ, t_{s^*})-continuous, then $(\pi(\mathfrak{A})[\mathsf{t}_{s^*}], \pi(\mathfrak{A}_0))$ is a locally convex quasi C*-algebra;*
2. *if π is $(\tau, \mathsf{t}_*^u(\mathcal{B}))$-continuous, then $(\pi(\mathfrak{A})[\mathsf{t}_*^u(\mathcal{B})], \pi(\mathfrak{A}_0))$ is a locally convex quasi C*-algebra.*

Proof Clearly $\pi(\mathfrak{A}_0)$ is a C*-algebra and

$$\pi : \mathfrak{A}[\tau] \to \pi(\mathfrak{A})[\mathsf{t}_{s^*}] \subset \widetilde{\pi(\mathfrak{A}_0)}[\mathsf{t}_{s^*}]$$

is a (τ, t_{s^*})-continuous *-representation of $(\widetilde{\mathfrak{A}[\tau], \mathfrak{A}_0}),$ where $(\pi(\mathfrak{A}), \pi(\mathfrak{A}_0))$ is a quasi *-algebra and $(\widetilde{\pi(\mathfrak{A}_0)[\mathsf{t}_{s^*}]}, \pi(\mathfrak{A}_0))$ (similarly $(\widetilde{\pi(\mathfrak{A}_0)[\mathsf{t}_*^u(\mathcal{B})]}, \pi(\mathfrak{A}_0))$ is a locally convex quasi C*-algebra, by Proposition 7.2.1. So, (1) and (2) follow from Definition 7.1.3. □

Now, recall that a sesquilinear form φ on $\mathfrak{A} \times \mathfrak{A}$ is called *positive*, resp. *invariant*, if and only if, $\varphi(a, a) \geq 0$, for each $a \in \mathfrak{A}$, resp. $\varphi(ax, y) = \varphi(x, a^*y)$, for all $a \in \mathfrak{A}$ and $x, y \in \mathfrak{A}_0$. Moreover, φ is called τ*-continuous*, if $|\varphi(a, b)| \leq p(a)p(b)$ for some τ-continuous seminorm p on $\mathfrak{A}$ and all $a, b \in \mathfrak{A}$ (see also Remark 2.3.1).

Further, let φ be a τ-continuous positive invariant sesquilinear form on $\mathfrak{A}_0 \times \mathfrak{A}_0$. Then, $\widetilde{\varphi}$ denotes the extension of φ to a τ-continuous positive invariant sesquilinear form on $\mathfrak{A} \times \mathfrak{A}$. Moreover, let $(\pi_\varphi, \lambda_\varphi, \mathcal{H}_\varphi)$ be the GNS construction for φ (see, Definition 2.4.2). Then, π_φ is extended on $\mathfrak{A}$, as follows:

$$\pi_\varphi(a)\lambda_\varphi(x) := \lim_\alpha \pi_\varphi(x_\alpha)\lambda_\varphi(x), \quad \forall\, x \in \mathfrak{A}_0, \tag{7.5.27}$$

where $\{x_\alpha\}$ is a net in $\mathfrak{A}[\tau]$ with $a = \tau\text{–}\lim_\alpha x_\alpha$. By the very definitions and the τ-continuity of φ, it follows that π_φ is a (τ, τ_{s^*})-continuous *-representation of $(\mathfrak{A}, \mathfrak{A}_0)$. Now, put

$$\mathfrak{P}(\mathfrak{A}_0) := \left\{\tau\text{-continuous positive invariant sesquilinear forms } \varphi \text{ on } \mathfrak{A}_0 \times \mathfrak{A}_0\right\}.$$

We shall say that the set $\mathfrak{P}(\mathfrak{A}_0)$ is *sufficient* (see also Definition 3.1.17), whenever

$$a \in \mathfrak{A} \text{ with } \widetilde{\varphi}(a, a) = 0, \quad \forall\, \varphi \in \mathfrak{P}(\mathfrak{A}_0), \text{ implies } a = 0.$$

- From the results that follow, *Theorems* 7.5.2, 7.5.4 (and, of course, Corollary 7.5.3) *give answers to the question* (a) stated at the beginning of this section. *These results* can be viewed as *analogues* of the Gelfand–Naimark theorem, *in the case of locally convex quasi C*-algebras.*

Theorem 7.5.2 *Let $(\mathfrak{A}[\tau], \mathfrak{A}_0)$ be a locally convex quasi C*-algebra. The following statements are equivalent:*

1. *there exists a faithful, (τ, τ_{s^*})-continuous *-representation π of $(\mathfrak{A}, \mathfrak{A}_0)$;*
2. *the set $\mathfrak{P}(\mathfrak{A}_0)$ is sufficient.*

Proof (1) ⇒ (2) For every $\xi \in \mathcal{D}(\pi)$ define

$$\varphi_\xi(x, y) := \langle \pi(x)\xi | \pi(y)\xi \rangle, \quad \forall\, x, y \in \mathfrak{A}_0.$$

Then, $\{\varphi_\xi : \xi \in \mathcal{D}(\pi)\} \subset \mathfrak{P}(\mathfrak{A}_0)$, so that from the preceding discussion it follows easily that $\mathfrak{P}(\mathfrak{A}_0)$ is sufficient.

(2) ⇒ (1) Let $\varphi \in \mathfrak{P}(\mathfrak{A}_0)$ and $(\pi_\varphi, \lambda_\varphi, \mathcal{H}_\varphi)$ the GNS construction for φ. Then, as we noticed before (see (7.5.27)), π_φ extends to a (τ, τ_{s^*})-continuous *-representation of $(\mathfrak{A}, \mathfrak{A}_0)$ with $\mathcal{D}(\pi_\varphi) = \lambda_\varphi(\mathfrak{A}_0)$. Now, take

$$\mathcal{D}(\pi) := \Big\{ (\lambda_\varphi(x_\varphi))_{\varphi \in \mathfrak{P}(\mathfrak{A}_0)} \in \bigoplus_{\varphi \in \mathfrak{P}(\mathfrak{A}_0)} \mathcal{H}_\varphi : x_\varphi \in \mathfrak{A}_0 \text{ and}$$

$$\lambda_\varphi(x_\varphi) = 0, \text{ except for a finite number of } \varphi\text{'s from } \mathfrak{P}(\mathfrak{A}_0) \Big\}$$

and define

$$\pi(a)(\lambda_\varphi(x_\varphi)) := (\lambda_\varphi(ax_\varphi)), \quad \forall\, a \in \mathfrak{A} \text{ and } (\lambda_\varphi(x_\varphi)) \in \mathcal{D}(\pi).$$

Then, it is easily seen that π is a faithful, (τ, τ_{s^*}) continuous *-representation of $(\mathfrak{A}, \mathfrak{A}_0)$. □

Results for (topological) quasi *-algebras $(\mathfrak{A}, \mathfrak{A}_0)$, with $\mathfrak{A}_0$ a unital C*-algebra, related to Theorem 7.5.2, have been considered in [44, Theorem 3.3] and [59, Theorem 3.2].

Now an application of Theorem 7.5.2 and Lemma 7.5.1, gives the following

Corollary 7.5.3 *Let $(\mathfrak{A}[\tau], \mathfrak{A}_0)$ be as in* Theorem 7.5.2. *Suppose that the set $\mathfrak{P}(\mathfrak{A}_0)$ is sufficient. Then, the locally convex quasi C*-algebra $(\mathfrak{A}[\tau], \mathfrak{A}_0)$ is continuously embedded in a locally convex quasi C*-algebra of operators.*

The next theorem gives further conditions under which a locally convex quasi C*-algebra $\mathfrak{A}[\tau]$ can be continuously embedded in a locally convex quasi C*-algebra of operators.

Theorem 7.5.4 *Let $(\mathfrak{A}[\tau], \mathfrak{A}_0)$ be a locally convex quasi C*-algebra. Suppose the multiplication of $\mathfrak{A}_0$ satisfies the following condition:*

For every τ-bounded subset B of $\mathfrak{A}_0$ and every $\lambda \in \Lambda$, there exist $\lambda' \in \Lambda$ and a positive constant c_B, such that

$$\sup_{y \in B} p_\lambda(xy) \le c_B\, p_{\lambda'}(x), \quad \forall\, x \in \mathfrak{A}_0.$$

Then, the next statements are equivalent:

(i) *there is a faithful $\big(\tau, \tau_*^u(\mathcal{B})\big)$-continuous *-representation π of $(\mathfrak{A}, \mathfrak{A}_0)$, where $\mathcal{B}$ is an admissible subset of $\mathcal{B}(\pi(\mathfrak{A}))$;*
(ii) *there is a faithful (τ, τ_{s^*})-continuous *-representation of $(\mathfrak{A}, \mathfrak{A}_0)$;*
(iii) *the set $\mathfrak{P}(\mathfrak{A}_0)$ is sufficient.*

Proof (i) ⇒ (ii) is trivial (see (7.2.10)).

(ii) ⇒ (iii) follows from Theorem 7.5.2.

(iii) ⇒ (i) Let $\varphi \in \mathfrak{P}(\mathfrak{A}_0)$ and $(\pi_\varphi, \lambda_\varphi, \mathcal{H}_\varphi)$ be the GNS construction for φ (see discussion before Theorem 7.5.2). Set

$$\mathcal{B}_\varphi := \{\lambda_\varphi(B) : B \text{ a } \tau\text{-bounded subset of } \mathfrak{A}_0\}.$$

Then, for each τ-bounded subset B of $\mathfrak{A}_0$, we have

$$\sup_{y \in B} \|\pi_\varphi(a)\lambda_\varphi(y)\| = \sup_{y \in B} \varphi(ay, ay)^{1/2} \leq \sup_{y \in B} p_\lambda(ay) \leq c_B p_{\lambda'}(a),$$

for all $a \in \mathfrak{A}$ and some $\lambda, \lambda' \in \Lambda$. It is clear now that $\lambda_\varphi(B) \in \mathcal{B}(\pi_\varphi(\mathfrak{A}))$ and that (see (7.2.9)) π_φ is $(\tau, \tau_*^u(\mathcal{B}_\varphi))$-continuous. Let now π be as in the proof of Theorem 7.5.2. Put

$$\mathcal{B}_\pi := \left\{ \bigoplus_{\varphi \in \mathfrak{P}(\mathfrak{A}_0)}^{\text{finite}} \lambda_\varphi(B_\varphi) : B_\varphi \text{ a } \tau\text{-bounded subset of } \mathfrak{A}_0 \right\}.$$

Then, it is easily seen that $\mathcal{B}_\pi$ is an admissible subset of $\mathcal{B}(\pi(\mathfrak{A}))$ and π a faithful, $(\tau, \tau_*^u(\mathcal{B}_\pi))$-continuous *-representation of $\mathfrak{A}$. □

An analogue of Corollary 7.5.3 is stated in the case of Theorem 7.5.4, too.

Taking again $(A[\tau], \mathfrak{A}_0)$ as in Theorem 7.5.2, we proceed to the study of a functional calculus for the commutatively positive elements of $\mathfrak{A}[\tau]$ (see (b) at the beginning of this section). So, let $a \in \mathfrak{A}_c^+$ (for this notation see discussion before Proposition 7.1.2). Then, from Proposition 7.1.2(1), the element $(e + a)^{-1}$ exists and belongs to $\mathcal{U}(\mathfrak{A}_0^+)$. Consider the maximal commutative C*-subalgebra $C^*(a)$ of $\mathfrak{A}_0$ containing the elements e, $(e + a)^{-1}$. Then,

- *$C^*(a)[\tau]$ satisfies the properties* (T$_1$)–(T$_4$) of Sect. 7.1. The properties (T$_1$)–(T$_3$) are trivially checked. To check (T$_4$), we must prove that $\mathcal{U}(C^*(a)^+)$ is τ-closed.

So, let $\{x_\alpha\}$ be a net in $\mathcal{U}(C^*(a)^+)$, such that $x_\alpha \underset{\tau}{\to} x$. But, $\mathcal{U}(C^*(a)^+) \subset \mathcal{U}(\mathfrak{A}_0^+)$ and since $\mathcal{U}(\mathfrak{A}_0^+)$ is τ-closed we have that $x \in \mathcal{U}(\mathfrak{A}_0^+)$. On the other hand,

$$xy \underset{\tau}{\leftarrow} x_\alpha y = y x_\alpha \underset{\tau}{\to} yx, \quad \forall\, y \in C^*(a).$$

Hence, $xy = yx$, which by the maximality of $C^*(a)$ means that $x \in C^*(a)$ and finally $x \in \mathcal{U}(C^*(a)^+)$. This completes the proof of (T$_4$) and all the above lead to the following

Proposition 7.5.5 *Let $(\mathfrak{A}[\tau], \mathfrak{A}_0)$ be a locally convex quasi C*-algebra. Let $a \in \mathfrak{A}_c^+$ and $C^*(a)$ the maximal commutative C*-subalgebra of $\mathfrak{A}_0$ containing $\{e, (e+a)^{-1}\}$. Then, $(\widetilde{C^*(a)[\tau]}, C^*(a))$ is a commutative locally convex quasi C*-algebra.*

Corollary 7.5.6 *The element a, as before, belongs to $\widetilde{C^*(a)[\tau]}^+$.*

Proof Since $a \in \mathfrak{A}_c^+$, Proposition 7.1.2(2) implies that

$$a(e+\varepsilon a)^{-1} = \frac{1}{\varepsilon}\left(e-(e+\varepsilon a)^{-1}\right) \in \mathfrak{A}_0^+, \quad \forall\, \varepsilon > 0.$$

Now, $(e+a)^{-1}$ commutes with every element $\omega \in C^*(a)$, therefore ω also commutes with $e+a$, hence with a, consequently with $(e+\varepsilon a)^{-1}$ too. Thus, $a(e+\varepsilon a)^{-1} \in C^*(a)$, for each $\varepsilon > 0$. Since moreover, $a = \tau\text{-}\lim_{\varepsilon\to 0} a(e+\varepsilon a)^{-1}$ (ibid.), Definition 7.1.1 gives that $a \in \widetilde{C^*(a)}[\tau]^+$. □

It is now clear from Corollary 7.5.6 that making use of Theorem 7.4.6 for $\widetilde{C^*(a)}[\tau]^+$, we can obtain the promised functional calculus for the commutatively positive elements of the noncommutative locally convex quasi C*-algebra $(\mathfrak{A}[\tau], \mathfrak{A}_0)$. That is, we have the following

Theorem 7.5.7 *Let $(\mathfrak{A}[\tau], \mathfrak{A}_0)$ be a noncommutative locally convex quasi C*-algebra. Let $a \in \mathfrak{A}_c^+$ such that a^n is well defined for some $n \in \mathbb{N}$. Then, there is a unique *-isomorphism $f \mapsto f(a)$ from $\bigcup_{k=1}^n C_k(\sigma_{C^*(a)}(a))$ into $\mathfrak{A}[\tau]$, such that*

1. *if $u_0 \in \bigcup_{k=1}^n C_k(\sigma_{C^*(a)}(a))$ with $u_0(\lambda) = 1$, for each $\lambda \in \sigma_{C^*(a)}(a)$, then $u_0(a) = e \in C^*(a) \hookrightarrow \mathfrak{A}[\tau]$;*
2. *if $u_1 \in \bigcup_{k=1}^n C_k(\sigma_{C^*(a)}(a))$ with $u_1(\lambda) = \lambda$, for each $\lambda \in \sigma_{C^*(a)}(a)$, then $u_1(a) = a \in \mathfrak{A}[\tau]$;*
3. *$\widehat{f(a)}(\varphi) = f(\widehat{a}(\varphi))$, for any $f \in \bigcup_{k=1}^n C_k(\sigma_{C^*(a)}(a))$ and $\varphi \in \mathfrak{M}(C^*(a))$;*
4. *$(f_1+f_2)(a) = f_1(a)+f_2(a)$, for any $f_1, f_2 \in \bigcup_{k=1}^n C_k(\sigma_{C^*(a)}(a))$, $(\lambda f)(a) = \lambda f(a)$, for any $f \in \bigcup_{k=1}^n C_k(\sigma_{C^*(a)}(a))$ and $\lambda \in \mathbb{C}$, $(f_1 f_2)(a) = f_1(a)f_2(a)$, for any $f_j \in C_{k_j}(\sigma_{C^*(a)}(a))$, $j = 1, 2$, with $k_1+k_2 \le n$;*
5. *restricted to $C_b(\sigma_{C^*(a)}(a))$ the map $f \mapsto f(a)$ is an isometric *-isomorphism of the C*-algebra $C_b(\sigma_{C^*(a)}(a))$ onto the closed *-subalgebra of the C*-algebra $C^*(a)$ generated by e and $(e+a)^{-1}$.*

Now, an application of Corollary 7.4.7 for the commutative locally convex quasi C*-algebra $(\widetilde{C^*(a)}[\tau], C^*(a))$ and Theorem 7.5.7 give the following

Corollary 7.5.8 *Let $(\mathfrak{A}[\tau], \mathfrak{A}_0)$ be as in* Theorem 7.5.7. *Let $a \in \mathfrak{A}_c^+$ and $n \in \mathbb{N}$. Then, there is a unique element $b \in \mathfrak{A}_c^+$, such that $a = b^n$. The element b is called commutatively nth-root of a and is denoted by $a^{\frac{1}{n}}$. If $n = 2$, the element $a^{\frac{1}{2}}$ is called commutatively square-root of a.*

7.6 Locally Convex Quasi C*-Algebras and Noncommutative Integration

In the case of Banach quasi *-algebras an important role has been played by the two sets of sesquilinear forms $\mathcal{S}_{\mathfrak{A}_0}(\mathfrak{A})$ and $\mathcal{T}_{\mathfrak{A}_0}(\mathfrak{A})$ (see Sects. 3.1.2 and 5.6.4). We extend these notions to the more general set-up of locally convex quasi *-algebras in a natural way.

Definition 7.6.1 Let $(\mathfrak{A}[\tau], \mathfrak{A}_0)$ be a locally convex quasi C*-algebra with unit e. We denote by $\mathcal{S}^{\tau}_{\mathfrak{A}_0}(\mathfrak{A})$ the set of all sesquilinear forms $\varphi \in \mathcal{Q}_{\mathfrak{A}_0}(\mathfrak{A})$ with the following additional properties:

(i) $|\varphi(a,b)| \leq p(a)p(b)$, for some τ-continuous seminorm p on $\mathfrak{A}$ and all $a, b \in \mathfrak{A}$;
(ii) $\varphi(e,e) \leq 1$.

The locally convex quasi C*-algebra $(\mathfrak{A}[\tau], \mathfrak{A}_0)$ is called **-semisimple* if $a \in \mathfrak{A}$, $\varphi(a,a) = 0$, for every $\varphi \in \mathcal{S}^{\tau}_{\mathfrak{A}_0}(\mathfrak{A})$, implies $a = 0$.

We denote by $\mathcal{T}^{\tau}_{\mathfrak{A}_0}(\mathfrak{A})$ the set of all sesquilinear forms φ from $\mathcal{S}^{\tau}_{\mathfrak{A}_0}(\mathfrak{A})$, with the following property

(iii) $\varphi(a,a) = \varphi(a^*, a^*), \quad \forall\, a \in \mathfrak{A}$.

Remark 7.6.2 Notice that

- By (iii) of Definition 7.6.1 and by polarization, we get

$$\varphi(a,b) = \varphi(b^*, a^*), \quad \forall\, a, b \in \mathfrak{A}.$$

- The set $\mathcal{T}^{\tau}_{\mathfrak{A}_0}(\mathfrak{A})$ is convex.

Let $\mathfrak{M}$ be a von Neumann algebra on a Hilbert space $\mathcal{H}$ and ϱ a normal faithful semifinite trace on $\mathfrak{M}^+$, then, as shown in Proposition 5.6.4, $(L^p(\varrho), L^\infty(\varrho) \cap L^p(\varrho))$ is a Banach quasi *-algebra and if ϱ is a finite trace, $(L^p(\varrho), \mathfrak{M})$ is a CQ*-algebra.

Example 7.6.3 Let $\mathfrak{M}$ be a von Neumann algebra and ϱ a normal faithful semifinite trace on $\mathfrak{M}^+$. Then, (see Sect. 5.6.2), $(L^p(\varrho), \mathcal{J}_p)$, $p \geq 2$, is a *-semisimple Banach quasi *-algebra. If ϱ is a finite trace (we assume $\varrho(e) = 1$), then $(L^p(\varrho), \mathfrak{M})$, with $p \geq 2$, is a *-semisimple locally convex quasi C*-algebra. If $p \geq 2$, L^p-spaces possess a sufficient family of positive sesquilinear forms. Indeed, in this case, since for every $W \in L^p(\varrho)$, $|W|^{p-2} \in L^{p/(p-2)}(\varrho)$, then the sesquilinear form φ_W defined by

$$\varphi_W(A,B) := \frac{\varrho\big(A(B|W|^{p-2})^*\big)}{\|W\|^{p-2}_{p,\varrho}}, \quad A, B \in L^p(\varrho)$$

satisfies the conditions of Definition 7.6.1, (see [52], for more details). Moreover,

$$\varphi_W(W, W) = \|W\|^p_{p,\varrho}.$$

Definition 7.6.4 Let $\mathfrak{M}$ be a von Neumann algebra and ϱ a normal faithful semifinite trace defined on $\mathfrak{M}^+$. We say that a measurable operator T belongs to $L^p_{\text{loc}}(\varrho)$ if $TP \in L^p(\varrho)$, for every central ϱ-finite projection P of $\mathfrak{M}$.

Remark 7.6.5 The von Neumann algebra $\mathfrak{M}$ is a subset of $L^p_{\text{loc}}(\varrho)$. Indeed, if $a \in \mathfrak{M}$, then for every ϱ-finite central projection P of $\mathfrak{M}$ the product XP belongs to the *-ideal $\mathcal{J}_p$ (see discussion at the beginning of Sect. 5.6.2).

Throughout this section we are given a von Neumann algebra $\mathfrak{M}$ on a Hilbert space $\mathcal{H}$ with a family $\{P_j\}_{j\in J}$ of ϱ-finite central projections of $\mathfrak{M}$, such that

- if $l, m \in J$, $l \neq m$, then $P_l P_m = 0$ (i.e., the P_j's are orthogonal);
- $\bigvee_{j\in J} P_j = e$; where $\bigvee_{j\in J} P_j$ denotes the projection onto the subspace generated by $\{P_j\mathcal{H}\}_{j\in J}$.

The previous two conditions are always realized in a von Neumann algebra $\mathfrak{M}$ with a faithful normal semifinite trace (see Lemma 5.6.1 and [15, 27], for more details).

If ϱ is a normal faithful semifinite trace on $\mathfrak{M}^+$, we define, for each $a \in \mathfrak{M}$, the following seminorms $q_j(a) := \|XP_j\|_{p,\varrho}$, $j \in J$, on $\mathfrak{M}$. The translation invariant locally convex topology defined by the system $\{q_j\}_{j\in J}$ is denoted by τ_p.

Definition 7.6.6 Let $\mathfrak{M}$ be a von Neumann algebra and ϱ a normal faithful semifinite trace defined on $\mathfrak{M}^+$. We denote by $\widetilde{\mathfrak{M}}^{\tau_p}$ the τ_p-completion of $\mathfrak{M}$.

Proposition 7.6.7 *Let $\mathfrak{M}$ be a von Neumann algebra and ϱ a normal faithful semifinite trace on $\mathfrak{M}^+$. Then, $L^p_{\text{loc}}(\varrho) \subseteq \widetilde{\mathfrak{M}}^{\tau_p}$. Moreover, if there exists a family $\{P_j\}_{j\in J}$ as above, where all P_j's are mutually equivalent, then $L^p_{\text{loc}}(\varrho) = \widetilde{\mathfrak{M}}^{\tau_p}$.*

Proof From Remark 7.6.5, $\mathfrak{M} \subseteq L^p_{\text{loc}}(\varrho)$. If $Y \in L^p_{\text{loc}}(\varrho)$, for every $j \in J$, we have $YP_j \in L^p(\varrho)$. Then, for every $j \in J$, there exists a sequence $\{X^j_n\}_{n=1}^{\infty} \subseteq \mathcal{J}_p$, such that $\|X^j_n - YP_j\|_{p,\varrho} \underset{n\to\infty}{\longrightarrow} 0$.

Let $\mathbb{F}_J$ be the family of finite subsets of J ordered by inclusion and $F \in \mathbb{F}_J$. We put

$$T_{n,F} := \sum_{j\in F} X^j_n P_j \in \mathfrak{M}.$$

Then, the net $\{T_{n,F}\}$ converges to Y with respect to τ_p. Indeed, for every $m \in J$,

$$q_m(T_{n,F} - Y) = \|(T_{n,F} - Y)P_m\|_{p,\varrho} = \|(X^m_n - Y)P_m\|_{p,\varrho},$$

for sufficiently large F. Thus, the inequality $\|(X_n^m - Y)P_m\|_{p,\varrho} \leq \|X_n^m - YP_m\|_{p,\varrho}$ implies that

$$q_m(T_{n,F} - Y) \underset{n,F}{\longrightarrow} 0.$$

Hence, $L^p_{\text{loc}}(\varrho) \subseteq \widetilde{\mathfrak{M}}^{\tau_p}$. On the other hand, assume that all P_j's are mutually equivalent. Then, if $Y \in \widetilde{\mathfrak{M}}^{\tau_p}$, there exists a net $\{X_\alpha\} \subseteq \mathfrak{M}$, such that $X_\alpha \to Y$, with respect to τ_p; consequently,

$$X_\alpha P_j \to YP_j \in L^p(\varrho), \quad \text{with respect to } \|\cdot\|_{p,\varrho}. \tag{7.6.28}$$

But, for each central ϱ-finite projection P, we have

$$\varrho(P) = \varrho\Big(P \sum_{j\in J} P_j\Big) = \sum_{j\in J} \varrho(PP_j). \tag{7.6.29}$$

By our assumption, for any $l, m \in J$, we may pick $U \in \mathfrak{M}$, such that $U^*U = P_l$ and $UU^* = P_m$, hence,

$$\varrho(PP_l) = \varrho(PU^*U) = \varrho(UPU^*) = \varrho(PUU^*) = \varrho(PP_m).$$

So, all terms on the right hand side of (7.6.29) are equal and since the above series converges, only a finite number of them can be nonzero. Thus, for some $s \in \mathbb{N}$ we may write $J = \{1, \ldots, s\}$ and then

$$P = P\sum_{j\in J} P_j = P\sum_{j=1}^{s} P_j = \sum_{j=1}^{s} PP_j \tag{7.6.30}$$

and thus

$$YP = \sum_{j=1}^{s} YPP_j = \sum_{j=1}^{s} YP_jP \in L^p(\varrho). \tag{7.6.31}$$

Therefore, if $Y \in \widetilde{\mathfrak{M}}^{\tau_p}$, for each central ϱ-finite projection P, we have $YP \in L^p(\varrho)$. Hence, $L^p_{\text{loc}}(\varrho) \supseteq \widetilde{\mathfrak{M}}^{\tau_p}$. □

Remark 7.6.8 In general, it is not guaranteed that a von Neumann algebra possesses an orthogonal family $\{P_j\}_{j\in J}$ of mutually equivalent finite central projections, such that $\bigvee_{j\in J} P_j = e$; but, if this is the case, then $L^p_{\text{loc}}(\varrho) = \widetilde{\mathfrak{M}}^{\tau_p}$.

Theorem 7.6.9 *Let $\mathfrak{M}$ be a von Neumann algebra on a Hilbert space $\mathcal{H}$ and ϱ a normal, faithful, semifinite trace on $\mathfrak{M}^+$. Then, $(\widetilde{\mathfrak{M}}^{\tau_p}, \mathfrak{M})$ is a locally convex quasi C^*-algebra, with respect to τ_p,, consisting of measurable operators.*

Proof The topology τ_p satisfies the properties (T$_1$)–(T$_4$). We just prove here (T$_3$) and (T$_4$).

- (T$_3$) For each $\lambda \in J$,

$$q_\lambda(XY) = \|P_\lambda XY\|_{p,\varrho} \leq \|X\|\|P_\lambda Y\|_{p,\varrho} = \|X\|q_\lambda(Y), \quad \forall\, X, Y \in \mathfrak{M};$$

- (T$_4$) The set $\mathcal{U}(\mathfrak{M}^+) := \{X \in \mathfrak{M}^+ : \|X\| \leq 1\}$ is τ_p-closed. To see this consider a net $\{F_\alpha\}$ in $\mathcal{U}(\mathfrak{M}^+)$, such that $F_\alpha \to F$, with respect to the topology τ_p. Then, for each $j \in J$, we have $\|(F_\alpha - F)P_j\|_{p,\varrho} \to 0$. By assumption on P_j, the trace ϱ is a normal faithful finite trace on the von Neumann algebra $(P_j\mathfrak{M})^+$ and by Proposition 5.6.4 $(L^p(\varrho), P_j\mathfrak{M})$ is a CQ*-algebra. Therefore, using (T$_4$) for $(L^p(\varrho), P_j\mathfrak{M})$, we have $FP_j \in \mathcal{U}((P_j\mathfrak{M})^+)$, for each $j \in J$. This, by definition, implies that $F \in \mathfrak{M}$. Indeed, for every $h = \sum_{j\in J} P_jh \in \mathcal{H} = \bigoplus_{j\in J} P_j\mathcal{H}$, we have

$$\|Fh\|^2 = \sum_{j\in J} \|FP_jh\|^2 = \sum_{j\in J} \|FP_jP_jh\|^2 \leq \sum_{j\in J} \|P_jh\|^2 = \|h\|^2.$$

Hence, $F \in \mathcal{U}(\mathfrak{M}^+)$. □

Remark 7.6.10 By Proposition 7.6.7, $(L^p_{\text{loc}}(\varrho), \mathfrak{M})$ itself is a *locally convex quasi C*-algebra*, with respect to τ_p.

7.7 The Representation Theorems

The results of this section generalize to locally convex quasi C*-algebras those obtained in Sect. 5.6.4 in the case of CQ*-algebras.

Let $(\mathfrak{A}[\tau], \mathfrak{A}_0)$ be a locally convex quasi C*-algebra with a unit e. For each $\varphi \in \mathcal{T}^\tau_{\mathfrak{A}_0}(\mathfrak{A})$, we define a linear functional ω_φ on $\mathfrak{A}_0$ by

$$\omega_\varphi(x) := \varphi(x, e), \quad \forall\, x \in \mathfrak{A}_0.$$

Then, we have

$$\omega_\varphi(x^*x) = \varphi(x^*x, e) = \varphi(x, x) = \varphi(x^*, x^*) = \omega_\varphi(xx^*) \geq 0$$

and this shows at once that ω_φ is positive and tracial.

We denote by π the universal *-representation of $\mathfrak{A}_0$ (Remark A.6.16) and for every $\varphi \in \mathcal{T}^\tau_{\mathfrak{A}_0}(\mathfrak{A})$ and $x \in \mathfrak{A}_0$, we put

$$\rho_\varphi(\pi(x)) := \omega_\varphi(x).$$

Then, for each $\varphi \in \mathcal{T}^{\tau}_{\mathfrak{A}_0}(\mathfrak{A})$, ρ_φ is a positive, bounded, linear functional on the operator algebra $\pi(\mathfrak{A}_0)$. Clearly,

$$\rho_\varphi(\pi(x)) = \omega_\varphi(x) = \varphi(x, e), \quad \forall\, x \in \mathfrak{A}_0.$$

If $\{p_\lambda\}$ is a directed family of seminorms defining the topology of $\mathfrak{A}$, by the fact that $\{p_\lambda\}$ is directed, we conclude that there exist $\gamma > 0$ and $\lambda \in \Lambda$, such that

$$|\,\rho_\varphi(\pi(x))\,| = |\,\omega_\varphi(x)\,| = |\,\varphi(x, e)\,| \leq \gamma^2 p_\lambda(xe) p_\lambda(e), \quad \forall\, x \in \mathfrak{A}_0.$$

Then, combining also with (T_3), we obtain

$$|\,\rho_\varphi(\pi(x))\,| \leq \gamma^2\, \|x\|_0\, p_{\lambda'}(e)^2, \quad \text{for some } \lambda' \in \Lambda \text{ and } \forall\, x \in \mathfrak{A}_0.$$

Thus, ρ_φ is continuous on $\pi(\mathfrak{A}_0)$.

By [64, Vol. 2, Proposition 10.1.1], ρ_φ is weakly continuous and so it extends uniquely to $\pi(\mathfrak{A}_0)''$, by the Hahn–Banach theorem. Moreover, since ρ_φ is a trace on $\pi(\mathfrak{A}_0)$, the extension $\widetilde{\rho_\varphi}$ is also a trace on the von Neumann algebra $\mathfrak{M} := \pi(\mathfrak{A}_0)''$ generated by $\pi(\mathfrak{A}_0)$.

Clearly, the set $\mathfrak{N}_{\mathcal{T}^{\tau}_{\mathfrak{A}_0}(\mathfrak{A})} \equiv \left\{\widetilde{\rho_\varphi} : \varphi \in \mathcal{T}^{\tau}_{\mathfrak{A}_0}(\mathfrak{A})\right\}$ is convex.

Definition 7.7.1 The locally convex quasi C*-algebra $(\mathfrak{A}[\tau], \mathfrak{A}_0)$ is called *strongly *-semisimple* if the following conditions are fulfilled:

(a) $x \in \mathfrak{A}_0$ and $\varphi(x, x) = 0$, for every $\varphi \in \mathcal{T}^{\tau}_{\mathfrak{A}_0}(\mathfrak{A})$, imply $x = 0$;
(b) the set $\mathfrak{N}_{\mathcal{T}^{\tau}_{\mathfrak{A}_0}(\mathfrak{A})}$ (see also Proposition 3.6.14 and discussion before it) is w*-closed.

Note that if $(\mathfrak{A}[\tau], \mathfrak{A}_0)$ is a CQ*-algebra, (b) is automatically satisfied, according to Proposition 5.6.14.

Example 7.7.2 Let $\mathfrak{M}$ be a von Neumann algebra and ϱ a normal, faithful, semifinite trace on $\mathfrak{M}^+$. Then, as seen in Example 7.6.3, if ϱ is a finite trace, $(L^p(\varrho), \mathfrak{M})$, with $p \geq 2$, is a *-semisimple, locally convex quasi C*-algebra. Moreover, the conditions (a) and (b) of Definition 7.7.1 are satisfied. Indeed, in this case, the set $\mathfrak{N}_{\mathcal{T}^{\tau}_{\mathfrak{A}_0}(\mathfrak{A})}$ is w*-closed by Proposition 5.6.14. Therefore, $(L^p(\varrho), \mathfrak{M})$, with ϱ finite, is a strongly *-semisimple, locally convex quasi C*-algebra.

Let $(\mathfrak{A}[\tau], \mathfrak{A}_0)$ be a locally convex quasi C*-algebra with unit e, π the universal representation of $\mathfrak{A}_0$ and $\mathfrak{M} = \pi(\mathfrak{A}_0)''$. Recall that $\|f\|^\star$ denotes the norm of a bounded functional f on $\mathfrak{M}$ and $\mathfrak{M}^\star$ the topological dual of $\mathfrak{M}$; then, the norm $\|\widetilde{\rho_\varphi}\|^\star$ of the linear functional $\widetilde{\rho_\varphi}$ on $\mathfrak{M}$, equals to the norm $\|\rho_\varphi\|$ of the linear functional ρ_φ on $\pi(\mathfrak{A}_0)$. By (ii) of Definition 7.6.1, $\|\widetilde{\rho_\varphi}\|^\star = \widetilde{\rho_\varphi}(\pi(e)) = \varphi(e, e) \leq 1$.

Hence, if (b) of Definition 7.7.1 is satisfied, the set $\mathfrak{N}_{\mathcal{T}^{\tau}_{\mathfrak{A}_0}(\mathfrak{A})}$, being a w*-closed subset of the unit ball of $\mathfrak{M}^\star$, is w*-compact.

Let $\mathfrak{EN}_{\mathcal{T}^\tau_{\mathfrak{A}_0}}(\mathfrak{A})$ be the set of extreme points of $\mathfrak{N}_{\mathcal{T}^\tau_{\mathfrak{A}_0}}(\mathfrak{A})$; then, $\mathfrak{N}_{\mathcal{T}^\tau_{\mathfrak{A}_0}}(\mathfrak{A})$ coincides with the w*-closure of the convex hull of $\mathfrak{EN}_{\mathcal{T}^\tau_{\mathfrak{A}_0}}(\mathfrak{A})$.

Moreover, $\mathfrak{EN}_{\mathcal{T}^\tau_{\mathfrak{A}_0}}(\mathfrak{A})$ is a family of normal finite traces on the von Neumann algebra $\mathfrak{M}$.

We put $\mathcal{F} := \{\varphi \in \mathcal{T}^\tau_{\mathfrak{A}_0}(\mathfrak{A}) : \widetilde{\rho_\varphi} \in \mathfrak{EN}_{\mathcal{T}^\tau_{\mathfrak{A}_0}}(\mathfrak{A})\}$ and denote by P_φ, the support projection, corresponding to the trace $\widetilde{\rho_\varphi}$. By [Lemma 5.6.12], $\{P_\varphi\}_{\varphi\in\mathcal{F}}$ consists of mutually orthogonal projections and if $Q := \bigvee_{\varphi\in\mathcal{F}} P_\varphi$, then

$$\mu := \sum_{\widetilde{\rho_\varphi}\in\mathfrak{EN}_{\mathcal{T}^\tau_{\mathfrak{A}_0}}(\mathfrak{A})} \widetilde{\rho_\varphi}$$

is a normal, faithful, semifinite trace defined on the direct sum (see [27] and [89])

$$Q\mathfrak{M} \equiv \bigoplus_{\varphi\in\mathcal{F}} P_\varphi\mathfrak{M}.$$

of von Neumann algebras.

For the topology τ_2 used in the following theorem, see discussion before Definition 7.6.6.

Theorem 7.7.3 *Let $(\mathfrak{A}[\tau], \mathfrak{A}_0)$ be a strongly *-semisimple locally convex quasi C^*-algebra with unit e and π the universal representation of $\mathfrak{A}_0$. Then, there exists a monomorphism*

$$\Psi : a \in \mathfrak{A} \to \Psi(a) := \widetilde{a} \in \widetilde{Q\mathfrak{M}}^{\tau_2},$$

with the following properties:

(i) *Ψ extends the isometry $\pi : \mathfrak{A}_0 \hookrightarrow \mathcal{B}(\mathcal{H})$ given by the Gelfand–Naimark theorem;*
(ii) *$\Psi(a^*) = \Psi(a)^*$, for every $a \in \mathfrak{A}$;*
(iii) *$\Psi(ab) = \Psi(a)\Psi(b)$, for all $a, b \in \mathfrak{A}$, such that either $a \in \mathfrak{A}_0$ or $b \in \mathfrak{A}_0$.*

Proof Let $\{p_\lambda\}_{\lambda\in\Lambda}$ be, as before, the family of seminorms defining the topology τ of $\mathfrak{A}$. Let $a \in \mathfrak{A}$ be fixed. Then, there exists a net $\{x_\alpha\}_{\alpha\in\Delta}$ of elements of $\mathfrak{A}_0$, such that $p_\lambda(x_\alpha - a) \to 0$, for each $\lambda \in \Lambda$. We put $a_\alpha = \pi(x_\alpha)$.

By (i) of Definition 7.6.1, for every $\varphi \in \mathcal{T}^\tau_{\mathfrak{A}_0}(\mathfrak{A})$, there exist $\gamma > 0$ and $\lambda' \in \Lambda$, such that, for each $\alpha, \beta \in \Delta$,

$$\begin{aligned}
\|P_\varphi(a_\alpha - a_\beta)\|_{2,\widetilde{\rho_\varphi}} &= \|P_\varphi\big(\pi(x_\alpha) - \pi(x_\beta)\big)\|_{2,\widetilde{\rho_\varphi}} \\
&= \big(\widetilde{\rho_\varphi}(|P_\varphi(\pi(x_\alpha) - \pi(x_\beta))|^2)\big)^{1/2} = \\
&= \big(\varphi\left((x_\alpha - x_\beta)^*(x_\alpha - x_\beta), e\right)\big)^{1/2} \\
&= \big(\varphi(x_\alpha - x_\beta, x_\alpha - x_\beta)\big)^{1/2} \le \gamma\, p_{\lambda'}(x_\alpha - x_\beta) \underset{\alpha,\beta}{\longrightarrow} 0.
\end{aligned}$$

Let $\widetilde{a_\varphi}$ be the $\|\cdot\|_{2,\widetilde{\rho_\varphi}}$-limit of the net $\{P_\varphi a_\alpha\}$ in $L^2(\widetilde{\rho_\varphi})$. Clearly, $\widetilde{a_\varphi} = P_\varphi \widetilde{a_\varphi}$. We define

$$\Psi(a) := \sum_{\varphi \in \mathcal{F}} P_\varphi \widetilde{a_\varphi} =: \widetilde{a}.$$

Evidently, $\widetilde{a} \in \widetilde{Q\mathfrak{M}}^{\tau_2}$. It is easy to see that the map $a \ni \mathfrak{A} \mapsto \widetilde{a} \in \widetilde{Q\mathfrak{M}}^{\tau_2}$ is well defined and injective. Indeed, if $x_\alpha \to 0$, there exist $\gamma > 0$ and $\lambda' \in \Lambda$, such that

$$\begin{aligned}
\|P_\varphi a_\alpha\|_{2,\widetilde{\rho_\varphi}} &= \|P_\varphi \pi(x_\alpha)\|_{2,\widetilde{\rho_\varphi}} \\
&= \left(\widetilde{\rho_\varphi}(|P_\varphi(\pi(x_\alpha)|^2)\right)^{1/2} = \\
&= \left(\varphi\left(x_\alpha^* x_\alpha, e\right)\right)^{1/2} \\
&= \left(\varphi(x_\alpha, x_\alpha)\right)^{1/2} \le \gamma\, p_{\lambda'}(x_\alpha) \to 0,
\end{aligned}$$

where $\|\cdot\|_{2,\widetilde{\rho_\varphi}}$ clearly denotes the norm in $L^2(\widetilde{\rho_\varphi})$. Thus, $P_\varphi(a_\alpha) = 0$, for every $\varphi \in \mathcal{T}^\tau_{\mathfrak{A}_0}(\mathfrak{A})$, therefore $\widetilde{a} = 0$. Moreover, if $P_\varphi \widetilde{a} = 0$, for each $\varphi \in \mathcal{F}$, then $\varphi(a, a) = 0$, for every $\varphi \in \mathcal{F}$. Since, every $\varphi \in \mathcal{T}^\tau_{\mathfrak{A}_0}(\mathfrak{A})$ is a w*-limit of convex combinations of elements from $\mathcal{F}$, we obtain $\varphi(a, a) = 0$, for every $\varphi \in \mathcal{T}^\tau_{\mathfrak{A}_0}(\mathfrak{A})$. Hence, by assumption, $a = 0$. □

Remark 7.7.4 In the same way one proves that

- If $(\mathfrak{A}[\tau], \mathfrak{A}_0)$ is a strongly *-semisimple, locally convex quasi C*-algebra and there exists a faithful $\varphi \in \mathcal{T}^\tau_{\mathfrak{A}_0}(\mathfrak{A})$ (i.e., the equality $\varphi(a, a) = 0$, implies $a = 0$), then there exists a monomorphism

$$\Psi : a \in \mathfrak{A} \to \Phi(a) := \widetilde{a} \in L^2(\widetilde{\rho_\varphi}),$$

  with the following properties:

  (i) Ψ extends the isometry $\pi : \mathfrak{A}_0 \hookrightarrow \mathcal{B}(\mathcal{H})$ given by the Gelfand–Naimark theorem;
  (ii) $\Psi(a^*) = \Psi(a)^*$, for every $a \in \mathfrak{A}$;
  (iii) $\Psi(ab) = \Psi(a)\Psi(b)$, for all $a, b \in \mathfrak{A}$, such that either $a \in \mathfrak{A}_0$ or $b \in \mathfrak{A}_0$.

- If the semifinite von Neumann algebra $\pi(\mathfrak{A}_0)''$ admits an orthogonal family of mutually equivalent projections $\{P'_i\}_{i \in I}$, such that $\sum_{i \in I} P'_i = e$, then it is easy to see that the map $a \in \mathfrak{A} \to \widetilde{a} \in L^2_{\text{loc}}(\varrho)$ (cf. Definition 7.6.4), is a monomorphism.

Appendix A
*-Algebras and Representations

This chapter is devoted to general aspects of the theory of *-algebras and their representations.

A.1 Algebras: Basic Definitions

A vector space $\mathfrak{A}$ is said to be an *algebra* if a map $(a, b) \in \mathfrak{A} \times \mathfrak{A} \mapsto ab \in \mathfrak{A}$ is defined and satisfies the following conditions:

(i) $a(bc) = (ab)c$,
(ii) $(a+b)c = ac + bc$ and $a(b+c) = ab + ac$,
(iii) $\alpha(ab) = (\alpha a)b = a(\alpha b)$,

for all $a, b, c \in \mathfrak{A}$ and $\alpha \in \mathbb{C}$. The element ab of $\mathfrak{A}$ is called the *product* of a and b. An element e of $\mathfrak{A}$ is called a *unit* of $\mathfrak{A}$ if $ea = ae = a$, for each $a \in \mathfrak{A}$; this element, if it exists, is necessarily unique.

An algebra $\mathfrak{A}$ is said to be a **-algebra* if there exists a conjugate linear map $a \in \mathfrak{A} \mapsto a^* \in \mathfrak{A}$ such that $(ab)^* = b^*a^*$ and $(a^*)^* = a$ for all $a, b \in \mathfrak{A}$. The map $a \in \mathfrak{A} \mapsto a^* \in \mathfrak{A}$ is called an *involution* of $\mathfrak{A}$. If $\mathfrak{A}$ has a unit e, then $e^* = e$.

Remark A.1.1 If the *-algebra $\mathfrak{A}$ has no unit, there is a standard procedure for embedding it in a *-algebra with unit $\mathfrak{A}^e$ called *unitization* of $\mathfrak{A}$. Indeed, let us consider the space $\mathfrak{A}^e := \mathfrak{A} \oplus \mathbb{C}$. For $(a, \lambda), (b, \mu) \in \mathfrak{A}^e$ and $\alpha \in \mathbb{C}$, define algebraic operations and involution as follows:

$$(a, \lambda) + (b, \mu) := (a + b, \lambda + \mu), \quad \alpha(a, \lambda) := (\alpha a, \alpha\lambda);$$

$$(a, \lambda) \cdot (b, \mu) := (ab + \mu a + \lambda b, \lambda\mu), \quad (a, \lambda)^* := (a^*, \overline{\lambda}).$$

M. Fragoulopoulou, C. Trapani, *Locally Convex Quasi *-Algebras and their Representations*, Lecture Notes in Mathematics 2257,
https://doi.org/10.1007/978-3-030-37705-2

Then, it is easily seen that, with these operations and involution, $\mathfrak{A}^e$ is a *-algebra with unit $(0, 1)$, such that $a \in \mathfrak{A} \mapsto (a, 0) \in \mathfrak{A}^e$, for all $a \in \mathfrak{A}$.

- From now on, for convenience, we put $e \equiv (0, 1)$ in $\mathfrak{A}^e$.

Example A.1.2 Very familiar examples of *-algebras are provided by certain spaces of complex valued functions. In all cases, the involution is defined by complex conjugation. Take, for instance, the space $C(X)$ of all continuous functions on a locally compact Hausdorff space X or the space $C_c^\infty(\mathbb{R})$ of all $\mathbb{C}$-valued C^∞-functions on $\mathbb{R}$, with compact support.

Example A.1.3 Let $\mathcal{D}$ be a dense subspace of a Hilbert space $\mathcal{H}$. We denote with $\mathcal{L}^\dagger(\mathcal{D})$ the space of all closable operators A in $\mathcal{H}$, such that $D(A) = \mathcal{D}$, $D(A^*) \supset \mathcal{D}$ (where $D(A)$ means domain of A, as we have noticed at the beginning of Sect. 2.1.3) and both A and A^* map $\mathcal{D}$ into itself. In this case, we define an involution $A \mapsto A^\dagger$ on $\mathcal{L}^\dagger(\mathcal{D})$ by putting $A^\dagger := A^* \restriction_{\mathcal{D}}$. It is easy to verify that with this involution and the natural algebraic operations $\mathcal{L}^\dagger(\mathcal{D})$ is a *-algebra.

A *normed algebra* $\mathfrak{A}$ is an algebra, which is also a normed space with norm $\|\cdot\|$, such that

$$\|ab\| \leq \|a\|\|b\|, \quad \forall\, a, b \in \mathfrak{A}.$$

If $\mathfrak{A}$ is a Banach space under the norm $\|\cdot\|$, we call it a *Banach algebra*. If $\mathfrak{A}$ has a unit e, we shall always suppose, without loss of generality, that $\|e\| = 1$. This follows from a theorem of Gelfand, according to which one can define a second norm $\|\cdot\|'$ on $\mathfrak{A}$ making it a Banach algebra and such that $\|\cdot\|'$ is equivalent to $\|\cdot\|$ with $\|e\|' = 1$.

A *normed *-algebra*, (respectively, *Banach *-algebra*) is a *-algebra $\mathfrak{A}$, which is also a normed (respectively, Banach algebra) with norm $\|\cdot\|$, such that

$$\|a^*\| = \|a\|, \quad \forall\, a \in \mathfrak{A}.$$

An involution with the previous property is called *isometric involution*.

A Banach *-algebra $\mathfrak{A}$ is said to be a *C*-algebra* if the given norm $\|\cdot\|$ satisfies the so-called *C*-condition*, namely

$$\|a^* a\| = \|a\|^2, \quad \forall\, a \in \mathfrak{A}.$$

Example A.1.4 The space $C(X)$ of $\mathbb{C}$-valued continuous functions on a compact Hausdorff space X provides the simplest commutative example of a C*-algebra with unit. The norm is defined, as usual, by

$$\|f\|_\infty := \sup_{x \in X} |f(x)|, \quad \forall\, f \in C(X).$$

It is easy to see that this norm satisfies the C*-condition.

Example A.1.5 The space $\mathcal{B}(\mathcal{H})$ of all bounded linear operators on a Hilbert space $\mathcal{H}$ is a *-algebra under the natural algebraic operations and the map $A \mapsto A^*$ as involution. As is already seen in Chap. 2, with the operator norm

$$\|A\| := \sup\left\{\|A\xi\| : \xi \in \mathcal{H},\ \|\xi\| = 1\right\}, \quad A \in \mathcal{B}(\mathcal{H}), \tag{A.1.1}$$

$\mathcal{B}(\mathcal{H})$ is a normed *-algebra with isometric involution. We now prove that $\mathcal{B}(\mathcal{H})$ is complete under the previous norm (A.1.1), thus it will be a Banach *-algebra. So let $\{A_n\}$ be a Cauchy sequence in $\mathcal{B}(\mathcal{H})$. For each $\xi \in \mathcal{H}$, the sequence $\{A_n\xi\}$ is a Cauchy sequence in $\mathcal{H}$. Then, it converges to a vector $\xi^* \in \mathcal{H}$. We put $A\xi = \xi^*$. It is easily checked that A is well-defined and linear. To see that A is bounded, we take into account that the sequence $\{\|A_n\|\}$ is bounded (i.e., there exists $M > 0$, such that $\|A_n\| \leq M$, for every $n \in \mathbb{N}$); thus

$$\|A\xi\| = \lim_{n\to\infty} \|A_n\xi\| \leq M\|\xi\|.$$

It remains to prove that $A_n \to A$, with respect to the operator norm. Since $\{A_n\}$ is Cauchy, for every $\epsilon > 0$, there exists n_ϵ, such that for $n, m > n_\epsilon$, $\|A_n - A_m\| < \epsilon$. Now fix $n > n_\epsilon$. Then, we have

$$\|(A_n - A)\xi\| = \lim_{m\to\infty} \|(A_n - A_m)\xi\| \leq \lim_{m\to\infty} \|A_n - A_m\|\|\xi\| \leq \epsilon\|\xi\|.$$

Hence,

$$\|A_n - A\| = \sup_{\|\xi\|=1} \|(A_n - A)\xi\| \leq \epsilon,$$

which proves the claim.

We now show that $\mathcal{B}(\mathcal{H})$ is actually a (non-commutative) C*-algebra, i.e., that the C*-condition is fulfilled. First we have $\|A^*A\| \leq \|A\|^2$, for all $A \in \mathcal{B}(\mathcal{H})$. On the other hand,

$$\begin{aligned}\|A^*A\| &= \sup_{\|\xi\|=\|\eta\|=1} |\langle A^*A\xi|\eta\rangle| \\ &= \sup_{\|\xi\|=\|\eta\|=1} |\langle A\xi|A\eta\rangle| \\ &\geq \sup_{\|\xi\|=1} |\langle A\xi|A\xi\rangle| = \|A\|^2, \quad \forall\, A \in \mathcal{B}(\mathcal{H}).\end{aligned}$$

Example A.1.6 An operator $B \in \mathcal{B}(\mathcal{H})$ is called compact if B maps bounded sets of $\mathcal{H}$ into relatively compact subsets of $\mathcal{H}$, or equivalently, if for any bounded sequence $\{\xi_n\}$ in $\mathcal{H}$, the sequence $\{B\xi_n\}$ contains a convergent subsequence. Let $\mathcal{C}(\mathcal{H})$ denote the space of all compact operators on $\mathcal{H}$. Then, $\mathcal{C}(\mathcal{H})$ is a closed two-sided ideal of $\mathcal{B}(\mathcal{H})$ and it is itself a C*-algebra. The quotient algebra $\mathcal{B}(\mathcal{H})/\mathcal{C}(\mathcal{H})$ is a Banach

*-algebra under the quotient norm

$$\|[A]\|_C := \inf\{\|A + B\| : B \in \mathcal{C}(\mathcal{H})\},$$

where $[A]$ denotes the equivalent class of A. This Banach *-algebra is called *Calkin algebra*.

Example A.1.7 Let $\mathfrak{M}$ be a subset of $\mathcal{B}(\mathcal{H})$. The *(von Neumann) commutant* of $\mathfrak{M}$ is defined as

$$\mathfrak{M}' = \{Y \in \mathcal{B}(\mathcal{H}) : XY = YX, \forall X \in \mathfrak{M}\}.$$

If $\mathfrak{M}$ is *-invariant (i.e., $X \in \mathfrak{M} \Leftrightarrow X^* \in \mathfrak{M}$), then $\mathfrak{M}'$ is a *-algebra. The *double commutant* (called also *bicommutant*) $\mathfrak{M}''$ is defined as $\mathfrak{M}'' := (\mathfrak{M}')'$. If $\mathfrak{M} = \mathfrak{M}''$, $\mathfrak{M}$ is said to be a *von Neumann algebra*. Due to von Neumann double commutant theorem [27, vol. I, Section 2.3], every von Neumann algebra is norm-closed in $\mathcal{B}(\mathcal{H})$ and so it is a C*-algebra.

A.2 Representations

Definition A.2.1 Let $\mathfrak{A}$ be an algebra. A *representation* of $\mathfrak{A}$ on a vector space $\mathcal{D}_\pi$ is a homomorphism π of $\mathfrak{A}$ into the algebra $L(\mathcal{D}_\pi)$ of all linear operators on $\mathcal{D}_\pi$; that is

(i) $\pi(a + b) = \pi(a) + \pi(b), \quad \forall\, a, b \in \mathfrak{A};$
(ii) $\pi(\alpha a) = \alpha\pi(a), \quad \forall\, a \in \mathfrak{A},\ \alpha \in \mathbb{C};$
(iii) $\pi(ab) = \pi(a)\pi(b), \quad \forall\, a, b \in \mathfrak{A}.$

A representation π on a vector space $\mathcal{D}_\pi$ is called *ultracyclic*, if there exists $\xi_0 \in \mathcal{D}_\pi$, such that $\mathcal{D}_\pi = \{\pi(a)\xi_0 : a \in \mathfrak{A}\}$.

A subspace $\mathcal{M}$ of $\mathcal{D}_\pi$ is said to be *invariant* under π if $\pi(a)\mathcal{M} \subseteq \mathcal{M}$, for every $a \in \mathfrak{A}$.

A representation π on a vector space $\mathcal{D}_\pi$ is called *algebraically irreducible*, if only the trivial subspaces $\{0\}$ and $\mathcal{D}_\pi$ are invariant subspaces of $\mathcal{D}_\pi$, under π.

A representation π of $\mathfrak{A}$ in a normed space $\mathcal{D}_\pi$ is called *bounded* if the operator $\pi(a)$ is bounded, for every $a \in \mathfrak{A}$.

If $\mathfrak{A}$ is a *-algebra and $\mathcal{D}_\pi$ a pre-Hilbert space, a *-representation of $\mathfrak{A}$ on $\mathcal{D}_\pi$ is a *-homomorphism of $\mathfrak{A}$ into $\mathcal{L}^\dagger(\mathcal{D}_\pi)$, that is a homomorphism satisfying, in addition to the properties listed in Definition A.2.1, the property $\pi(a^*) = \pi(a)^\dagger$, for every $a \in \mathfrak{A}$.

Let $\mathcal{H}_\pi$ denote the Hilbert space completion of $\mathcal{D}_\pi$. If $\mathcal{D}_{\pi_1}$ is another dense subspace of $\mathcal{H}_\pi$ with $\mathcal{D}_\pi \subseteq \mathcal{D}_{\pi_1}$ and π_1 is a *-representation of $\mathfrak{A}$ defined on $\mathcal{D}_{\pi_1}$, then π is said to be a *sub *-representation* of π_1, writing $\pi \subseteq \pi_1$, if $\pi_1(a)\xi = \pi(a)\xi$, for any $a \in \mathfrak{A}$ and $\xi \in \mathcal{D}_\pi$.

Remark A.2.2 Let π be a *-representation of a *-algebra $\mathfrak{A}$ without unit, defined on $\mathcal{D}_\pi$. Then, π can be extended to a *-representation π^e of the unitization $\mathfrak{A}^e$ of $\mathfrak{A}$ (see Remark A.1.1) by putting

$$\pi^e\big((a,\lambda)\big) = \pi(a) + \lambda\mathbb{I}, \quad \forall\,(a,\lambda) \in \mathfrak{A}^e,$$

where $\mathbb{I}$ denotes the identity operator of $\mathcal{D}_\pi$. It is easily seen that π^e is a *-representation on the same domain $\mathcal{D}_\pi$, with the property $\pi^e\big(e \equiv (0,1)\big) = \mathbb{I}$.

Remark A.2.3 If π is a *-representation of $\mathfrak{A}$ on $\mathcal{D}_\pi$, then $\mathcal{D}_\pi$ can be endowed with a topology, denoted by t_π, linked to the representation π itself. More precisely, t_π is the *locally convex topology* on $\mathcal{D}_\pi$ defined, for a given $a \in \mathfrak{A}$, by the family of seminorms

$$\xi \mapsto \|\xi\|_{\pi(a)} := \|\xi\| + \|\pi(a)\xi\|, \quad \forall\,\xi \in \mathcal{D}_\pi, \tag{A.2.1}$$

where $\|\cdot\|$ is the Hilbert norm. The topology t_π is known as the *graph topology* on $\mathcal{D}_\pi$.

Let π be a *-representation of the *-algebra $\mathfrak{A}$ on the pre-Hilbert space $\mathcal{D}_\pi$. We put

$$\mathcal{D}_{\widetilde{\pi}} = \bigcap_{a\in\mathfrak{A}} D(\overline{\pi(a)}),$$

where $\overline{\pi(a)}$ denotes the closure of the operator $\pi(a)$ (see beginning of Sect. 2.1.3). Then, we define

$$\widetilde{\pi}(a) := \overline{\pi(a)}\restriction_{\mathcal{D}_{\widetilde{\pi}}}, \quad a \in \mathfrak{A}.$$

Lemma A.2.4 *Let $\{a_1,\dots,a_n\}$ be a finite subset of $\mathfrak{A}$ and π a *-representation of $\mathfrak{A}$ on $\mathcal{D}_\pi$. Then, there exists $b \in \mathfrak{A}$, such that*

$$\sum_{i=1}^n \|\pi(a_i)\xi\| \le \|\pi(b)\xi\|, \quad \forall\,\xi \in \mathcal{D}_\pi.$$

Proof First add a unit, if necessary. Then, put $b = e + \frac{n}{2}\sum_{i=1}^n a_i^* a_i$. If $b_1 = \frac{n}{2}\sum_{i=1}^n a_i^* a_i$, we obtain $\langle\pi(b_1)\xi|\xi\rangle = \frac{n}{2}\sum_{i=1}^n\langle\pi(a_i^* a_i)\xi|\xi\rangle$, hence

$$\|\pi(b)\xi\|^2 = \|\xi\|^2 + 2\langle\pi(b_1)\xi|\xi\rangle + \|\pi(b_1)\xi\|^2 \ge n\sum_{i=1}^n\langle\pi(a_i^* a_i)\xi|\xi\rangle$$

$$= n\sum_{i=1}^n \|\pi(a_i)\xi\|^2 \ge \left(\sum_{i=1}^n \|\pi(a_i)\xi\|\right)^2.$$

□

The latter inequality is nothing but the Cauchy–Schwarz inequality.

Theorem A.2.5 *For any *-representation π of $\mathfrak{A}$ on $\mathcal{D}_\pi$, $\widetilde{\pi}$ is a *-representation of $\mathfrak{A}$ on $\mathcal{D}_{\widetilde{\pi}}$.*

Proof We may suppose, without loss of generality, that $\mathfrak{A}$ has a unit e. We show first that $\widetilde{\pi}(b)\mathcal{D}_{\widetilde{\pi}} \subset \mathcal{D}_{\widetilde{\pi}}$, for all $b \in \mathfrak{A}$. Let $a, b \in \mathfrak{A}$. From Lemma A.2.4, there exists $c \in \mathfrak{A}$, such that

$$\|\pi(c)\xi\| \geq \|\pi(ab)\xi\| + \|\pi(b)\xi\| + \|\xi\|, \quad \forall \xi \in \mathcal{D}_{\widetilde{\pi}}. \tag{A.2.2}$$

If $\xi \in \mathcal{D}_{\widetilde{\pi}}$, then $\xi \in D(\overline{\pi(c)})$; thus, there exists a sequence $\{\xi_n\} \subset \mathcal{D}_\pi$, such that $\xi_n \to \xi$ and $\pi(c)\xi_n \to \overline{\pi(c)}\xi$. From (A.2.2) the following hold

$$\begin{cases} \|\pi(b)(\xi_n - \xi_m)\| \to 0, \\ \|\pi(a)\pi(b)(\xi_n - \xi_m)\| = \|\pi(ab)(\xi_n - \xi_m)\| \to 0, \end{cases}$$

that imply $\xi \in D(\overline{\pi(b)})$, $\pi(b)\xi_n \to \overline{\pi(b)}\xi$ and $\overline{\pi(b)}\xi \in D(\overline{\pi(a)})$. Since $a \in \mathfrak{A}$ is arbitrary, it follows that $\overline{\pi(b)}\mathcal{D}_{\widetilde{\pi}} \subset \mathcal{D}_{\widetilde{\pi}}$. Moreover, $\overline{\pi(ab)}\xi = \overline{\pi(a)}\,\overline{\pi(b)}\xi$. Therefore, $\widetilde{\pi}$ is a representation of $\mathfrak{A}$ in $\mathcal{D}_{\widetilde{\pi}}$. To prove that it is a *-representation, we take $\xi, \eta \in \mathcal{D}_{\widetilde{\pi}}$. Then, there exist two sequences $\{\xi_n\}, \{\eta_n\}$ in $\mathcal{D}(\pi)$, such that $\xi_n \to \xi$, $\eta_n \to \eta$ and $\pi(a)\xi_n \to \overline{\pi(a)}\xi$, $\pi(a^*)\eta_n \to \pi(a^*)\eta$. Hence,

$$\langle \widetilde{\pi}(a)\xi|\eta\rangle = \lim_{n\to\infty} \langle \pi(a)\xi_n|\eta_n\rangle = \lim_{n\to\infty} \langle \xi_n|\pi(a^*)\eta_n\rangle = \langle \xi|\widetilde{\pi}(a^*)\eta\rangle$$

and this concludes the proof. □

The *-representation $\widetilde{\pi}$ is called the *closure* of π (see discussion after Definition 2.2.5). If $\mathcal{D}_{\widetilde{\pi}} = \mathcal{D}_\pi$, then $\widetilde{\pi} = \pi$ and π is called *closed* (ibid.). More generally, it can be proved that $\widetilde{\pi}$ is the minimal closed extension of π and this fact motivates its name.

Remark A.2.6 From the construction itself of $\mathcal{D}_{\widetilde{\pi}}$ it follows that $\mathcal{D}_\pi$ is dense in $\mathcal{D}_{\widetilde{\pi}}$ with respect to the locally convex topology defined by (A.2.1). Moreover, it can be proved that $\mathcal{D}_{\widetilde{\pi}}$ can be identified with the completion of $\mathcal{D}_\pi$ in this topology.

If π is a bounded *-representation, then for each $a \in \mathfrak{A}$, $\widetilde{\pi}(a)$ is an everywhere defined bounded operator on the Hilbert space $\mathcal{H}_\pi$, completion of $\mathcal{D}_\pi$, with respect to the norm defined by the inner product. More precisely, we have

Theorem A.2.7 *Let π be a *-representation of $\mathfrak{A}$ on $\mathcal{D}_\pi \subset \mathcal{H}_\pi$. The following statements are equivalent:*

(i) *π is bounded on $\mathcal{D}_\pi$;*
(ii) *π is the restriction to $\mathcal{D}_\pi$ of a bounded *-representation π_1 on $\mathfrak{A}$, with $\mathcal{D}_{\pi_1} = \mathcal{H}_\pi$;*
(iii) *$\mathcal{D}_{\widetilde{\pi}} = \mathcal{H}_\pi$ and $\widetilde{\pi}$ is bounded.*

Proof (i) $\Rightarrow$ (ii) Let π be bounded. Then, for every $a \in \mathfrak{A}$, the operator $\pi(a)$ is bounded and has a bounded everywhere defined closure $\overline{\pi(a)}$ on $\mathcal{H}_\pi$. Define

$$\pi_1(a) := \overline{\pi(a)}, \quad a \in \mathfrak{A}.$$

Then, π_1 is a *-representation of $\mathfrak{A}$ with domain $\mathcal{D}_{\pi_1} = \mathcal{H}_\pi$ and clearly $\pi \subseteq \pi_1$.

(ii) $\Rightarrow$ (iii) Since $\pi \subseteq \pi_1$, with π_1 bounded, π itself is bounded and so $\pi(a)$ is a bounded everywhere defined operator on $\mathcal{H}_\pi$, for every $a \in \mathfrak{A}$. Hence,

$$\mathcal{D}_{\widetilde{\pi}} = \bigcap_{a \in \mathfrak{A}} D(\overline{\pi(a)}) = \mathcal{H}_\pi$$

and $\widetilde{\pi}$ is bounded.

(iii) $\Rightarrow$ (i) This is obvious. □

Definition A.2.8 Let $\mathfrak{A}$ be a *-algebra, $\mathcal{H}_\pi, \mathcal{H}_\rho$ two Hilbert spaces, π and ρ two *-representations of $\mathfrak{A}$ with domains $\mathcal{D}_\pi \subseteq \mathcal{H}_\pi$ and $\mathcal{D}_\rho \subseteq \mathcal{H}_\rho$, respectively. We say that π and ρ are *unitarily equivalent* (and we write $\pi \approx \rho$) if there exists a unitary operator $U : \mathcal{H}_\pi \to \mathcal{H}_\rho$, such that

$$U\mathcal{D}_\pi = \mathcal{D}_\rho \text{ and}$$
$$\rho(a)U\xi = U\pi(a)\xi, \quad \forall\, a \in \mathfrak{A},\ \xi \in \mathcal{D}_\pi.$$

Lemma A.2.9 *Let π and ρ be two *-representations of $\mathfrak{A}$ with domains $\mathcal{D}_\pi \subseteq \mathcal{H}_\pi$ and $\mathcal{D}_\rho \subseteq \mathcal{H}_\rho$, respectively. If $\pi \approx \rho$, then $\widetilde{\pi} \approx \widetilde{\rho}$.*

Proof Let $U : \mathcal{H}_\pi \to \mathcal{H}_\rho$ be a unitary operator satisfying the two conditions of Definition A.2.8. Let $\xi \in D(\overline{\pi(a)})$, $a \in \mathfrak{A}$. Then, there exists a sequence $\{\xi_n\}$ of elements of $\mathcal{D}_\pi$, such that $\xi_n \to \xi$ and $\pi(a)\xi_n \to \overline{\pi(a)}\xi$. Moreover, $U\xi_n \to U\xi$ and

$$\|\rho(U\xi_n - U\xi_m)\| = \|U\pi(a)(\xi_n - \xi_m)\| = \|\pi(a)(\xi_n - \xi_m)\| \to 0, \text{ as } n, m \to \infty.$$

Hence, $U\xi \in \mathcal{D}(\overline{\rho(a)})$ and

$$\overline{\rho(a)}U\xi = \lim_{n\to\infty} \rho(a)U\xi_n = \lim_{n\to\infty} U\pi(a)\xi_n = U\overline{\pi(a)}\xi. \tag{A.2.3}$$

From the arbitrariness of $a \in A$, it follows that $U\mathcal{D}_{\widetilde{\pi}} \subset \mathcal{D}(\widetilde{\rho})$; interchanging the roles of π and ρ and using U^{-1} instead of U, one can prove the converse inclusion. Finally, by (A.2.3), one easily obtains

$$\widetilde{\rho}(a)U\xi = U\widetilde{\pi}(a)\xi, \quad \forall\, a \in \mathfrak{A},\ \xi \in \mathcal{D}_{\widetilde{\pi}}.$$ □

The definition of an ultracyclic representation given at the beginning of this section is of purely algebraic nature. In what follows, one takes into account the topological structure of the domains involved.

Definition A.2.10 A *-representation π of a *-algebra $\mathfrak{A}$ with domain $\mathcal{D}_\pi$ is called *cyclic* if there exists $\xi_0 \in \mathcal{D}_\pi$, such that $\{\pi(a)\xi_0 : a \in \mathfrak{A}\}$ is norm-dense in $\mathcal{H}_\pi$; it is called *strongly-cyclic* if there exists $\xi_0 \in \mathcal{D}_\pi$, such that $\{\pi(a)\xi_0 : a \in \mathfrak{A}\}$ is t_π-dense in $\mathcal{D}_{\tilde{\pi}}$.

The vector ξ_0 is called *cyclic*, or *strongly-cyclic* for π, respectively.

Theorem A.2.11 *Let π and ρ be two closed *-representations of $\mathfrak{A}$ with domains $\mathcal{D}_\pi \subseteq \mathcal{H}_\pi$ and $\mathcal{D}_\rho \subseteq \mathcal{H}_\rho$, respectively. Assume that π is cyclic* (respectively, strongly-cyclic)*, with cyclic* (respectively, strongly-cyclic) *vector ξ_0. Then, ρ is unitarily equivalent to π, if and only if, ρ is cyclic* (respectively, strongly-cyclic) *and there exists a cyclic* (respectively, strongly-cyclic) *vector $\eta_0 \in \mathcal{D}_\rho$, such that*

$$\langle \pi(a)\xi_0|\xi_0\rangle = \langle \rho(a)\eta_0|\eta_0\rangle, \quad \forall\, a \in \mathfrak{A}. \tag{A.2.4}$$

Proof Assume that π and ρ are unitarily equivalent and let U be the unitary operator, which establishes the equivalence. We shall prove that $\eta_0 = U\xi_0$ is strongly-cyclic for ρ. Let $\eta \in \mathcal{D}_\rho$. Then, there exists $\xi \in \mathcal{D}_\pi$ and a net $\{a_\alpha\}$ in $\mathfrak{A}$, such that $U\xi = \eta$ and $\xi = t_\pi\text{-}\lim_\alpha \pi(a_\alpha)\xi_0$. This is equivalent to

$$\pi(b)\xi = t_\pi\text{-}\lim_\alpha \pi(b)\pi(a_\alpha)\xi_0, \quad \forall\, b \in \mathfrak{A}.$$

Then, we have

$$\begin{aligned}\rho(b)\eta = \rho(b)U\xi = U\pi(b)\xi &= U t_\pi\text{-}\lim_\alpha \pi(b)\pi(a_\alpha)\xi_0\\ &= U t_\pi\text{-}\lim_\alpha \pi(ba_\alpha)\xi_0 = t_\pi\text{-}\lim_\alpha \rho(ba_\alpha)U\xi_0\\ &= t_\pi\text{-}\lim_\alpha \rho(b)\rho(a_\alpha)U\xi_0 = t_\pi\text{-}\lim_\alpha \rho(b)\rho(a_\alpha)\eta_0.\end{aligned}$$

This implies that η_0 is strongly-cyclic for ρ. Moreover,

$$\langle \rho(a)\eta_0|\eta_0\rangle = \langle \rho(a)U\xi_0|U\xi_0\rangle = \langle U\pi(a)\xi_0|U\xi_0\rangle = \langle \pi(a)\xi_0|\xi_0\rangle, \quad \forall\, a \in \mathfrak{A}.$$

Conversely, assume that ρ is strongly-cyclic with strongly-cyclic vector η_0 satisfying (A.2.4). We begin with defining

$$U\xi_0 := \eta_0 \text{ and } U\pi(a)\xi_0 := \rho(a)\eta_0, \quad a \in \mathfrak{A}.$$

The equality

$$\begin{aligned}\|\rho(a)\eta_0\|^2 = \langle \rho(a)\eta_0|\rho(a)\eta_0\rangle &= \langle \rho(a^*a)\eta_0|\eta_0\rangle\\ &= \langle \pi(a^*a)\xi_0|\xi_0\rangle = \langle \pi(a)\xi_0|\pi(a)\xi_0\rangle\\ &= \|\pi(a)\xi_0\|^2\end{aligned}$$

proves at once that U is well-defined and isometric. Since $\{\pi(a)\xi_0 : a \in \mathfrak{A}\}$ is dense in $\mathcal{H}_\pi$, U extends to an isometric operator $\overline{U}$ from $\mathcal{H}_\pi$ into $\mathcal{H}_\rho$. Since U is clearly invertible and the inverse is also isometric, $\overline{U}$ is unitary. By an argument similar to that used in the first part of the proof, one can easily show that $\overline{U}\mathcal{D}_\pi = \mathcal{D}_\rho$. Finally, we have

$$\begin{aligned} U\pi(a)\pi(b)\xi_0 &= U\pi(ab)\xi_0 = \rho(ab)U\xi_0 = \rho(a)\rho(b)\eta_0 \\ &= \rho(a)U\pi(b)\xi_0, \quad \forall\, a, b \in \mathfrak{A}. \end{aligned}$$

Hence $U\pi(a)\xi = \rho(a)U\xi$, for every $\xi \in \{\pi(b)\xi_0 : b \in \mathfrak{A}\}$. This easily implies that $\overline{U}\pi(a)\xi = \rho(a)\overline{U}\xi$, for every $a \in \mathfrak{A}$ and $\xi \in \mathcal{D}_\pi$. □

A.2.1 Hermitian and Positive Linear Functionals

Let now π be a *-representation of $\mathfrak{A}$ with domain $\mathcal{D}_\pi \subseteq \mathcal{H}_\pi$. Fix an element $\xi \in \mathcal{D}_\pi$ and define

$$\omega_\xi(a) := \langle \pi(a)\xi|\xi\rangle, \quad a \in \mathfrak{A}.$$

Then, ω_ξ is a linear functional on $\mathfrak{A}$ with the properties:

- $\omega_\xi(a^*) = \langle \pi(a^*)\xi|\xi\rangle = \langle \xi|\pi(a)\xi\rangle = \overline{\langle \pi(a)\xi|\xi\rangle} = \overline{\omega_\xi(a)}, \quad \forall\, a \in \mathfrak{A}$;
- $\omega_\xi(a^*a) = \langle \pi(a^*a)\xi|\xi\rangle = \langle \pi(a^*)\pi(a)\xi|\xi\rangle = \langle \pi(a)\xi|\pi(a)\xi\rangle \geq 0$, $\forall\, a \in \mathfrak{A}$;
- $|\omega_\xi(a)|^2 = |\langle \pi(a)\xi|\xi\rangle|^2 \leq \|\pi(a)\xi\|^2\, \|\xi\|^2 = \|\xi\|^2\omega_\xi(a^*a), \quad \forall\, a \in \mathfrak{A}$.

These properties of ω_ξ suggest the following

Definition A.2.12 Let $\mathfrak{A}$ be a *-algebra. A linear functional ω on $\mathfrak{A}$ is called *hermitian* if $\omega(a^*) = \overline{\omega(a)}$, for every $a \in \mathfrak{A}$; *positive* if $\omega(a^*a) \geq 0$, for all $a \in \mathfrak{A}$. A positive linear functional ω is called

- *faithful* if $a \in \mathfrak{A}$ and $\omega(a^*a) = 0$ imply $a = 0$.
- *Hilbert bounded* if there exists $\gamma > 0$, such that $|\omega(a)|^2 \leq \gamma\, \omega(a^*a)$, for every $a \in \mathfrak{A}$.
- If $\mathfrak{A}$ has a unit e, ω is called a *state* if $\omega(e) = 1$.

Clearly, if ω is Hilbert bounded, then it is automatically positive. Moreover, in this case, one can put (see [19, Definition 9.4.2])

$$\|\omega\|_H = \sup\left\{|\omega(a)|^2 : a \in \mathfrak{A},\, \omega(a^*a) = 1\right\}.$$

Every positive linear functional ω on $\mathfrak{A}$ satisfies the equality,

$$\omega(b^*a) = \overline{\omega(a^*b)}, \quad \forall\, a, b \in \mathfrak{A}, \tag{A.2.5}$$

as well as the Cauchy–Schwarz inequality

$$|\omega(b^*a)|^2 \le \omega(a^*a)\omega(b^*b), \quad \forall\, a, b \in \mathfrak{A}. \tag{A.2.6}$$

To prove (A.2.5), let $\alpha \in \mathbb{C}$. Then,

$$0 \le \omega\big((\alpha a + b)^*(\alpha a + b)\big) = |\alpha|^2\omega(a^*a) + \alpha\omega(b^*a) + \overline{\alpha}\omega(a^*b) + \omega(b^*b).$$

Since $|\alpha|^2\omega(a^*a) + \omega(b^*b)$ is real, $\alpha\omega(b^*a) + \overline{\alpha}\omega(a^*b)$ must also be real, for every $\alpha \in \mathbb{C}$. This implies that $\omega(b^*a) = \overline{\omega(a^*b)}$, for all a, b in $\mathfrak{A}$.

As for the Cauchy–Schwarz inequality, for $\alpha \in \mathbb{R}$ and $a, b \in \mathfrak{A}$, we have

$$\begin{aligned} 0 &\le \omega\big((\alpha a + b)^*(\alpha a + b)\big) \\ &= \alpha^2\omega(a^*a) + 2\alpha\,\Re\omega(b^*a) + \omega(b^*b) \\ &\le \alpha^2\omega(a^*a) + 2\alpha|\omega(b^*a)| + \omega(b^*b), \end{aligned}$$

where $\Re$ means 'real part'. For the last term to be nonnegative for any $\alpha \in \mathbb{R}$, the discriminant cannot be positive, i.e.,

$$|\omega(b^*a)|^2 - \omega(a^*a)\omega(b^*b) \le 0.$$

As a direct consequence of (A.2.5) and (A.2.6), we have

Proposition A.2.13 *If $\mathfrak{A}$ has a unit e and ω is a positive linear functional of $\mathfrak{A}$, then*

$$\omega(a^*) = \overline{\omega(a)} \ \text{ and } \ |\omega(a)|^2 \le \omega(e)\omega(a^*a), \quad \forall\, a \in \mathfrak{A};$$

i.e., ω is hermitian and Hilbert bounded with $\gamma = \omega(e)$.

Remark A.2.14 If $\mathfrak{A}$ has no unit, then ω may fail to be hermitian. If ω is hermitian, then clearly (A.2.5) holds.

Remark A.2.15 If ω is a linear functional on $\mathfrak{A}$, then ω has a natural extension ω^e to the unitization $\mathfrak{A}^e$ of $\mathfrak{A}$ (Remark A.1.1), obtained by setting:

$$\omega^e\big((a, \lambda)\big) := \omega(a) + \lambda, \quad \forall\, (a, \lambda) \in \mathfrak{A}^e.$$

It is easily seen that if ω is hermitian on $\mathfrak{A}$, then ω^e is also hermitian on $\mathfrak{A}^e$; but, if ω is positive on $\mathfrak{A}$, then ω^e need not be positive on $\mathfrak{A}^e$. Nevertheless, an extension of a positive linear functional ω on $\mathfrak{A}$ can be achieved under certain conditions and such conditions are given in Theorems A.2.19 and A.2.22.

Since the functional ω_ξ defined, at the beginning of this section from a *-representation π of $\mathfrak{A}$, is the prototype of a positive linear functional on $\mathfrak{A}$, it is natural to pose the *question* as to whether *any* positive linear functional can be realized as a *vector* functional ω_ξ; more precisely, does there exist a *-representation π_ω and a vector $\xi_\omega \in \mathcal{D}(\pi_\omega)$, such that $\omega(a) = \langle \pi_\omega(a)\xi_\omega|\xi_\omega\rangle$, for every $a \in \mathfrak{A}$? We shall discuss this *question* in the next subsections.

A.2.2 The Gelfand–Naimark–Segal Theorem

Let ω be a positive linear functional on a *-algebra $\mathfrak{A}$. Let

$$\mathfrak{I}_\omega = \{a \in \mathfrak{A} : \omega(a^*a) = 0\}.$$

Then, $\mathfrak{I}_\omega$ is a *left-ideal* of $\mathfrak{A}$. Indeed, if $a \in \mathfrak{I}_\omega$ and $b \in \mathfrak{A}$, the Cauchy–Schwarz inequality gives

$$\omega(a^*b^*ba)^2 \le \omega(a^*a)\omega\big((b^*ba)^*b^*ba\big) = 0.$$

Thus, $ba \in \mathfrak{I}_\omega$. We denote by $\mathcal{D}_\omega$ the quotient space $\mathfrak{A}/\mathfrak{I}_\omega$ and with $\lambda_\omega(a)$ the equivalent class of $a \in \mathfrak{A}$. Then, $\mathcal{D}_\omega$ is a pre-Hilbert space with inner product defined by

$$\langle \lambda_\omega(a)|\lambda_\omega(b)\rangle := \omega(b^*a), \quad a, b \in \mathfrak{A}.$$

Let $\mathcal{H}_\omega$ denote the Hilbert space completion of $\mathcal{D}_\omega$. It is easily checked that $a \in \mathfrak{A} \mapsto \lambda_\omega(a) \in \mathcal{H}_\omega$ is a linear map. We now put

$$\pi_\omega^\circ(a)\lambda_\omega(b) := \lambda_\omega(ab), \quad a, b \in \mathfrak{A}.$$

Then, for each $a \in \mathfrak{A}$, $\pi_\omega^\circ(a)$ is well-defined. Indeed if $\lambda_\omega(b) = 0$, then clearly $\lambda_\omega(ab) = 0$, since $\mathfrak{I}_\omega$ is a left-ideal. Moreover, we have

$$\langle \pi_\omega^\circ(a)\lambda_\omega(b)|\lambda_\omega(c)\rangle = \omega\big(c^*(ab)\big) = \omega\big((a^*c)^*b\big) = \langle \lambda_\omega(b)|\pi_\omega^\circ(a^*)\lambda_\omega(c)\rangle,$$

with $c \in \mathfrak{A}$. This implies that $\pi_\omega^\circ(a^*) = \pi_\omega^\circ(a)^\dagger$, for all $a \in \mathfrak{A}$. Therefore, π_ω° is a *-representation of $\mathfrak{A}$ in $\mathcal{H}_\omega$. *Its closure π_ω is a closed *-representation of* $\mathfrak{A}$, which is called the *Gelfand–Naimark–Segal* (for short, *GNS*) *representation* of $\mathfrak{A}$ constructed by ω. Of course, π_ω satisfies the relation

$$\omega(c^*ab) = \langle \pi_\omega(a)\lambda_\omega(b)|\lambda_\omega(c)\rangle, \quad \forall\, a, b, c \in \mathfrak{A}.$$

Suppose now that $\mathfrak{A}$ has a unit e. Then, we have that $\{\pi_\omega^\circ(a)\lambda_\omega(e) : a \in \mathfrak{A}\} = \mathcal{D}_\omega$; hence, π_ω° is algebraically cyclic. The vector $\xi_\omega := \lambda_\omega(e)$ is a *strongly-cyclic* vector for π_ω; i.e., $\{\pi_\omega(a)\lambda_\omega(e) : a \in \mathfrak{A}\}$ is dense in $\mathcal{D}_{\pi_\omega}$, with respect to the topology defined by the seminorms (A.2.1). In this case, one has

$$\omega(a) = \langle \pi(a)\xi_\omega | \xi_\omega \rangle.$$

Summing up, we have the following

Theorem A.2.16 *For every positive linear functional ω on a *-algebra $\mathfrak{A}$, there exists a Hilbert space $\mathcal{H}_\omega$, a linear map $\lambda_\omega : \mathfrak{A} \to \mathcal{H}_\omega$ and a closed *-representation π_ω of $\mathfrak{A}$ with domain $\mathcal{D}_{\pi_\omega} \subseteq \mathcal{H}_\omega$, such that*

$$\omega(c^* ab) = \langle \pi_\omega(a)\lambda_\omega(b) | \lambda_\omega(c) \rangle, \quad \forall\, a, b, c \in \mathfrak{A}.$$

The triple $(\pi_\omega, \lambda_\omega, \mathcal{H}_\omega)$ is called the GNS construction for ω. If $\mathfrak{A}$ has a unit e, then π_ω is strongly-cyclic with strongly-cyclic vector $\xi_\omega = \lambda_\omega(e)$ and

$$\omega(a) = \langle \pi_\omega(a)\xi_\omega | \xi_\omega \rangle, \quad \forall\, a \in \mathfrak{A}.$$

*Moreover, if ρ is another closed strongly-cyclic *-representation of $\mathfrak{A}$ with domain $\mathcal{D}_\rho \subseteq \mathcal{H}_\rho$ and η_ω a strongly-cyclic vector in $\mathcal{D}_\rho$, such that $\omega(a) = \langle \rho(a)\eta_\omega | \eta_\omega \rangle$, then ρ is unitarily equivalent to π_ω.*

The statement about the essential uniqueness of the GNS construction for *-algebras with unit follows from Theorem A.2.11.

As seen above, there is some difference in the GNS theorem when considering algebras with unit or without unit. As we know, every *-algebra $\mathfrak{A}$ without unit has a unitization $\mathfrak{A}^e$; but a positive linear functional ω does not extend in a natural way to $\mathfrak{A}^e$, in general. We shall now examine, in more details, this problem.

Definition A.2.17 Let $\mathfrak{A}$ be a *-algebra. A positive linear functional ω on $\mathfrak{A}$ is called *extensible* if ω is the restriction to $\mathfrak{A}$ of some positive linear functional ω' on $\mathfrak{A}^e$, the unitization of $\mathfrak{A}$.

It is clear that if ω is extensible, then it is Hilbert bounded. We will show that the converse also holds. To begin with, let us give the following

Lemma A.2.18 *Let ω be a positive linear functional on $\mathfrak{A}$. Let $(\pi_\omega, \lambda_\omega, \mathcal{H}_\omega)$ be the GNS construction for ω. Then, ω is Hilbert bounded, if and only if, there exists a vector $\zeta_\omega \in \mathcal{H}_\omega$, such that*

$$\omega(a) = \langle \lambda_\omega(a) | \zeta_\omega \rangle, \quad \forall\, a \in \mathfrak{A}. \tag{A.2.7}$$

In this case, the vector ζ_ω is uniquely determined and $\|\zeta_\omega\| = \|\omega\|_H^{1/2}$.

Proof Assume that ω is Hilbert bounded. Let us define on $\mathcal{D}_\omega \equiv \{\lambda_\omega(a) : a \in \mathfrak{A}\}$ a linear functional f, as follows

$$f(\lambda_\omega(a)) := \omega(a), \quad a \in \mathfrak{A}.$$

The functional f is well-defined since if $\omega(a^*a) = 0$, its Hilbert boundedness implies that $\omega(a) = 0$. On the other hand,

$$|f(\lambda_\omega(a))| = |\omega(a)| \leq \|\omega\|_H^{1/2}\omega(a^*a)^{1/2} = \|\omega\|_H^{1/2}\|\lambda_\omega(a)\|, \quad a \in \mathfrak{A}.$$

Thus, f is bounded on $\mathcal{D}_\omega$. Therefore, it can be uniquely extended to a bounded linear functional on $\mathcal{H}_\omega$. By the Riesz lemma, there exists a vector $\zeta_\omega \in \mathcal{H}_\omega$, such that

$$f(\lambda_\omega(a)) = \omega(a) = \langle\lambda_\omega(a)|\zeta_\omega\rangle, \quad \forall\, a \in \mathfrak{A}.$$

Conversely, assume that there exists ζ_ω, such that (A.2.7) holds. Then,

$$|\omega(a)| = |\langle\lambda_\omega(a)|\zeta_\omega\rangle| \leq \|\lambda_\omega(a)\|\,\|\zeta_\omega\| = \|\zeta_\omega\|\,\omega(a^*a)^{1/2}, \quad \forall\, a \in \mathfrak{A}.$$

Hence, ω is Hilbert bounded. The uniqueness of ζ_ω and the equality $\|\zeta_\omega\| = \|\omega\|_H^{1/2}$ also follow from the Riesz lemma. □

Theorem A.2.19 *Let ω be a positive linear functional on $\mathfrak{A}$ and let $(\pi_\omega, \lambda_\omega, \mathcal{H}_\omega)$ be the GNS construction for ω. The following statements are equivalent:*

(i) *ω is extensible;*
(ii) *ω is hermitian and Hilbert bounded;*
(iii) *ω is Hilbert bounded and $\omega^e : \mathfrak{A}^e \to \mathbb{C}$ defined by*

$$\omega^e\big((a,\lambda)\big) = \omega(a) + \lambda\gamma, \quad \forall\,(a,\lambda) \in \mathfrak{A}^e,$$

is a positive linear functional on $\mathfrak{A}^e$, for every $\gamma \geq \|\omega\|_H$;
(iv) *there exists a *-representation π of $\mathfrak{A}$ with domain $\mathcal{D}_\pi$ and a vector $\xi \in \mathcal{D}_\pi$, such that*

$$\omega(a) = \langle\pi(a)\xi|\xi\rangle, \quad \forall\, a \in \mathfrak{A};$$

(v) *there exists a vector $\zeta_\omega \in \mathcal{H}_\omega$, such that*

$$\omega(a) = \langle\lambda_\omega(a)|\zeta_\omega\rangle, \quad \forall\, a \in \mathfrak{A}. \tag{A.2.8}$$

Proof (i) $\Rightarrow$ (ii) Assume that ω' is a positive linear functional that extends ω to $\mathfrak{A}^e$. Then, since $\mathfrak{A}^e$ has a unit, ω is hermitian and

$$|\omega(a)|^2 = |\omega'(a)|^2 \leq \omega'(e)\omega'(a^*a) = \omega'(e)\omega(a^*a), \quad \forall\, a \in \mathfrak{A}.$$

(ii) ⇒ (iii) Let ω be hermitian and Hilbert bounded. Let $\gamma \geq \|\omega\|_H$. Then, for every $a \in \mathfrak{A}$ and $\lambda \in \mathbb{C}$, we have

$$\begin{aligned}\omega^e\big((a,\lambda)^*(a,\lambda)\big) &= \omega(a^*a) + \overline{\lambda}\omega(a) + \lambda\omega(a^*) + \gamma|\lambda|^2 \\ &\geq \omega(a^*a) + \overline{\lambda}\omega(a) + \lambda\omega(a^*) + \frac{|\omega(a)|^2}{\omega(a^*a)}|\lambda|^2 \\ &= \big(\omega(a^*a)^{\frac{1}{2}} + \omega(a^*a)^{-\frac{1}{2}}\overline{\lambda}\omega(a)\big)\big(\omega(a^*a)^{\frac{1}{2}} + \omega(a^*a)^{-\frac{1}{2}}\lambda\overline{\omega(a)}\big) \\ &= \left|\omega(a^*a)^{\frac{1}{2}} + \omega(a^*a)^{-\frac{1}{2}}\overline{\lambda}\omega(a)\right|^2 \geq 0.\end{aligned}$$

Consequently, ω^e is positive.

To prove that (iii) ⇒ (iv) it suffices to consider the GNS representation of $\mathfrak{A}^e$ constructed from ω^e and restrict it to $\mathfrak{A}$.

(iv) ⇒ (v) If π is a *-representation as described in (iv), then ω is Hilbert bounded, therefore the result follows from Lemma A.2.18.

(v) ⇒ (i) From Lemma A.2.18 it follows that ω is Hilbert bounded and that $\|\omega\|_H = \|\zeta_\omega\|^2$. We define

$$\omega^e\big((a,\lambda)\big) = \omega(a) + \lambda\|\zeta_\omega\|^2, \quad \forall\, a \in \mathfrak{A},\ \lambda \in \mathbb{C}.$$

Then, with the same computation made in (ii) ⇒ (iii) one shows that ω^e is positive on $\mathfrak{A}^e$. Hence, ω is extensible. □

Let ω be a positive linear functional on $\mathfrak{A}$. Let $b \in \mathfrak{A}$ and put

$$\omega_b(a) := \omega(b^*ab), \quad a \in \mathfrak{A}.$$

Then, it is easily seen that ω_b is a positive linear functional on $\mathfrak{A}$, for every $b \in \mathfrak{A}$.

Proposition A.2.20 *For every $b \in \mathfrak{A}$, ω_b is extensible and $\|\omega_b\|_H \leq \omega(b^*b)$. Moreover,*

$$\omega_b(a) = \langle \pi_\omega(a)\lambda_\omega(b)|\lambda_\omega(b)\rangle, \quad a \in \mathfrak{A}.$$

Proof By the Cauchy–Schwarz inequality we have

$$|\omega_b(a)|^2 = |\omega(b^*ab)|^2 \leq \omega(b^*b)\omega(b^*a^*ab) = \omega(b^*b)\omega_b(a^*a), \quad \forall\, a \in \mathfrak{A}.$$

This shows at once that ω_b is Hilbert bounded and that $\|\omega_b\|_H \leq \omega(b^*b)$. The statement then follows from Theorem A.2.19. □

Definition A.2.21 Let $\mathfrak{A}$ be a *-algebra and ω a positive linear functional on $\mathfrak{A}$. We say that ω is *representable* if there exists a closed strongly-cyclic *-representation

π, with strongly-cyclic vector ξ_0, such that

$$\omega(a) = \langle \pi(a)\xi_0 | \xi_0 \rangle, \quad \forall a \in \mathfrak{A}.$$

From the GNS theorem it follows easily that every positive linear functional on a *-algebra $\mathfrak{A}$ with unit e is representable. In the general case we have the following.

Theorem A.2.22 *Let $\mathfrak{A}$ be a *-algebra and ω a positive linear functional on $\mathfrak{A}$. The following statements are equivalent:*

(i) *ω is representable;*
(ii) *ω is Hilbert bounded;*
(iii) *ω is extensible.*

Proof It is an easy consequence of Theorems A.2.19 and A.2.16. □

A.2.3 Boundedness of the GNS Representation; Admissibility

Proposition A.2.23 *Let $\mathfrak{A}$ be a *-algebra with unit e. Let ω be a positive linear functional on $\mathfrak{A}$. The following statements are equivalent:*

(1) *the GNS representation π_ω constructed from ω is bounded;*
(2) *for every $a \in \mathfrak{A}$, there exists a constant C_a, such that*

$$\omega(b^* a^* ab) \leq C_a^2 \omega(b^* b), \quad \forall b \in \mathfrak{A}. \tag{A.2.9}$$

Proof The boundedness of the GNS representation is equivalent to the following condition: For every $a \in \mathfrak{A}$, there exists a constant C_a, such that

$$\|\pi_\omega(a)\lambda_\omega(b)\|^2 \leq C_a^2 \|\lambda_\omega(b)\|^2, \quad \forall b \in \mathfrak{A}.$$

The statement then follows from the equalities

$$\|\pi_\omega(a)\lambda_\omega(b)\|^2 = \omega(b^* a^* ab) \text{ and } \|\lambda_\omega(b)\|^2 = \omega(b^* b).$$ □

Definition A.2.24 Let $\mathfrak{A}$ be a *-algebra with unit e and ω a positive linear functional on $\mathfrak{A}$. We say that ω is *admissible* if condition (2) of Proposition A.2.23 holds.

Remark A.2.25 If ω is admissible, the smallest constant C_a satisfying (A.2.9) is easily determined. Indeed, since, for every $a \in \mathfrak{A}$, $\pi_\omega(a)$ is bounded, its closure $\overline{\pi_\omega(a)}$ is a bounded linear operator on $\mathcal{H}_\omega$. Hence,

$$\omega(b^* a^* ab) = \|\pi_\omega(a)\lambda_\omega(b)\|^2 \leq \|\overline{\pi_\omega(a)}\|^2 \|\lambda_\omega(b)\|^2 = \|\overline{\pi_\omega(a)}\|^2 \omega(b^* b).$$

Hence, $\|\overline{\pi_\omega(a)}\|$ is the best constant for (A.2.9).

Lemma A.2.26 (Kaplansky's Inequality) *Let ω be a positive linear functional on $\mathfrak{A}$. Then, for every $n \in \mathbb{N}$,*

$$\omega(b^*a^*ab) \leq \omega(b^*b)^{1-2^{-n}} \left(\omega(b^*(a^*a)^{2^n} b)\right)^{2^{-n}}, \quad \forall\, a, b \in \mathfrak{A}.$$

Proof Using the Cauchy–Schwarz inequality, we get

$$\omega(b^*a^*ab) \leq \omega(b^*b)^{1/2} \omega\big(b^*(a^*a)^2 b\big)^{1/2}.$$

Thus the statement is true for $n = 1$. Assume that it holds for $n \in \mathbb{N}$. Since

$$\omega\big(b^*(a^*a)^{2^n} b\big) \leq \omega(b^*b)^{1/2} \omega\big(b^*(a^*a)^{2^{n+1}} b\big)^{1/2},$$

we have

$$\begin{aligned}
\omega(b^*a^*ab) &\leq \omega(b^*b)^{1-2^{-n}} \left(\omega\big(b^*(a^*a)^{2^n} b\big)\right)^{2^{-n}} \\
&\leq \omega(b^*b)^{1-2^{-n}} \left(\omega(b^*b)^{1/2} \omega(b^*(a^*a)^{2^{n+1}} b)^{1/2}\right)^{2^{-n}} \\
&= \omega(b^*b)^{1-2^{-(n+1)}} \left(\omega\big(b^*(a^*a)^{2^{n+1}} b\big)\right)^{2^{-(n+1)}}.
\end{aligned}$$

□

Definition A.2.27 Let p be a seminorm on $\mathfrak{A}$. We say that ω is *relatively p-bounded* if

$$\forall\, b \in \mathfrak{A},\ \exists\, \gamma_b > 0, \text{ such that } |\omega_b(a)| \leq \gamma_b\, p(a), \quad \forall\, a \in \mathfrak{A}.$$

We say that ω is *p-bounded* if

$$\exists\, \gamma > 0, \text{ such that } |\omega(a)| \leq \gamma\, p(a), \quad \forall\, a \in \mathfrak{A}.$$

Remark A.2.28 If p is submultiplicative (i.e., $p(ab) \leq p(a)p(b)$, for every $a, b \in \mathfrak{A}$), then p-boundedness implies relative p-boundedness. If $\mathfrak{A}$ has a unit e, relative p-boundedness implies p-boundedness. So, if $\mathfrak{A}$ has a unit and p is submultiplicative, relative p-boundedness and p-boundedness are equivalent.

Proposition A.2.29 *Let p be a seminorm on $\mathfrak{A}$, such that*

$$C_a^2 := \liminf_{n\to\infty} p\big((a^*a)^{2^n}\big)^{2^{-n}} < \infty, \quad \forall\, a \in \mathfrak{A}.$$

Then, if ω is relatively p-bounded, ω is admissible.

Proof By Kaplansky's inequality, for every $a, b \in \mathfrak{A}$,

$$\omega_b(a^*a) \leq \omega(b^*b)^{1-2^{-n}} \left(\omega_b\big((a^*a)^{2^n}\big)\right)^{2^{-n}}$$

$$\leq \omega(b^*b)^{1-2^{-n}} \gamma_b^{2^{-n}} p\big((a^*a)^{2^n}\big)^{2^{-n}}.$$

Taking the lim inf of the right hand side, we obtain

$$\omega(b^*a^*ab) \leq C_a^2 \omega(b^*b)$$

and this completes the proof. □

Corollary A.2.30 *If p is submultiplicative, $p(a^*) \leq p(a)$, for every a in $\mathfrak{A}$ and ω is relatively p-bounded, then*

$$\omega(b^*a^*ab) \leq p(a)^2\omega(b^*b), \quad \forall\, a, b \in \mathfrak{A}.$$

Proof In this case, we have, in fact

$$\liminf_{n\to\infty} p\big((a^*a)^{2^n}\big)^{2^{-n}} \leq \liminf_{n\to\infty} p(a^*a) \leq p(a)^2, \quad \forall\, a \in \mathfrak{A}.$$

□

Definition A.2.31 Let $\mathfrak{A}$ be a *-algebra. A seminorm p on $\mathfrak{A}$ is called a *C*-seminorm* if

(1) $p(ab) \leq p(a)p(b)$, for all $a, b \in \mathfrak{A}$;
(2) $p(a^*a) = p(a)^2$, for each $a \in \mathfrak{A}$.

Remark A.2.32 We remark that by a beautiful result of Sebestyén, (2) always implies (1). We also notice that (1) and (2) imply $p(a^*) = p(a)$, for each $a \in \mathfrak{A}$. Indeed, one has that $p(a)^2 = p(a^*a) \leq p(a^*)p(a)$, therefore $p(a) \leq p(a^*)$; the result then follows by interchanging the roles of a and a^*.

Proposition A.2.33 *Suppose that $\mathfrak{A}$ admits a bounded *-representation π. Then, $p(a) := \|\overline{\pi(a)}\|$, for every $a \in \mathfrak{A}$, is a C*-seminorm on $\mathfrak{A}$.*

The proof is straightforward.

Proposition A.2.34 *Let ω be a positive linear functional on a *-algebra $\mathfrak{A}$ with unit e. Then, ω is admissible, if and only if, there exists a C*-seminorm p on $\mathfrak{A}$, such that ω is relatively p-bounded.*

Proof If there exists a C*-seminorm p on $\mathfrak{A}$, such that ω is relatively p-bounded, then by Corollary A.2.30 we get the assertion. Conversely, assume that ω is admissible. Then, by Remark A.2.25,

$$\omega(b^*a^*ab) \leq \|\overline{\pi_\omega(a)}\|^2\omega(b^*b), \quad \forall\, a \in \mathfrak{A}.$$

Thus, taking also into account Proposition A.2.13 , we obtain

$$|\omega_b(a)| \le \omega_b(e)^{1/2}\omega_b(a^*a)^{1/2} \le \omega_b(e)^{1/2}\omega(b^*b)^{1/2}\|\overline{\pi_\omega(a)}\| = \gamma_b p(a), \quad \forall\, a \in \mathfrak{A},$$

where $\gamma_b \equiv \omega_b(e)^{1/2}\omega(b^*b)^{1/2}$ and $p(a) := \|\overline{\pi_\omega(a)}\|$, $a \in \mathfrak{A}$. The assertion now follows from Definition A.2.27 and the fact that $p(\cdot)$ is a C*-seminorm. □

It is quite clear that the notions of admissibility and representability for a positive linear functional are independent.

Example A.2.35 Consider the *-algebra, say $\mathfrak{A}$, of complex polynomials with pointwise algebraic operations and involution by the complex conjugation. Define

$$\omega(p) = \int_0^\infty p(x)e^{-x}dx, \; p \in \mathfrak{A}.$$

Then, ω is positive and representable, since $\mathfrak{A}$ has a unit. On the other hand, if we take $p(x) = x$ and $q(x) = x^n$, we get

$$\frac{\omega(p^2q^2)}{\omega(q^2)} = \frac{(2n+2)!}{(2n)!} = (2n+2)(2n+1).$$

Hence, ω is not admissible.

In Sect. A.4, we will study in more details the admissibility of positive linear functionals for normed *-algebras. Before doing this, we need, however, a deeper knowledge of the structure of normed or Banach *-algebras.

A.3 Spectral Radius, Spectrum and All That

Definition A.3.1 Let $\mathfrak{A}$ be a normed algebra. For a in $\mathfrak{A}$ define

$$\nu(a) := \limsup_{n\to\infty} \|a^n\|^{1/n}.$$

Proposition A.3.2 *Let a, b be elements in a normed algebra $\mathfrak{A}$, and $\alpha \in \mathbb{C}$. Then,*

(1) $\nu(a) = \inf\{\|a^n\|^{1/n} : n \in \mathbb{N}\}$;
(2) $0 \le \nu(a) \le \|a\|$;
(3) $\nu(\alpha a) = |\alpha|\nu(a)$;
(4) $\nu(ab) = \nu(ba)$ *and* $\nu(a^k) = \nu(a)^k$, $k \in \mathbb{N}$;
(5) *If* $ab = ba$, *then* $\nu(a+b) \le \nu(a) + \nu(b)$ *and* $\nu(ab) \le \nu(a)\nu(b)$.

Proof We shall prove only (1).

Put $\nu = \inf\{\|a^n\|^{1/n} : n \in \mathbb{N}\}$; we shall prove that $\nu = \lim_{n\to\infty} \|a^n\|^{1/n}$. Let $\epsilon > 0$; then, there exists $m \in \mathbb{N}$, such that $\|a^m\|^{1/m} < \nu + \epsilon$. For $n > m$ we can write $n = pm + q$ with $0 \le q \le m-1$. Since $q/m \to 0$, $pm/n \to 1$. Then, we have

$$\|a^n\|^{1/n} = \|a^{pm+q}\|^{1/n} \le \|a^m\|^{p/n}\|a\|^{q/n} < (\nu+\epsilon)^{pm/n}\|a\|^{q/n} \to \nu+\epsilon.$$

This implies that $\limsup_{n\to\infty}\|a^n\|^{1/n} < \nu+\epsilon$; from the arbitrariness of ϵ, we get

$$\limsup_{n\to\infty}\|a^n\|^{1/n} \le \nu.$$

On the other hand, $\nu \le \|a^n\|^{1/n}$; therefore, $\nu \le \liminf_{n\to\infty}\|a^n\|^{1/n}$. In conclusion,

$$\lim_{n\to\infty}\|a^n\|^{1/n} = \nu.$$

□

Proposition A.3.3 *Let a be an element of a normed algebra $\mathfrak{A}$, such that $\|a^2\| = \|a\|^2$. Then, $\nu(a) = \|a\|$.*

Proof If $\|a^2\| = \|a\|^2$, then $\|a^{2^k}\| = \|a\|^{2^k}$, $k \in \mathbb{N}$. Therefore, $\nu(a) = \lim_{k\to\infty}\|a^{2^k}\|^{1/2^k} = \|a\|$. □

Proposition A.3.4 *The equality $\nu(a) = \|a\|$ holds, for every $a \in \mathfrak{A}$, if and only if, $\|a^2\| = \|a\|^2$, for every $a \in \mathfrak{A}$.*

Proof The sufficiency follows from Proposition A.3.3. As for the necessity, if $\nu(a) = \|a\|$, for every $a \in \mathfrak{A}$, we have

$$\|a^2\| = \nu(a^2) = \nu(a)^2 = \|a\|^2.$$

□

In a C*-algebra every hermitian element a, i.e., $a = a^*$, satisfies the condition $\|a^2\| = \|a\|^2$. Thus, one has the following

Corollary A.3.5 *Let $\mathfrak{A}$ be a C*-algebra. For every $a \in \mathfrak{A}$, such that $a = a^*$, $\nu(a) = \|a\|$.*

Remark A.3.6 If $ab = ba$, then $\nu(a+b) \le \nu(a) + \nu(b)$ and $\nu(ab) \le \nu(a)\nu(b)$. Therefore, if $\mathfrak{A}$ is commutative, the function $\nu(\cdot)$ is a seminorm. If, in addition, $\|a^2\| = \|a\|^2$, for every $a \in \mathfrak{A}$, then $\nu(\cdot)$ is a norm. It can be proved that a normed algebra $\mathfrak{A}$, where $\|a^2\| = \|a\|^2$, for every $a \in \mathfrak{A}$, is *necessarily* commutative (see, for instance, [8, p. 345, (B.6.16) Theorem]).

Let $\mathfrak{A}$ be a *-algebra with unit e. We say that an element $a \in \mathfrak{A}$ has a *right inverse* if there exists $b \in \mathfrak{A}$, such that $ab = e$. Similarly, a has a *left inverse* if there exists $c \in \mathfrak{A}$, such that $ca = e$. An element $a \in \mathfrak{A}$ is called *invertible* if it has both left and right inverse. In this case, the left inverse c and the right inverse b of a coincide. Indeed, $c = ce = c(ab) = (ca)b = eb = b$. The inverse of a is denoted, as usual, by a^{-1}. The set $\mathfrak{G}$ of all invertible elements of $\mathfrak{A}$ constitutes a multiplicative group.

Proposition A.3.7 *Let $\mathfrak{A}$ be a Banach algebra with unit e. If $a \in \mathfrak{A}$ with $\nu(e-a) < 1$, then a is invertible and $a^{-1} = e + \sum_{n=1}^{\infty}(e-a)^n$.*

Proof If $\nu(e-a) < 1$, then $\lim_{n\to\infty} \|(e-a)^n\|^{1/n} < 1$. Thus, the numerical series $\sum_{n=0}^{\infty} \|(e-a)^n\|$ is convergent. This implies that the series $\sum_{n=0}^{\infty}(e-a)^n$ converges absolutely to an element $b \in \mathfrak{A}$. We prove that $a^{-1} = e + b$. Indeed,

$$a(e+b) = a+\sum_{n=1}^{\infty} a(e-a)^n = a+\sum_{n=1}^{\infty}(a-e)(e-a)^n+\sum_{n=1}^{\infty}(e-a)^n = a+(e-a) = e.$$

In a similar way one also proves that $(e+b)a = e$. □

Since, for $b \in \mathfrak{A}$, one has $\nu(b) \leq \|b\|$, we get

Corollary A.3.8 *If $a \in \mathfrak{A}$, where $\mathfrak{A}$ is a Banach algebra with unit, and $\|e-a\| < 1$, then a is invertible.*

With an obvious substitution in Proposition A.3.7, one also has the following

Corollary A.3.9 *Let $\mathfrak{A}$ be a Banach algebra with unit e. If $a \in \mathfrak{A}$ with $\nu(a) < 1$, then $e-a$ is invertible and $(e-a)^{-1} = e + \sum_{n=1}^{\infty} a^n$.*

Theorem A.3.10 *If $\mathfrak{A}$ is a Banach algebra with unit e, then the group $\mathfrak{G}$ of all invertible elements of $\mathfrak{A}$ is open in $\mathfrak{A}$. Moreover, the map $a \mapsto a^{-1}$ is continuous on $\mathfrak{G}$.*

Proof Let $a \in \mathfrak{G}$. The set $\mathcal{U}_a = \left\{b \in \mathfrak{A} : \|a-b\| < \frac{1}{\|a^{-1}\|}\right\}$ is obviously a neighborhood of a. We prove that $\mathcal{U}_a \subset \mathfrak{G}$. Indeed, if $b \in \mathcal{U}_a$, one has

$$\|e - a^{-1}b\| = \|a^{-1}(a-b)\| \leq \|a^{-1}\|\,\|a-b\| < 1.$$

From Corollary A.3.8, $a^{-1}b$ is invertible and this implies that b is also invertible.

As for the continuity of the inversion, we can use the identity

$$b^{-1} - a^{-1} = a^{-1}(a-b)a^{-1} + (b^{-1} - a^{-1})(a-b)a^{-1}$$

which implies

$$\|b^{-1} - a^{-1}\| \leq \|a^{-1}\|^2\|b-a\| + \|b^{-1} - a^{-1}\|\|b-a\|\|a^{-1}\|.$$

Thus

$$\|b^{-1} - a^{-1}\| \leq \frac{\|a^{-1}\|^2}{1 - \|b-a\|\|a^{-1}\|}\|b-a\|$$

and the right hand side converges to 0 if $b \to a$. □

One of the most important and useful concepts in the theory of Banach algebras is the notion of the spectrum of an element.

Definition A.3.11 Let a be an element of an algebra $\mathfrak{A}$ with unit e. The *resolvent set* of a is the following subset of the complex plane $\mathbb{C}$

$$\rho_{\mathfrak{A}}(a) = \{\lambda \in \mathbb{C} : a - \lambda e \text{ is invertible in } \mathfrak{A}\}.$$

The *spectrum* of a is the set $\sigma_{\mathfrak{A}}(a) = \mathbb{C} \setminus \rho_{\mathfrak{A}}(a)$.

In general, $\sigma_{\mathfrak{A}}(\cdot)$ is neither non-empty, nor bounded (see, e.g., [9, 4.3 Examples]). Theorems A.3.14 and A.3.15 give that context, where both properties always hold.

Remark A.3.12 Clearly if $\mathfrak{B}$ is a subalgebra of $\mathfrak{A}$, we may have $\rho_{\mathfrak{B}}(a) \neq \rho_{\mathfrak{A}}(a)$ (and consequently $\sigma_{\mathfrak{B}}(a) \neq \sigma_{\mathfrak{A}}(a)$). For this reason, we have explicitly mentioned the dependence on $\mathfrak{A}$ in the notation.

The following statement can be easily proved.

Proposition A.3.13 *If $\mathfrak{B} \subset \mathfrak{A}$ is a subalgebra containing the unit of $\mathfrak{A}$ then, for each $a \in \mathfrak{B}$, $\sigma_{\mathfrak{A}}(a) \subseteq \sigma_{\mathfrak{B}}(a)$.*

Theorem A.3.14 *If a is an element of a normed algebra $\mathfrak{A}$ with unit e, then $\sigma_{\mathfrak{A}}(a)$ is nonempty.*

Proof When it is not so, the inverse $(a - \lambda e)^{-1}$ would exist, for every $\lambda \in \mathbb{C}$. In particular, the element a^{-1} exists. By the Hahn–Banach theorem there is a bounded linear functional Φ on $\mathfrak{A}$, such that $\Phi(a^{-1}) = 1$. Let $a(\lambda) = (a - \lambda e)^{-1}$ and $f(\lambda) = \Phi(a(\lambda))$. Since $\Phi(a^{-1}) = 1$, $f(0) = 1$. We shall prove that f is analytic on the whole plane $\mathbb{C}$. Indeed, the identity $a(\lambda) - a(\mu) = (\lambda - \mu)a(\lambda)a(\mu)$, $\lambda, \mu \in \mathbb{C}$, implies that

$$\lim_{\lambda \to \mu} \frac{f(\lambda) - f(\mu)}{\lambda - \mu} = \lim_{\lambda \to \mu} \Phi(a(\lambda)a(\mu)) = \Phi(a(\mu)^2).$$

Since,

$$|f(\lambda)| \leq \|\Phi\| \|a(\lambda)\|, \quad \lambda \in \mathbb{C}$$

and $a(\lambda) = \lambda^{-1}\left(\lambda^{-1}a - e\right)^{-1}$, with $a(\lambda) \to 0$, as $|\lambda| \to \infty$ (take into account the continuity of the inverse map), it follows that $|f(\lambda)|$ is bounded. On the other hand, by Liouville's theorem the function f is constant; hence $f = 0$, which is impossible since $f(0) = 1$. □

Theorem A.3.15 *Let $\mathfrak{A}$ be a Banach algebra with unit and $a \in \mathfrak{A}$. Then,*

(1) *$|\lambda| \leq \|a\|$, for every $\lambda \in \sigma_{\mathfrak{A}}(a)$;*
(2) *$\sigma_{\mathfrak{A}}(a)$ is a nonempty compact subset of $\mathbb{C}$.*

Proof

1. Let $\lambda \in \sigma_{\mathfrak{A}}(a)$ with $|\lambda| > \|a\|$. Then, $\|\lambda^{-1}a\| < 1$, hence by Corollary A.3.8 the element $e - \lambda^{-1}a$ is invertible and clearly the same is also true for the element $\lambda e - a$. But, this means that $\lambda \notin \sigma_{\mathfrak{A}}(a)$, a contradiction.
2. We already know that $\sigma_{\mathfrak{A}}(a)$ is nonempty and bounded. We now prove that it is closed. The map $f : \lambda \in \mathbb{C} \to a - \lambda e \in \mathfrak{A}$ is continuous. Since the set $\mathfrak{G}$ of invertible elements is open, the set $f^{-1}(\mathfrak{G} \cap \operatorname{Im} f)$ is also open, therefore its complement is closed. But, this is exactly $\sigma_{\mathfrak{A}}(a)$. □

Theorem A.3.16 (Mazur–Gelfand) *Let $\mathfrak{A}$ be a normed algebra with unit e. Assume that each nonzero element of $\mathfrak{A}$ is invertible* (i.e., $\mathfrak{A}$ is a division algebra). *Then, $\mathfrak{A}$ is isometrically isomorphic to $\mathbb{C}$.*

Proof It suffices to show that the map $\lambda \in \mathbb{C} \to \lambda e \in \mathfrak{A}$ is surjective. Let $a \in \mathfrak{A}$ be nonzero. By Theorem A.3.14, there exists $\lambda \in \mathbb{C}$, such that $a - \lambda e$ is not invertible. Since each nonzero element in $\mathfrak{A}$ has an inverse, it must be $a - \lambda e = 0$, i.e., $a = \lambda e$. □

Remark A.3.17 We warn the reader that not too much regularity can be imposed to a Banach algebra without trivializing it. The Mazur–Gelfand theorem already implies the existence of non invertible elements in a non trivial normed algebra. Similarly, in a Banach algebra $\mathfrak{A}$ with unit, the following two facts hold (see, for instance, [8, (B.4.7) Proposition])

1. If $\|a^{-1}\| = \|a\|^{-1}$, for every $a \in \mathfrak{G}$, then $\mathfrak{A}$ is topologically isomorphic to $\mathbb{C}$.
2. If $\|ab\| = \|a\|\|b\|$, for all $a, b \in \mathfrak{A}$, then $\mathfrak{A}$ is topologically isomorphic to $\mathbb{C}$.

Lemma A.3.18 *Let $\mathfrak{A}$ be a Banach algebra with unit e and $\sum_{n=0}^{\infty} \beta_n z^n$ be a power series with radius of convergence R. Let $f(z)$ be the sum of the series. Then, for each $a \in \mathfrak{A}$, with $\|a\| \le r < R$, the series $\sum_{n=0}^{\infty} \beta_n a^n$ converges to an element of $\mathfrak{A}$ that we call $f(a)$. Moreover, if $\lambda \in \sigma_{\mathfrak{A}}(a)$ and $|\lambda| < r$, then $f(\lambda) \in \sigma_{\mathfrak{A}}(f(a))$.*

Proof For $M > N$, we have

$$\left\| \sum_{n=N}^{M} \beta_n a^n \right\| \le \sum_{n=N}^{M} |\beta_n| \|a^n\| \le \sum_{n=N}^{M} |\beta_n| r^n \to 0$$

as $N, M \to \infty$. To prove the second statement we proceed as follows:

$$\begin{aligned} f(a) - f(\lambda)e &= \sum_{n=0}^{\infty} \beta_n (a^n - \lambda^n e) \\ &= \sum_{n=0}^{\infty} \beta_n (a - \lambda e) \sum_{k=0}^{n-1} \lambda^k a^{n-k-1} \\ &= (a - \lambda e) \sum_{n=0}^{\infty} \beta_n \sum_{k=0}^{n-1} \lambda^k a^{n-k-1}. \end{aligned}$$

Since, $\left\|\sum_{k=0}^{n-1}\lambda^k a^{n-k-1}\right\| \leq nr^{n-1}$, the latter series converges to some $b \in \mathfrak{A}$. Thus, $f(a) - f(\lambda)e = (a - \lambda e)b$. If $f(a) - f(\lambda)e$ has an inverse, then $a - \lambda e$ also has an inverse. This completes the proof. □

In the next Lemma, we list some easy relations, of purely algebraic nature, that hold for the spectrum.

Lemma A.3.19 *Let $\mathfrak{A}$ be a *-algebra with unit e. For $a, b \in \mathfrak{A}$ and $\beta \in \mathbb{C}$, we have*

(1) $\sigma_{\mathfrak{A}}(a - \beta e) = \{\lambda - \beta : \lambda \in \sigma_{\mathfrak{A}}(a)\}$;
(2) $\sigma_{\mathfrak{A}}(a^*) = \{\overline{\lambda} : \lambda \in \sigma_{\mathfrak{A}}(a)\}$;
(3) $\sigma_{\mathfrak{A}}(ab) \cup \{0\} = \sigma_{\mathfrak{A}}(ba) \cup \{0\}$;
(4) *If a^{-1} exists, then* $\sigma_{\mathfrak{A}}(a^{-1}) = \{\lambda^{-1} : \lambda \in \sigma_{\mathfrak{A}}(a)\}$.

Proof For proving (3), we take into account that if $\lambda \notin \sigma_{\mathfrak{A}}(ba)$, then

$$(ab - \lambda e)\big(a(ba - \lambda e)^{-1}b - e\big) = \lambda e = \big(a(ba - \lambda e)^{-1}b - e\big)(ab - \lambda e).$$

This implies that $(ab - \lambda e)$ is invertible, at least if $\lambda \neq 0$. Thus, $\sigma_{\mathfrak{A}}(ab) \cup \{0\} \subseteq \sigma_{\mathfrak{A}}(ba) \cup \{0\}$. Interchanging the roles of a and b one obtains the opposite inclusion □

The statement (3) has the following very important consequence:

Theorem A.3.20 (Wielandt–Wintner) *Let $\mathfrak{A}$ be a Banach algebra with unit e. Then, there are no elements $a, b \in \mathfrak{A}$, such that $ab - ba = e$.*

Proof Assume that two such elements exist. From Theorem A.3.15 $\sigma_{\mathfrak{A}}(ab)$ and $\sigma_{\mathfrak{A}}(ba)$ are non empty subsets of $\mathbb{C}$. By (3) of Lemma A.3.19, $\sigma_{\mathfrak{A}}(ab)$ and $\sigma_{\mathfrak{A}}(ba)$ differ at most by 0. But, this is impossible, since the assumption and (1) of Lemma A.3.19 imply that $\sigma_{\mathfrak{A}}(ab) = 1 + \sigma_{\mathfrak{A}}(ba)$. □

Theorem A.3.20 shows the existence of algebras that do not have *bounded* representations. Indeed, let us consider on $C_c^\infty(\mathbb{R})$ the operators $a = d/dx$ and $b : f \to id_{\mathbb{R}} f, x \in \mathbb{R}$, where $id_{\mathbb{R}}$ is the identity map of $\mathbb{R}$. Then, it is easily seen that $(ab - ba)f(x) = f(x)$; i.e., $ab - ba = e$. Thus, the algebra $\mathfrak{A}$ of operators on $C_c^\infty(\mathbb{R})$ generated by a, b has no bounded representations.

Definition A.3.21 Let $\mathfrak{A}$ be an algebra. The *spectral radius* of an element $a \in \mathfrak{A}$ is defined as

$$|a|_\sigma = \sup\{|\lambda| : \lambda \in \sigma_{\mathfrak{A}}(a)\}.$$

In general, one has that $0 \leq |a|_\sigma \leq +\infty$. The next theorem, due to Beurling and Gelfand, shows that in a Banach algebra the spectral radius of an element is always finite; for $\nu(\cdot)$, see Definition A.3.1.

Theorem A.3.22 *Let $\mathfrak{A}$ be a Banach algebra with unit e and $a \in \mathfrak{A}$. Then,*

$$|a|_\sigma = \lim_{n\to\infty} \|a^n\|^{1/n} = \nu(a).$$

Proof If $a \in \mathfrak{A}$ and $\lambda \in \sigma_{\mathfrak{A}}(a)$, then by Theorem A.3.15, $|\lambda| \leq \|a\|$ and by Lemma A.3.18, $\lambda^n \in \sigma_{\mathfrak{A}}(a^n)$. This implies

$$|a|_\sigma^n \leq |a^n|_\sigma.$$

Then,

$$|a|_\sigma \leq (|a^n|_\sigma)^{1/n} \leq \|a^n\|^{1/n},$$

and so

$$|a|_\sigma \leq \lim_{n\to\infty} \|a^n\|^{1/n} = \nu(a).$$

To prove the converse, we make use of an argument of complex variables.

If $r > |a|_\sigma$, then the function $f(\lambda) = (\lambda e - a)^{-1}$ is analytic in the exterior of the circle γ_r of radius r. In this region, it has the following Laurent's expansion

$$f(\lambda) = \sum_{n=0}^{\infty} \frac{a^n}{\lambda^{n+1}}, \quad \text{with } a^n = \frac{1}{2\pi i}\int_{\gamma_r} \lambda^n f(\lambda)d\lambda.$$

Now, put $M(r) = \sup_{0\leq\theta\leq 2\pi} \|f(re^{i\theta}\|$. Then,

$$\|a^n\| \leq r^{n+1}M(r) = r^n(rM(r)) \text{ and so } \lim_{n\to\infty} \|a^n\|^{1/n} \leq r.$$

Since this is true for any $r > |a|_\sigma$, we finally get

$$\lim_{n\to\infty} \|a^n\|^{1/n} \leq |a|_\sigma.$$

□

Definition A.3.23 Let $\mathfrak{A}$ be a Banach *-algebra with unit e. An element $a \in \mathfrak{A}$ is called *positive* if $a^* = a$ and $\sigma_{\mathfrak{A}}(a) \subset \mathbb{R}^+ \cup \{0\}$.

Positive elements, as well as positive linear functionals, play a crucial role in the theory of Banach *-algebras.

Proposition A.3.24 *Let $\mathfrak{A}$ be a C*-algebra with unit e and let $a = a^* \in \mathfrak{A}$. The following statements are equivalent:*

(1) $\sigma_{\mathfrak{A}}(a) \subset \mathbb{R}^+ \cup \{0\}$ (i.e., a is positive)*;*
(2) $a = b^*b$*, for some* $b \in \mathfrak{A}$*;*
(3) $a = h^2$*, for some* $h \in \mathfrak{A}$ *with* $h^* = h$*;*
(4) *if* $\|a\| \leq \alpha$*, then* $\|\alpha e - a\| \leq \alpha$.

Remark A.3.25 Generally, in a Banach *-algebra there is no equivalence between the four conditions above. We only have (1) $\Rightarrow$ (2), (1) $\Rightarrow$ (4).

A.4 Admissible Positive Linear Functionals

Lemma A.4.1 *Let $\mathfrak{A}$ be a Banach *-algebra with unit e. For any $a \in \mathfrak{A}$ with $a = a^*$ and $\nu(a) < 1$, there exists an element $b \in \mathfrak{A}$ with $b = b^*$, such that $2b - b^2 = a$.*

Proof The basic idea of the proof is to try to define the element $b = e + \sqrt{e-a}$, which formally solves the given equation. As it is known from the elementary calculus, if $|\lambda| < 1$, the series

$$-\sum_{k=1}^{\infty}\binom{1/2}{n}(-\lambda)^n,$$

converges to a solution ζ of the equation $2z - z^2 = \lambda$. We can apply Lemma A.3.18, taking into account that $\nu(a) = \|a\|$, since $a = a^*$. Thus, the series

$$-\sum_{k=1}^{\infty}\binom{1/2}{n}(-a)^n,$$

converges to an element $b \in \mathfrak{A}$, which satisfies the desired equation. The fact that $b = b^*$ follows from the continuity of the involution. □

Theorem A.4.2 *Let $\mathfrak{A}$ be a Banach *-algebra with unit e and ω a positive linear functional on $\mathfrak{A}$. Then,*

$$|\omega(b^*hb)| \le \nu(h)\omega(b^*b), \quad \forall\, b \in \mathfrak{A},\ h = h^* \in \mathfrak{A}.$$

Proof Assume that $\nu(h) < 1$. By Lemma A.4.1, there exist elements $r, s \in \mathfrak{A}$ with $r = r^*$, $s = s^*$, such that $2r - r^2 = h$, $2s - s^2 = -h$. For $b \in \mathfrak{A}$ we put $x = (e-r)b$, $y = (e-s)b$. Then, it is easily seen that

$$x^*x = b^*(e-r)^2b = b^*(e-h)b$$
$$y^*y = b^*(e-s)^2b = b^*(e+h)b.$$

Then, $\omega(b^*(e-h)b) \ge 0$ and $\omega(b^*(e+h)b) \ge 0$. These inequalities imply that

$$|\omega(b^*hb)| \le \omega(b^*b), \tag{A.4.1}$$

in the case $\nu(h) < 1$. For an arbitrary $h = h^* \in \mathfrak{A}$ and for $\epsilon > 0$, we put $k = h/(\nu(h)+\epsilon)$; thus $\nu(k) < 1$. Applying the inequality (A.4.1), we obtain

$$|\omega(b^*hb)| \leq (\nu(h)+\epsilon)\omega(b^*b).$$

Since ϵ is arbitrary, we get the result. □

From Definition A.2.24 and the previous theorem it follows immediately that

Corollary A.4.3 *Every positive linear functional on a Banach *-algebra with unit is admissible.*

But there is something more!

Theorem A.4.4 *Every positive linear functional ω on a Banach *-algebra $\mathfrak{A}$ with unit e is bounded and $\|\omega\|^\star = \omega(e)$.*

Proof Indeed, applying Theorem A.4.2 with $b = e$ and $h = a^*a$, $a \in \mathfrak{A}$, we obtain

$$\omega(a^*a) \leq \omega(e)\nu(a^*a).$$

Then, a simple application of the Cauchy–Schwarz inequality gives

$$|\omega(a)| \leq \omega(e)\nu(a^*a)^{1/2}, \ a \in \mathfrak{A}.$$

Taking into account that $\nu(a^*a) = \|a^*a\|$, we obtain

$$|\omega(a)| \leq \omega(e)\|a^*a\|^{1/2} \leq \omega(e)\|a\|, \ a \in \mathfrak{A};$$

thus ω is bounded and $\|\omega\|^\star \leq \omega(e)$, but evidently $\omega(e) \leq \|\omega\|^\star$ too, therefore the equality holds. □

The following theorem gives the converse of the previous one for C*-algebras.

Theorem A.4.5 *Let $\mathfrak{A}$ be a C*-algebra with unit e. Every continuous linear functional ω on $\mathfrak{A}$, such that $\|\omega\|^\star = \omega(e)$ is positive.*

Proof First of all, we prove that $\omega(a^*a)$ is real, for every $a \in \mathfrak{A}$. If $\omega : \mathfrak{A} \to \mathbb{C}$ is a linear functional, define $\omega^* : \mathfrak{A} \to \mathbb{C}$, such that $\omega^*(a) := \overline{\omega(a^*)}$, for all $a \in \mathfrak{A}$. Then, ω^* is a linear functional on $\mathfrak{A}$, called *adjoint* of ω. If $\omega^* = \omega$, that is $\omega(a^*) = \overline{\omega(a)}$, for all $a \in \mathfrak{A}$, then ω is called *hermitian* (see also Sect. A.2.1). Each linear functional ω can be written as $\omega = \omega_1 + i\omega_2$, with ω_1, ω_2 hermitian. It suffices to put $\omega_1(a) = \frac{1}{2}\big(\omega(a) + \overline{\omega(a^*)}\big)$ and $\omega_2(a) = \frac{1}{2i}\big(\omega(a) - \overline{\omega(a^*)}\big)$, $a \in \mathfrak{A}$. We assume that $\omega(e) = 1$. Then, $\omega_1(e) = 1$ and $\omega_2(e) = 0$. Moreover, $\omega_1(a^*a)$ is real, for every $a \in \mathfrak{A}$. We shall show that ω_2 is on the whole $\mathfrak{A}$ zero. Let $h = h^* \in \mathfrak{A}$; define $x = \alpha e - ih$, where α is a real number. Then,

$$\|x\|^2 = \|\alpha^2 e + h^2\| \leq \alpha^2 + \|h\|^2.$$

The equality

$$|\omega(x)|^2 = \alpha^2 + 2\alpha\omega_2(h) + \omega_1(h)^2 + \omega_2(h)^2$$

implies that $\alpha^2 \leq |\omega(x)|^2 - 2\alpha\omega_2(h)$. Hence,

$$\|x\|^2 \leq |\omega(x)|^2 - 2\alpha\omega_2(h) + \|h\|^2 \leq \|x\|^2 - 2\alpha\omega_2(h) + \|h\|^2.$$

In conclusion,

$$2\alpha\omega_2(h) \leq \|h\|^2.$$

This inequality holds for any $\alpha \in \mathbb{R}$. Thus, $\omega_2(h) = 0$, for all $h = h^* \in \mathfrak{A}$. This easily implies that ω_2 is identically zero. Hence, $\omega(a^*a)$ is real, for every $a \in \mathfrak{A}$.

To prove the positivity of ω, we assume that there exists $a \in \mathfrak{A}$, such that $\|a^*a\| < 1$ and $\omega(a^*a) < 0$. By Lemma A.4.1, there exists $b = b^* \in \mathfrak{A}$, such that $a^*a = 2b - b^2$; thus $e - a^*a = (e - b)^2$, i.e., $\omega(e - a^*a)$ is positive. Therefore, by our assumption and Proposition A.3.24, we have

$$1 = \omega(e) = \omega(e - a^*a) + \omega(a^*a) < \omega(e - a^*a) \leq \|e - a^*a\| \leq 1.$$

But, this is a contradiction. □

Remark A.4.6 One may wonder if any *hermitian* linear functional on a C*-algebra is also continuous. The answer is negative as the following example shows.

Let $\mathfrak{A}$ be a C*-algebra and $\{x_n\}$ a sequence of linearly independent elements of the real vector subspace $\mathfrak{A}_h := \{x \in \mathfrak{A} : x = x^*\}$. Assume $\|x_n\| = 1$, for every natural number n. Let $E = \{y_\alpha : \alpha \in \Delta\}$ be a Hamel basis for $\mathfrak{A}_h$ containing the sequence $\{x_n\}$. Clearly E is a Hamel basis for $\mathfrak{A}$, too. Now define

$$f(x) = \begin{cases} n, & x = x_n \\ 0, & x \in E \setminus \{x_n\}. \end{cases}$$

Extend f by linearity to the whole algebra $\mathfrak{A}$. Then, f is a linear functional on $\mathfrak{A}$ and it is hermitian. Indeed, let $x = \sum_{\alpha \in F} \lambda_\alpha y_\alpha + \sum_{k=1}^n \mu_k x_k$, where F is a finite subset of Δ and $y_\alpha \neq x_k$, for every $\alpha \in F$ and $k \in \{1, \ldots, n\}$. Then, $x^* = \sum_{\alpha \in F} \overline{\lambda_\alpha} y_\alpha + \sum_{k=1}^n \overline{\mu_k} x_k$. Hence,

$$f(x) = \sum_{k=1}^n \mu_k \cdot k \ \text{ and } \ f(x^*) = \sum_{k=1}^n \overline{\mu_k} \cdot k.$$

Thus, $f(x^*) = \overline{f(x)}$, $\forall\, x \in \mathfrak{A}$; i.e., f is hermitian. The functional f is clearly not bounded, since $\|x_n\| = 1$, but $f(x_n) = n \to \infty$.

A.5 C*-Seminorms on Banach *-Algebras

If $\mathfrak{A}$ is a Banach *-algebra with unit e, Theorem A.4.4 allows us to construct a seminorm on $\mathfrak{A}$ by putting

$$p(a) = \sup_{\omega \in \mathcal{S}(\mathfrak{A})} \omega(a^* a)^{1/2}, \quad a \in \mathfrak{A},$$

where $\mathcal{S}(\mathfrak{A})$ denotes the set of all states on $\mathfrak{A}$; that is, of all positive linear functionals ω on $\mathfrak{A}$ with $\omega(e) = 1$. Then, $p(a) \leq \|a\|$, for every $a \in \mathfrak{A}$.

We notice that if ω is a positive linear functional on $\mathfrak{A}$ and $b \in \mathfrak{A}$, then the linear functional ω_b defined before Proposition A.2.20 is also positive. Moreover, if $\omega(b^* b) = 1$, then $\omega_b \in \mathcal{S}(\mathfrak{A})$.

Proposition A.5.1 *If $\mathfrak{A}$ is a Banach *-algebra with unit e, then*

$$p(a) := \sup_{\omega \in \mathcal{S}(\mathfrak{A})} \omega(a^* a)^{1/2},$$

defines a C-seminorm on $\mathfrak{A}$.*

Proof We prove (1) of Definition A.2.31. We follow [95]. Let $a, b \in \mathfrak{A}$ and $c = ab$. Then, if $\omega \in \mathcal{S}(\mathfrak{A})$, $\omega(c^* c) = \omega_b(a^* a)$. If $\omega(b^* b) = 0$, then by the Cauchy–Schwarz inequality, $\omega(c^* c) = 0$, so that inequality (1) trivially holds. Now suppose that $\omega(b^* b) > 0$ and put $x = b/\omega(b^* b)^{1/2}$. It follows that $\omega_x \in \mathcal{S}(\mathfrak{A})$ and

$$\omega(c^* c) = \omega_x(a^* a)\omega(b^* b),$$

therefore $\omega(c^* c) \leq p(a)^2 p(b)^2$. Taking sup on the left hand side, we get the result.

As for (2) of Definition A.2.31, if $a \in \mathfrak{A}$ and $\omega \in \mathcal{S}(\mathfrak{A})$, we have

$$\omega(a^* a)^2 \leq \omega(a^* a a^* a) \leq p(a^* a)^2,$$

where the first inequality is due to the Cauchy–Schwarz inequality (see, in particular, Proposition A.2.13) and the second to the definition of p. This implies that $p(a)^2 \leq p(a^* a)$. The statement then follows from submultiplicativity of p that we have already proven. □

Note that, in general, p is not a norm; i.e., it may happen that $p(a) = 0$, with $0 \neq a \in \mathfrak{A}$. Let

$$\mathcal{N}_p = \{b \in \mathfrak{A} : p(b) = 0\}.$$

It is easily seen that $\mathcal{N}_p$ is a closed *-ideal of $\mathfrak{A}$. Then, the quotient $\mathfrak{A}/\mathcal{N}_p$ is a C*-algebra, under the norm $\|[a]\| = p(a)$, where $a \in \mathfrak{A}$ and $[a] \equiv a + \mathcal{N}_p$.

Definition A.5.2 The **-radical* $\mathsf{R}(\mathfrak{A})$ of a Banach *-algebra $\mathfrak{A}$ is the intersection of the kernels of all *-representations of $\mathfrak{A}$ on some Hilbert space. The Banach *-algebra $\mathfrak{A}$ is called **-semisimple* if its *-radical $\mathsf{R}(\mathfrak{A})$ is $\{0\}$ (in this regard, see also Definitions 3.1.21 and 3.1.23).

Proposition A.5.3 *The *-radical* $\mathsf{R}(\mathfrak{A})$ *of a Banach *-algebra* $\mathfrak{A}$ *with unit* e *coincides with* $\mathcal{N}_p$.

Proof If $a \in \mathsf{R}(\mathfrak{A})$ and π is a *-representation of $\mathfrak{A}$ on a Hilbert space, then $\pi(a) = 0$. This is, in particular true, for the GNS representation π_ω constructed by starting from an element $\omega \in \mathcal{S}(\mathfrak{A})$. Thus, $\|\pi_\omega(a)\lambda_\omega(e)\| = 0$ and this implies that $p(a) = 0$. Conversely, we prove that if $\pi(a) \neq 0$, for some *-representation π, then $a \notin \mathcal{N}_p$. Indeed, if $\pi(a) \neq 0$, there exists a vector $\xi_0 \in \mathcal{H}$, $\|\xi_0\| = 1$, such that $\pi(a)\xi_0 \neq 0$. We define

$$\omega(a) := \langle \pi(a)\xi_0|\xi_0\rangle, \quad a \in \mathfrak{A}.$$

Since $\omega(a^*a) = \|\pi(a)\xi_0\|^2 \neq 0$ and ω is positive, it follows that $a \notin \mathcal{N}_p$. □

In what we have done so far, there is a *question* that remains up to now unsolved. Can we be sure that the set of positive linear functionals on $\mathfrak{A}$ does not reduce only to $\{0\}$? *We shall answer this question* in the case where $\mathfrak{A}$ is a C*-algebra.

Proposition A.5.4 *Let* $\mathfrak{A}$ *be a C*-algebra with unit* e. *For each* $a \in \mathfrak{A}$, *there exists a positive linear functional* ω *on* $\mathfrak{A}$, *with* $\omega(e) = 1$ *and* $\omega(a^*a) = \|a\|^2$.

Proof Let $a \in \mathfrak{A}$ and $\mathfrak{B} := \{\alpha e + \beta a^*a : \alpha, \beta \in \mathbb{C}\}$. Then, $\mathfrak{B}$ is a vector subspace of $\mathfrak{A}$. We define a linear functional f on $\mathfrak{B}$ by $f(\alpha e + \beta a^*a) := \alpha + \beta\|a\|^2$. Since $|a^*a|_\sigma = \|a^*a\|$, $a \in \mathfrak{A}$, we have that

$$|\alpha + \beta\|a\|^2| \leq \sup\{|\alpha + \beta\lambda| : \lambda \in \sigma_{\mathfrak{A}}(a^*a)\} \leq \|\alpha e + \beta a^*a\|.$$

Therefore, f is continuous and $\|f\|^* \leq 1$; but $f(e) = 1$, hence $\|f\|^* = 1$. By the Hahn–Banach theorem, f has an extension ω to the whole $\mathfrak{A}$ with $\|\omega\|^* = 1$. By Proposition A.4.5 it follows that ω is positive. □

The following theorem, which is one of the crucial points of the theory of C*-algebras, is now a simple consequence of A.5.4 (recall also that $p(a) =: \|[a]\|, a \in \mathfrak{A}$).

Theorem A.5.5 *Let* $\mathfrak{A}$ *be a C*-algebra with unit. Then,*

$$\|a\| = p(a) = \sup_{\omega \in \mathcal{S}(\mathfrak{A})} \omega(a^*a)^{1/2}, \quad \forall\, a \in \mathfrak{A}.$$

Remark A.5.6 The previous theorem, which also shows that any C*-algebra is *-semisimple, was first proved by Gelfand and Naimark; for this reason p is often called in the literature *Gelfand–Naimark seminorm*.

A.6 Gelfand Theory

We recall some basic definitions on ideals.

Let $\mathfrak{A}$ be an algebra. A vector subspace $\mathfrak{I}$ of $\mathfrak{A}$ is called a *left ideal* of $\mathfrak{A}$ if $ax \in \mathfrak{I}$, for any $a \in \mathfrak{A}$ and $x \in \mathfrak{I}$. A right ideal is similarly defined. A two-sided ideal, or simply *ideal* is a left ideal that is also a right ideal.

If $\mathfrak{A}$ is unital, a left ideal $\mathfrak{I}$ in $\mathfrak{A}$ is called *maximal* if $\mathfrak{I}$ is not contained in any other proper left ideal of $\mathfrak{A}$. With the help of Zorn's lemma it can be proved that each left ideal of $\mathfrak{A}$ is contained in a maximal one.

If $\mathfrak{A}$ has no unit, then a left ideal $\mathfrak{I}$ in $\mathfrak{A}$ is called *modular* if there exists $u \in \mathfrak{A}$, such that $a - au \in \mathfrak{I}$, for all $a \in \mathfrak{A}$. In this case, u is called a *right unit modulo* $\mathfrak{I}$.

The (Jacobson) *radical* of $\mathfrak{A}$, denoted by $\mathrm{rad}(\mathfrak{A})$ is the intersection of all left maximal modular ideals of $\mathfrak{A}$. For more details, see [8, pp. 124, 125]). If $\mathrm{rad}(\mathfrak{A}) = \{0\}$, then $\mathfrak{A}$ is called *semisimple*.

A.6.1 The Commutative Case

Let $\mathfrak{A}$ be a *commutative* Banach algebra. A *multiplicative linear functional* or *character* of $\mathfrak{A}$ is a non-zero linear functional ω on $\mathfrak{A}$ that preserves multiplication, i.e.,

$$\omega(ab) = \omega(a)\omega(b), \quad \forall\, a, b \in \mathfrak{A}.$$

Since ω is non-zero, it is surjective. Indeed, if $0 \neq \alpha \in \mathbb{C}$ and $a \in \mathfrak{A}$, with $\omega(a) \neq 0$, then the element $b = \alpha a/\omega(a)$ of $\mathfrak{A}$ gives $\omega(b) = \alpha$. Summing up, each character ω of $\mathfrak{A}$ is a non-trivial homomorphism of $\mathfrak{A}$ onto $\mathbb{C}$.

➧ For the sake of simplicity, *the set of all characters of* $\mathfrak{A}$ *will be denoted by* $\widehat{\mathfrak{A}}$, instead of $\mathfrak{M}(\mathfrak{A})$, as we did in Sect. 7.3.

Furthermore, note that *if* $\mathfrak{A}$ *has a unit e, then a* $\mathbb{C}$*-valued homomorphism* ω *on* $\mathfrak{A}$ *is a character*, if and only if, $\omega(e) = 1$.

Proposition A.6.1 *Let* $\mathfrak{A}$ *be a commutative Banach algebra and* $\omega \in \widehat{\mathfrak{A}}$. *Then,* $\mathcal{K} = \mathrm{Ker}\,\omega$ *is a maximal modular ideal of* $\mathfrak{A}$. *In particular,* $\mathfrak{A}$ *is generated by* $\mathcal{K}$ *and an element* $b \notin \mathcal{K}$.

Proof As we have seen, ω is surjective. Thus, $\mathfrak{A}/\mathcal{K}$ is isomorphic to $\mathbb{C}$; hence $\mathfrak{A}/\mathcal{K}$ has a unit, i.e., there exists $u \in \mathfrak{A}$, such that $(a + \mathcal{K})(u + \mathcal{K}) = a + \mathcal{K}$, for all $a \in \mathfrak{A}$. This implies that $a - au \in \mathcal{K}$. Therefore, $\mathcal{K}$ is modular. Now we prove that $\mathfrak{A}$ is generated by $\mathcal{K}$ and by an element $b \notin \mathcal{K}$. If $b \notin \mathcal{K}$, then $\omega(b) \neq 0$. For $a \in \mathfrak{A}$, put $c \equiv a - \frac{\omega(a)}{\omega(b)}b$. Then, $\omega(c) = 0$ and clearly $a = \frac{\omega(a)}{\omega(b)}b + c$ is the desired decomposition. □

Theorem A.6.2 *Let $\mathfrak{A}$ be a commutative Banach algebra with unit e and $\omega \in \widehat{\mathfrak{A}}$. Then, $|\omega(a)| \leq \|a\|$, for all $a \in \mathfrak{A}$. Therefore, ω is continuous and $\|\omega\|^\star = 1$.*

Proof Assume that there is $a \in \mathfrak{A}$, such that $|\omega(a)| > \|a\|$; put $\lambda = \omega(a)$. Then, $\|a/\lambda\| < 1$, and so $(e - a/\lambda)$ has an inverse b in $\mathfrak{A}$. Then, we have

$$1 = \omega(e) = \omega(b(e - a/\lambda)) = \omega(b)\omega(e - a/\lambda) = 0,$$

a contradiction. □

Remark A.6.3 By Proposition A.2.13, each character on a C*-algebra with unit is a positive linear functional.

Proposition A.6.4 *Let $\mathfrak{A}$ be a commutative Banach algebra. Then, every maximal modular ideal of $\mathfrak{A}$ is the kernel of a unique character of $\mathfrak{A}$.*

Proof Every maximal modular ideal $\mathfrak{I}$ of $\mathfrak{A}$ has codimension 1 and it is closed (for the latter, see [8, (B.5.2) Proposition]); thus, $\mathfrak{A}/\mathfrak{I}$ is a Banach algebra, which has no proper ideals. The modularity of $\mathfrak{I}$ implies that $\mathfrak{A}/\mathfrak{I}$ has a unit $u + \mathfrak{I}$, with $u \in \mathfrak{A}$. We now prove that each non-zero element of $\mathfrak{A}/\mathfrak{I}$ has an inverse. Indeed, if $b + \mathfrak{I} \neq 0$, then $\mathfrak{M} = \{ab + \mathfrak{I} : a \in \mathfrak{A}\}$ is a non-zero ideal (it contains $b + \mathfrak{I}$) in $\mathfrak{A}/\mathcal{J}$; hence, $\mathfrak{M} = \mathfrak{A}/\mathfrak{I}$. Thus, $u + \mathfrak{I} \in \mathfrak{M}$, so that there exists $a_0 \in \mathfrak{A}$, such that $a_0 b + \mathfrak{I} = u + \mathfrak{I}$; but, $a_0 b + \mathfrak{I} = (a_0 + \mathfrak{I})(b + \mathfrak{I})$ and so $a_0 + \mathfrak{I}$ is the inverse of $b + \mathfrak{I}$. Now Theorem A.3.16, implies that $\mathfrak{M}$ is isometrically isomorphic to $\mathbb{C}$. The quotient map $\Phi : \mathfrak{A} \to \mathfrak{A}/\mathfrak{I}$, whose kernel is $\mathfrak{I}$, is an algebra homomorphism with $\Phi(u) = u + \mathfrak{I} \neq 0$, therefore it is a character of $\mathfrak{A}$. As for the uniqueness, if there were two characters ω and ω', with kernel $\mathfrak{I}$, since u and $\mathfrak{I}$ generate $\mathfrak{A}$ (Proposition A.6.1) and $\omega(u) = \omega'(u) = 1$, they necessarily coincide. □

The following theorem allows to describe the spectrum of an element in terms of characters.

Theorem A.6.5 *Let $\mathfrak{A}$ be a commutative Banach algebra with unit e. Let $a \in \mathfrak{A}$. Then, $\lambda \in \sigma_{\mathfrak{A}}(a)$, if and only if, there exists $\omega \in \widehat{\mathfrak{A}}$, such that $\omega(a) = \lambda$.*

Proof If $\lambda \in \sigma_{\mathfrak{A}}(a)$, the element $a - \lambda e$ is not invertible. This implies that $\mathfrak{I} = \{(a - \lambda e)x : x \in \mathfrak{A}\}$ is a proper ideal of $\mathfrak{A}$ containing $a - \lambda e$. This ideal is contained in a maximal ideal $\mathcal{J}$. This ideal is clearly modular, since $e + \mathcal{J}$ is the unit of $\mathfrak{A}/\mathcal{J}$. By Proposition A.6.4, $\mathcal{J}$ is the kernel of a character ω. Then, $\omega(a - \lambda e) = 0$; thus $\omega(a) = \lambda\omega(e) = \lambda$. Conversely, if $\lambda \notin \sigma_{\mathfrak{A}}(a)$, then $\omega(a - \lambda e) \neq 0$ for every $\omega \in \widehat{\mathfrak{A}}$, otherwise, we would have

$$1 = \omega(e) = \omega\big((a - \lambda e)(a - \lambda e)^{-1})\big) = \omega(a - \lambda e)\omega\big((a - \lambda e)^{-1}\big) = 0,$$

a contradiction. □

Before going forth, we recall some facts of duality theory that will be needed in what follows.

If $\mathfrak{A}$ is a Banach space, as we have already seen, its topological dual space $\mathfrak{A}^\star$, i.e., the space of all bounded linear functionals on $\mathfrak{A}$, is a Banach space under the norm

$$\|\omega\|^\star = \sup_{\|a\|\le 1} |\omega(a)|, \quad \omega \in \mathfrak{A}^\star.$$

The unit ball $\mathcal{U}_{\mathfrak{A}^\star}$ is not compact for the norm topology (unless $\mathfrak{A}$ is finite-dimensional). There is, however, another natural topology that can be defined on $\mathfrak{A}^\star$, namely the *weak* topology*, which is a locally convex topology defined by the seminorms

$$\mathfrak{A}^\star \ni \omega \mapsto |\omega(a)|, \quad a \in \mathfrak{A},$$

where according to the celebrated Banach–Alaoglou theorem, $\mathcal{U}_{\mathfrak{A}^\star}$ is weakly*-compact in $\mathfrak{A}^\star$. But, $\mathcal{U}_{\mathfrak{A}^\star}$ is also convex; then, the Krein–Milman theorem states that $\mathcal{U}_{\mathfrak{A}^\star}$ has *extreme points* i.e., elements that cannot be expressed as convex combinations of two other elements of $\mathcal{U}_{\mathfrak{A}^\star}$ and that $\mathcal{U}_{\mathfrak{A}^\star}$ is the weak*-closure of the set of all combinations of its extreme points. One of the consequences of this fact is that for the Gelfand–Naimark seminorm introduced in Sect. A.5, one has that

$$p(a) = \sup_{\omega\in\mathcal{S}(\mathfrak{A})} \omega(a^*a)^{1/2} = \sup_{\omega\in E\mathcal{S}(\mathfrak{A})} \omega(a^*a)^{1/2} \tag{A.6.1}$$

where $E\mathcal{S}(\mathfrak{A})$ is the set of extreme points of $\mathcal{S}(\mathfrak{A})$.

Theorem A.6.6 *If $\mathfrak{A}$ is a commutative Banach algebra with unit, then the space $\widehat{\mathfrak{A}}$ of characters of $\mathfrak{A}$ is weakly*-compact in $\mathfrak{A}^\star$.*

Proof By Theorem A.6.2, $\widehat{\mathfrak{A}} \subset \mathcal{U}_{\mathfrak{A}^\star}$. It suffices to show that $\widehat{\mathfrak{A}}$ is weakly*-closed in $\mathcal{U}_{\mathfrak{A}^\star}$ First, we notice (but we omit the easy proof) that the limit ω of a net $\{\omega_\gamma\}$ of characters is still a multiplicative linear functional. There are two possibilities: either $\omega = 0$ or $\omega \in \widehat{\mathfrak{A}}$. The case $\omega = 0$ is excluded by the fact that $\mathfrak{A}$ has a unit and for each γ, $\omega_\gamma(e) = 1$. □

We are now ready to define the Gelfand transform.

Let $\mathfrak{A}$ be a commutative Banach algebra with unit e and $\widehat{\mathfrak{A}}$ the space of characters of $\mathfrak{A}$. For $a \in \mathfrak{A}$ we define a complex valued function $\widehat{a}$ on $\widehat{\mathfrak{A}}$, by $\widehat{a}(\omega) := \omega(a)$, for $\omega \in \widehat{\mathfrak{A}}$. The function $\widehat{a}$ is continuous on $\widehat{\mathfrak{A}}$ endowed with the weak* topology. Indeed, one has

$$|\widehat{a}(\omega) - \widehat{a}(\omega')| = |\omega(a) - \omega'(a)|,$$

where on the right hand side, seminorms defining the weak* topology of $\mathfrak{A}^\star$ appear. Therefore, $\widehat{a} \in C(\widehat{\mathfrak{A}})$, the Banach *-algebra of continuous functions on $\widehat{\mathfrak{A}}$, whose supremum norm will be denoted as $\|\cdot\|_\infty$. The function $\widehat{a}$ is called the *Gelfand* *transform* of a and the homomorphism

$$\mathfrak{A} \to C(\widehat{\mathfrak{A}}) : a \mapsto \widehat{a}$$

is called *Gelfand map*.

Theorem A.6.7 (Gelfand) *Let $\mathfrak{A}$ be a commutative Banach algebra with unit e. The Gelfand map is a homomorphism of $\mathfrak{A}$ into $C(\widehat{\mathfrak{A}})$, with the property $\|\widehat{a}\|_\infty \leq \|a\|$, for all $a \in \mathfrak{A}$. Hence, the preceding map is continuous and moreover,*

$$\|\widehat{a}\|_\infty = |a|_\sigma = \nu(a), \quad \forall\, a \in \mathfrak{A}.$$

Proof By the very definitions one has that

$$\|\widehat{a}\|_\infty = \sup_{\omega \in \widehat{\mathfrak{A}}} |\omega(a)| \leq \sup_{\omega \in \widehat{\mathfrak{A}}} \|\omega\|^* \|a\| \leq \|a\|, \quad \forall\, a \in \mathfrak{A}.$$

Moreover, by Theorem A.6.5,

$$\|\widehat{a}\|_\infty = \sup_{\omega \in \widehat{\mathfrak{A}}} |\omega(a)| = \sup\left\{|\lambda| : \lambda \in \sigma_{\mathfrak{A}}(a)\right\} = |a|_\sigma, \quad \forall\, a \in \mathfrak{A}.$$

□

Remark A.6.8 We notice that, if $\mathfrak{A}$ is a commutative Banach *-algebra with unit, the Gelfand map is not, in general, *-preserving (see Definition A.6.12(1) below), unless each character is hermitian, like in the case of commutative C*-algebras. In this situation one has

$$\widehat{a^*}(\omega) = \omega(a^*) = \overline{\omega(a)} = \overline{\widehat{a}(\omega)}, \quad a \in \mathfrak{A}.$$

The hermiticity of a character in the aforementioned case follows from the fact that the spectrum of a hermitian element in an arbitrary C*-algebra is real [8, (8.1) Proposition]). Therefore, $\widehat{a^*} = \overline{\widehat{a}}$, for each $a \in \mathfrak{A}$.

The kernel of the Gelfand map

$$\left\{a \in \mathfrak{A} : \widehat{a} = 0\right\} = \left\{a \in \mathfrak{A} : \nu(a) = 0\right\}$$

coincides, according to Propositions A.6.1 and A.6.4, with the intersection of all maximal modular ideals of $\mathfrak{A}$, i.e., the (Jacobson) radical of $\mathfrak{A}$. *This set can be proved to be equal to the* 'radical' *of the algebra $\mathfrak{A}$, defined as the intersection of the kernels of all characters of $\mathfrak{A}$.* Thus, *if $\mathfrak{A}$ is semisimple, the Gelfand map is injective*.

We now consider the case where $\mathfrak{A}$ is a C*-algebra. If $\mathfrak{A}$ is a Banach *-algebra with unit, taking into account the definitions (see discussion before Proposition A.5.3), *the radical of $\mathfrak{A}$ is contained in the *-radical of $\mathfrak{A}$* (cf., e.g., [8, (30.2) Proposition]). Thus, if $\mathfrak{A}$ is *-semisimple, it is also semisimple and, if $\mathfrak{A}$

is commutative, the Gelfand map is injective. By Theorem A.5.5 *any C*-algebra is *-semisimple*.

Theorem A.6.9 *Let $\mathfrak{A}$ be a commutative C*-algebra with unit. Then, the Gelfand map $a \mapsto \widehat{a}$ is an isometric *-isomorphism of $\mathfrak{A}$ onto the C*-algebra $C(\widehat{\mathfrak{A}})$ of continuous functions on the weakly*-compact space $\widehat{\mathfrak{A}}$ of characters of $\mathfrak{A}$.*

We need the following lemma (see, e.g., [8, (29.5) Theorem])

Lemma A.6.10 *Let $\mathfrak{A}$ be a commutative C*-algebra with unit e and ω a positive linear functional with $\omega(e) = 1$, i.e., $\omega \in \mathcal{S}(\mathfrak{A})$. Then, the following statements are equivalent:*

(1) *ω is an extreme point of $\mathcal{S}(\mathfrak{A})$;*
(2) *ω is a character.*

Proof of Theorem A.6.9 We prove that the Gelfand map $a \mapsto \widehat{a}$ is isometric. Indeed, using Theorem A.5.5 and (A.6.1), we have

$$\begin{aligned}\|\widehat{a}\|_\infty^2 &= \sup_{\omega\in\widehat{\mathfrak{A}}} |\widehat{a}(\omega)|^2 = \sup_{\omega\in\widehat{\mathfrak{A}}} |\omega(a)|^2 \\ &= \sup_{\omega\in\widehat{\mathfrak{A}}} \omega(a^*)\omega(a) = \sup_{\omega\in\widehat{\mathfrak{A}}} \omega(a^*a) = \sup_{\omega\in E\mathcal{S}(\mathfrak{A})} \omega(a^*a) = \|a\|^2.\end{aligned}$$

It remains only to prove that the Gelfand map is surjective. Let X denote the image of $\mathfrak{A}$ under the map $a \mapsto \widehat{a}$. Then, X is norm-closed in $C(\widehat{\mathfrak{A}})$ and contains the unit function. Clearly, for any pair $\omega_1, \omega_2 \in \widehat{\mathfrak{A}}$, such that $\omega_1 \neq \omega_2$, there exists $a \in A$ with $\omega_1(a) \neq \omega_2(a)$ or equivalently, $\widehat{a}(\omega_1) \neq \widehat{a}(\omega_2)$. So X separates the points of $\widehat{\mathfrak{A}}$. Thus, by the Stone–Weierstrass theorem, it follows that $X = C(\widehat{\mathfrak{A}})$. □

With the help of the Gelfand map, some very important properties of non commutative C*-algebras, can be showed. As an instance, we prove the following

Proposition A.6.11 *Let $\mathfrak{A}$ be a C*-algebra with unit and $a = a^*$ in $\mathfrak{A}$. Then, there exist two positive elements $a_+, a_- \in \mathfrak{A}$, such that $a_+a_- = a_-a_+ = 0$ and $a = a_+ - a_-$*

Proof Let $\mathfrak{M}(a)$ denote the commutative unital C*-subalgebra of $\mathfrak{A}$ generated by a. Then, $\mathfrak{M}(a)$ is isometrically isomorphic to $C(X)$, for some compact Hausdorff space X. Let $\widehat{a}$ be the Gelfand transform of a and $\widehat{a}_+, \widehat{a}_-$ the respective positive and negative parts of $\widehat{a}$ in $C(X)$. Taking the inverse images of these functions, we find the requested elements of $\mathfrak{A}$. □

A.6.2 The Noncommutative Case

Definition A.6.12 Let $\mathfrak{A}$, $\mathfrak{B}$ be *-algebras. A map $\tau : \mathfrak{A} \to \mathfrak{B}$ is a *-homomorphism if

(1) $\tau(a^*) = \tau(a)^*$, for all $a \in \mathfrak{A}$;
(2) $\tau(\alpha a + \beta b) = \alpha\tau(a) + \beta\tau(b)$, for all $\alpha, \beta \in \mathbb{C}$ and $a, b \in \mathfrak{A}$;
(3) $\tau(ab) = \tau(a)\tau(b)$, for all $a, b \in \mathfrak{A}$.

If $\mathfrak{A}$ and $\mathfrak{B}$ have units, denoted by $e_{\mathfrak{A}}$, $e_{\mathfrak{B}}$ respectively, we may suppose without loss of generality that τ *preserves units*; i.e., $\tau(e_{\mathfrak{A}}) = e_{\mathfrak{B}}$. If it is not so we replace $\mathfrak{B}$ with $\tau(e_{\mathfrak{A}})\mathfrak{B}\tau(e_{\mathfrak{A}})$.

Proposition A.6.13 *The following hold:*

(1) *if $\mathfrak{A}$, $\mathfrak{B}$ are Banach algebras with units and τ a homomorphism between them, then $\sigma_{\mathfrak{B}}(\tau(a)) \subseteq \sigma_{\mathfrak{A}}(a)$, for all $a \in \mathfrak{A}$;*
(2) *if $\mathfrak{A}$, $\mathfrak{B}$ are Banach *-algebras with units and τ a *-homomorphism between them, then for $a \in \mathfrak{A}$, positive* (i.e., $a \geq 0$), *one has that $\tau(a)$ is positive in $\mathfrak{B}$, too;*
(3) *if $\mathfrak{A}$, $\mathfrak{B}$ are C*-algebras and τ a *-homomorphism between them, then τ is continuous and, in particular, $\|\tau(a)\| \leq \|a\|$, for all $a \in \mathfrak{A}$.*

Proof

1. Let $\lambda \in \rho_{\mathfrak{A}}(a)$. Then, $(a - \lambda e_{\mathfrak{A}})^{-1}$ exists in $\mathfrak{A}$. This implies that also $\tau(a) - \lambda e_{\mathfrak{B}}$ has an inverse in $\mathfrak{B}$, since
$$(\tau(a) - \lambda e_{\mathfrak{B}})\tau\big((a - \lambda e_{\mathfrak{A}})^{-1}\big) = \tau(a - \lambda e_{\mathfrak{A}})\tau\big((a - \lambda e_{\mathfrak{A}})^{-1}\big) = \tau(e_{\mathfrak{A}}) = e_{\mathfrak{B}}.$$
Thus, $\lambda \in \rho_{\mathfrak{B}}\big(\tau(a)\big)$.
2. Since $a \geq 0$, we have that $a^* = a$ and $\sigma_{\mathfrak{A}}(a) \subset \mathbb{R}^+ \cup \{0\}$ (cf. Definition A.3.23), therefore $\tau(a)^* = \tau(a^*) = \tau(a)$, for all $a \in \mathfrak{A}$. The assertion now follows from (1).
3. Add units $e_{\mathfrak{A}}$, $e_{\mathfrak{B}}$, in $\mathfrak{A}$ and $\mathfrak{B}$ respectively, if necessary and suppose that $\tau(e_{\mathfrak{A}}) = e_{\mathfrak{B}}$. Then, for any $a \in \mathfrak{A}$, we have
$$\|\tau(a)\|^2 = \|\tau(a)^*\tau(a)\| = \|\tau(a^*a)\| = \nu(\tau(a^*a)).$$
Now, by (1) and Corollary A.3.5, it follows that $\nu(\tau(a^*a)) \leq \nu(a^*a) = \|a^*a\| = \|a\|^2$ and this completes the proof of (3). □

Corollary A.6.14 *Each* (algebraic) **-isomorphism of C*-algebras is isometric.*

The foregoing statements apply in particular to *-representations of a C*-algebra that are, by definition, *-homomorphisms of $\mathfrak{A}$ into $\mathcal{B}(\mathcal{H})$. In particular, any *faithful* *-representation π (i.e., $\pi(a) = 0$ implies $a = 0$) is norm-preserving and $\pi(\mathfrak{A})$ is a C*-algebra of bounded operators.

Before proving the main theorem of this section, we need the notion of *direct sum* of *-representations.

Let $\mathfrak{A}$ be a C*-algebra. For each $j \in I$, where I is a set of indices, let π_j be a *-representation of $\mathfrak{A}$ on the Hilbert space $\mathcal{H}_j$. First, we construct the direct sum

$$\mathcal{H} = \bigoplus_{j \in I} \mathcal{H}_j$$

of the Hilbert spaces $\mathcal{H}_j$ in the following way. The finite subsets F of I form a directed set with the set inclusion as order. An element ξ of $\mathcal{H}$ is a family $\xi = \{\xi_j\}$ with $\xi_j \in H_j$, such that

$$\lim_F \sum_{j \in F} \|\xi_j\|_j^2 < \infty,$$

where $\|\cdot\|_j$ denotes the norm in $\mathcal{H}_j$ derived from the inner product $\langle \cdot|\cdot \rangle_j$, $j \in J$. The inner product in $\mathcal{H}$ is then defined by

$$\langle \xi | \eta \rangle = \lim_F \sum_{j \in F} \langle \xi_j | \eta_j \rangle_j, \quad \xi, \eta \in \mathcal{H}.$$

The *direct sum *-representation* $\pi \equiv \bigoplus_{j \in I} \pi_j$ is given by

$$\pi(a)\{\xi_j\} = \{\pi_j(a)\xi_j\}, \quad a \in \mathfrak{A}, \ \{\xi_j\} \in \mathcal{H}.$$

Since $\|\pi_j(a)\| \leq \|a\|$, $j \in J, a \in \mathfrak{A}$, we have that $\sup_{j \in J} \|\pi_j(a)\| < \infty$. Moreover, $\|\pi(a)\| = \sup_{j \in J} \|\pi_j(a)\| \leq \|a\|$, therefore $\pi(a)$ is a bounded operator on $\mathcal{H}$, for all $a \in \mathfrak{A}$.

Theorem A.6.15 (Gelfand–Naimark) *Each C*-algebra $\mathfrak{A}$ with unit has a faithful *-representation. It follows that $\mathfrak{A}$ is isometrically *-isomorphic to a C*-algebra of bounded linear operators on a Hilbert space.*

Proof For each $\omega \in \mathcal{S}(\mathfrak{A})$, the GNS construction yields a bounded *-representation π_ω of $\mathfrak{A}$ on a Hilbert space $\mathcal{H}_\omega$ (see Theorem A.2.16). Let π be the *-representation direct sum of the family $\{\pi_\omega\}_{\omega \in \mathcal{S}(\mathfrak{A})}$, acting on the Hilbert space $\mathcal{H} = \bigoplus_{\omega \in \mathcal{S}(\mathfrak{A})} \mathcal{H}_\omega$. By Proposition A.5.4, for each non-zero $a \in \mathfrak{A}$ there exists $\omega \in \mathcal{S}(\mathfrak{A})$, such that $\|\pi_\omega(a)\| = \|a\|$. Then, $\|a\| = \|\pi_\omega(a)\| \leq \|\pi(a)\|$. By (3) of Proposition A.6.13, we finally get $\|\pi(a)\| = \|a\|$, $a \in \mathfrak{A}$, thus π is a faithful *-representation. In conclusion, $\mathfrak{A}$ is isometrically *-isomorphic to the C*-algebra $\pi(\mathfrak{A})$. □

Remark A.6.16 The *-representation π constructed in the proof of Theorem A.6.15 is usually called *universal *-representation* of $\mathfrak{A}$. Thus, one can formulate Theorem A.6.15 by saying that π is an isometric *-isomorphism of $\mathfrak{A}$ onto a C*-algebra of bounded linear operators on the carrier space $\mathcal{H}$.

Appendix B
Operators in Hilbert Spaces

In this chapter after recalling shortly some basic facts on Hilbert spaces, we list, mainly without proofs, the fundamental aspects of the theory of bounded and unbounded operators in Hilbert spaces that have been used throughout this book.

B.1 Hilbert Spaces

Definition B.1.1 Let $\mathcal{D}$ be a complex vector space. A map associating to an ordered pair (ξ, η) of elements of $\mathcal{D} \times \mathcal{D}$ the complex number $\langle \xi | \eta \rangle$ is called an *inner product* if

(i) $\langle \alpha\xi + \beta\eta | \zeta \rangle = \alpha\langle \xi | \zeta \rangle + \beta\langle \eta | \zeta \rangle$;
(ii) $\langle \xi | \eta \rangle = \overline{\langle \eta | \xi \rangle}$;
(iii) $\langle \xi | \xi \rangle \geq 0$;
(iv) $\langle \xi | \xi \rangle = 0 \;\Leftrightarrow\; \xi = 0$;

where $\xi, \eta, \zeta \in \mathcal{D}$ and $\alpha, \beta \in \mathbb{C}$.

A vector space $\mathcal{D}$ endowed with an inner product is called an *inner product space* or *pre-Hilbert space*

Now, put

$$\|\xi\| := \langle \xi | \xi \rangle^{1/2}, \quad \xi \in \mathcal{D}; \tag{B.1.1}$$

then, it is easily seen that the triangle inequality

$$\|\xi + \eta\| \leq \|\xi\| + \|\eta\|, \quad \forall\, \xi, \eta \in \mathcal{D},$$

M. Fragoulopoulou, C. Trapani, *Locally Convex Quasi *-Algebras and their Representations*, Lecture Notes in Mathematics 2257,
https://doi.org/10.1007/978-3-030-37705-2

holds true. This fact, together with (iii) and (iv) of Definition B.1.1, implies that (B.1.1) actually defines a norm on $\mathcal{D}$. Moreover, the following Cauchy–Schwarz inequality holds

$$|\langle \xi | \eta \rangle| \leq \|\xi\| \|\eta\|, \quad \forall \xi, \eta \in \mathcal{D}. \tag{B.1.2}$$

Therefore, every pre-Hilbert space is a normed space.

Remark B.1.2 From the Cauchy–Schwarz inequality it follows easily that

$$\|\xi\| = \sup_{\|\eta\| \leq 1} |\langle \xi | \eta \rangle|. \tag{B.1.3}$$

Definition B.1.3 Two vectors, ξ and η of $\mathcal{D}$ are said to be *orthogonal* if $\langle \xi | \eta \rangle = 0$. A set ξ_i of vectors of $\mathcal{D}$ is called *orthonormal* if $\langle \xi_i | \xi_i \rangle = 1$ and $\langle \xi_i | \xi_j \rangle = 0$, for $i \neq j$.

Remark B.1.4 For two orthogonal vectors ξ and η of a pre-Hilbert space $\mathcal{D}$ the *Pythagoras theorem* holds; i.e., $\|\xi + \eta\|^2 = \|\xi\|^2 + \|\eta\|^2$.

Definition B.1.5 A pre-Hilbert space $\mathcal{H}$ which is complete with respect to the norm (B.1.1) is said to be a *Hilbert space*.

If $\mathcal{M} \subset \mathcal{H}$, then $\mathcal{M}^\perp \equiv \left\{\xi \in \mathcal{H} : \langle \xi | \eta \rangle = 0, \forall \eta \in \mathcal{M}\right\}$ is a closed subspace of $\mathcal{H}$. If $\mathcal{M}$ itself is a closed subspace of $\mathcal{H}$, then $\mathcal{M}^\perp$ is called the *orthogonal complement* of $\mathcal{M}$. It can be proved that, if $\mathcal{M}$ is a closed subspace of $\mathcal{H}$, then $\mathcal{M} \oplus \mathcal{M}^\perp = \mathcal{H}$. Similarly, $\mathcal{M}$ is dense in $\mathcal{H}$, if and only if, $\mathcal{M}^\perp = \{0\}$

Example B.1.6

(a) For $n \in \mathbb{N}$ fixed, the space $\mathbb{C}^n$ of all n-tuples of complex numbers

$$z = (z_1, z_2, \dots, z_n)$$

is a Hilbert space if the inner product of z and $w = (w_1, w_2, \dots, w_n)$ is defined by

$$\langle z | w \rangle = \sum_{j=1}^{n} z_j \bar{w}_j.$$

(b) The space $L^2(\mathbb{R})$ of Lebesgue square integrable functions (modulo the set of all almost everywhere null functions) is a Hilbert space if the inner product of two elements f, g in $L^2(\mathbb{R})$ is defined by

$$\langle f | g \rangle = \int_{\mathbb{R}} f(x)\overline{g(x)}\, dx, \; x \in \mathbb{R}. \tag{B.1.4}$$

We notice that (B.1.4) is well defined because of Hölder inequality. The completeness of $L^2(\mathbb{R})$ constitutes the content of the Riesz–Fisher theorem.

(c) The space $C[0, 1]$ of all complex valued continuous functions in $[0, 1]$ is an inner product space if the inner product is defined by

$$\langle f|g\rangle = \int_0^1 f(x)\overline{g(x)}\,dx,\ x \in [0, 1],$$

but it is not a Hilbert space. Indeed, let us consider the sequence of functions

$$f_n(x) = \begin{cases} 0, & \text{if } 0 \le x \le \frac{1}{2} - \frac{1}{n} \\ \frac{n}{2}\left(x - \frac{1}{2}\right) + \frac{1}{2}, & \text{if } \frac{1}{2} - \frac{1}{n} \le x \le \frac{1}{2} + \frac{1}{n} \\ 1, & \text{if } \frac{1}{2} + \frac{1}{n} \le x \le 1 \end{cases}$$

for $n > 2$. It is easy to see that if f is the discontinuous function

$$f(x) = \begin{cases} 0, & \text{if } 0 \le x \le \frac{1}{2} \\ 1, & \text{if } \frac{1}{2} < x \le 1 \end{cases}$$

one has

$$\lim_{n\to\infty} \int_0^1 |f_n(x) - f(x)|^2\,dx = 0,\ x \in [0, 1].$$

Thus (f_n) is a Cauchy sequence in $C[0, 1]$ but $f \notin C[0, 1]$.

Let $E[\|\cdot\|_E]$, $F[\|\cdot\|_F]$ be normed spaces. We remind that a linear map $f : E \to F$, i.e.,

$$f(\alpha x + \beta y) = \alpha f(x) + \beta f(y), \quad \forall x, y \in E,\ \alpha, \beta \in \mathbb{C},$$

is said to be *bounded* if there exists a constant $C > 0$, such that

$$\|f(x)\|_F \le C\|x\|_E, \quad \forall x \in E.$$

It is well known that boundedness of f is equivalent to its continuity at 0 (and then at every point of E). The *norm* of a continuous linear map $f : E[\|\cdot\|_E] \to F[\|\cdot\|_F]$ is defined as

$$\|f\| := \sup\left\{\|f(x)\|_F : \|x\|_E \le 1\right\},\ x \in E.$$

In the case $F = \mathbb{C}$, a linear map from E into $\mathbb{C}$ is called a *linear functional* on E. The space of all bounded (continuous) linear functionals on E is denoted by $E^\star$ and called the *dual* space of E. The space $E^\star$ is a Banach space under the norm

$$\|f\|^\star := \sup\left\{|f(x)| : \|x\|_E \leq 1\right\}, \; f \in E^\star, \; x \in E.$$

The following theorem, known as *Riesz Lemma*, is one of the main results of the theory of Hilbert spaces. It is due to Riesz and Fréchet and characterizes bounded linear functionals on a Hilbert space $\mathcal{H}$.

Theorem B.1.7 *Let $\mathcal{H}$ be a Hilbert space and $\eta \in \mathcal{H}$. Let*

$$f_\eta(\xi) = \langle \xi | \eta \rangle, \quad \xi \in \mathcal{H}.$$

Then, f_η is a continuous linear functional on $\mathcal{H}$ and $\|f_\eta\|^\star = \|\eta\|$.

Conversely, if f is a continuous linear functional on $\mathcal{H}$, then there exists a unique $\eta \in \mathcal{H}$, such that $f = f_\eta$.

Proof The fact that f_η is continuous, follows immediately by (B.1.2). The same inequality shows that $\|f_\eta\|^\star \leq \|\eta\|$. On the other hand,

$$\|f_\eta\|^\star = \sup\left\{|f_\eta(\xi)| : \|\xi\| = 1\right\} \geq \left|f_\eta\left(\|\eta\|^{-1}\eta\right)\right| = \left\langle \|\eta\|^{-1}\eta | \eta \right\rangle = \|\eta\|$$

and this concludes the proof of first part.

Conversely, let f be a continuous linear functional on $\mathcal{H}$. Put $\mathcal{M} = \operatorname{Ker} f$; then, $\mathcal{M}$ is a closed vector subspace of $\mathcal{H}$, which does not coincide with $\mathcal{H}$. Then, $\mathcal{M}^\perp \neq \{0\}$. Taking $u \in \mathcal{M}^\perp$ with $\|u\| = 1$, we have $f(u)f(\xi) - f(\xi)f(u) = 0$, therefore $f(u)\xi - f(\xi)u \in \mathcal{M}$. Since $u \in \mathcal{M}^\perp$, we obtain

$$0 = \langle f(u)\xi - f(\xi)u | u \rangle = f(u)\langle \xi | u \rangle - f(\xi),$$

that is

$$f(\xi) = f(u)\langle \xi | u \rangle.$$

If we put $\eta = u\overline{f(u)}$, we obtain $f = f_\eta$.

As for the uniqueness, let $\zeta \in \mathcal{H}$ be another vector in $\mathcal{H}$, such that $f = f_\zeta$. Then,

$$\|\eta - \zeta\| = \|f_{\eta - \zeta}\| = \|f_\eta - f_\zeta\| = \|f - f\| = 0,$$

whence $\eta = \zeta$. □

B.2 Bounded Operators

In the discussion before Theorem B.1.7, the definition of a bounded linear map between Banach spaces and its norm are given. Since every Hilbert space $\mathcal{H}$ is a Banach space, a bounded linear map $T : \mathcal{H} \to \mathcal{H}$ is called a *bounded operator* and its norm $\|T\|$ is exactly the norm of a bounded linear map T between Banach spaces; i.e.,

$$\|T\| := \sup\left\{\|T\xi\| : \|\xi\| \leq 1\right\},\ \xi \in \mathcal{H}.$$

The set of all bounded operators on a Hilbert space $\mathcal{H}$ will be denoted by $\mathcal{B}(\mathcal{H})$.

Remark B.2.1 It is easy to prove that the sum, scalar multiple and product of bounded operators are again bounded operators, such that

$$\|T+S\| \leq \|T\| + \|S\|, \quad \|\alpha T\| = |\alpha|\|T\|, \quad \|TS\| \leq \|T\|\,\|S\|,$$

for any $T, S \in \mathcal{B}(\mathcal{H})$ and $\alpha \in \mathbb{C}$.

An interesting application of Riesz lemma is the following

Theorem B.2.2 *Let $\Omega(\ ,\ )$ be a bounded sesquilinear form on $\mathcal{H}$, i.e., a map from $\mathcal{H} \times \mathcal{H}$ into $\mathbb{C}$ satisfying the following conditions:*

(i) $\Omega(\alpha\xi + \beta\eta, \zeta) = \alpha\Omega(\xi, \zeta) + \beta\Omega(\eta, \zeta)$;
(ii) $\Omega(\xi, \alpha\eta + \beta\zeta) = \overline{\alpha}\Omega(\xi, \eta) + \overline{\beta}\Omega(\xi, \zeta)$;
(iii) *there exists a constant $C > 0$, such that $|\Omega(\xi, \eta)| \leq C\|\xi\|\|\eta\|$,*

for any $\xi, \eta, z \in \mathcal{H}$ $\alpha, \beta \in \mathbb{C}$. Then, there exists a unique bounded linear operator T from $\mathcal{H}$ into $\mathcal{H}$, such that

$$\Omega(\xi, \eta) = \langle \xi | T\eta \rangle, \quad \forall\, \xi, \eta \in \mathcal{H}$$

and

$$\|T\| = \sup\left\{|\Omega(\xi, \eta)| : \|\xi\| = \|\eta\| = 1\right\}. \tag{B.2.1}$$

Proof Fix $\eta \in \mathcal{H}$; then $\Omega_\eta(\xi) = \Omega(\xi, \eta)$ is a bounded linear functional on $\mathcal{H}$. By Riesz lemma, there exists $\zeta \in \mathcal{H}$, such that

$$\Omega_\eta(\xi) = \Omega(\xi, \eta) = \langle \xi | \zeta \rangle, \quad \forall\, \xi \in \mathcal{H}.$$

Put $A\eta = \zeta$. One defines in this way a map A from $\mathcal{H}$ into itself. It is easy to prove that A is a linear operator. To see that it is bounded, we compute $\|A\eta\|^2$, for $\eta \in \mathcal{H}$; indeed,

$$\|A\eta\|^2 = \langle A\eta | A\eta \rangle = \Omega(A\eta, \eta) \leq C\|A\eta\|\,\|\eta\|.$$

It remains to prove uniqueness. Let A' be another linear operator, such that $\Omega(\xi, \eta) = \langle \xi | A\eta \rangle$, for all $\xi, \eta \in \mathcal{H}$. Then, $\langle \xi | A'\eta - A\eta \rangle = 0$, for all $\xi \in \mathcal{H}$; but $\mathcal{H}^\perp = \{0\}$. The equality (B.2.1) follows easily from (B.1.3). □

B.3 The Adjoint of a Bounded Operator

Theorem B.3.1 *Let $T \in \mathcal{B}(\mathcal{H})$. Then, there exists an operator $T^* \in \mathcal{B}(\mathcal{H})$* (called the *adjoint of T*), *such that*

$$\langle T\xi|\eta\rangle = \langle \xi|T^*\eta\rangle, \quad \forall\, \xi, \eta \in \mathcal{H}.$$

Proof For fixed $\eta \in \mathcal{H}$, put

$$\omega_{T,\eta}(\xi) \equiv \langle T\xi|\eta\rangle, \quad \xi \in \mathcal{H}.$$

Then, $\omega_{T,\eta}$ is a linear functional on $\mathcal{H}$ and since

$$|\omega_{T,\eta}(\xi)| = |\langle T\xi|\eta\rangle| \leq \|T\|\|\xi\|\|\eta\|, \quad \xi \in \mathcal{H},$$

it is bounded. By the Riesz lemma, there exists $\zeta \in \mathcal{H}$, such that

$$\omega_{T,\eta}(\xi) = \langle T\xi|\eta\rangle = \langle \xi|\zeta\rangle, \quad \xi \in \mathcal{H}.$$

Now put $T^*\eta = \zeta$. The map defined in this way is linear. We prove that it is bounded. We have

$$\|T^*\eta\|^2 = \langle T^*\eta|T^*\eta\rangle = \langle TT^*\eta|\eta\rangle \leq \|T\|\,\|T^*\eta\|\,\|\eta\|.$$

This proves that $T^* \in \mathcal{B}(\mathcal{H})$ and also that $\|T^*\| \leq \|T\|$. □

Proposition B.3.2 *The map $T \mapsto T^*$ in $\mathcal{B}(\mathcal{H})$ has the following properties:*

(1) $T^{**} = T$, *for every $T \in \mathcal{B}(\mathcal{H})$;*
(2) $(\alpha T + \beta S)^* = \overline{\alpha}A^* + \overline{\beta}S$, *for all $T, S \in \mathcal{B}(\mathcal{H})$ and $\alpha, \beta \in \mathbb{C}$;*
(3) $(TS)^* = S^*T^*$, *for all $T, S \in \mathcal{B}(\mathcal{H})$.*

Remark B.3.3 From (1) of Proposition B.3.2 and the last part of the proof of Theorem B.3.1 it follows easily that $\|T^*\| = \|T\|$, for every $T \in \mathcal{B}(\mathcal{H})$.

Before going forth, we define some special class of bounded operators.

B.3.1 Symmetric Operators

A bounded operator T in $\mathcal{H}$ is called *selfadjoint* if $T = T^*$, i.e., if

$$\langle T\xi|\eta\rangle = \langle \xi|T\eta\rangle, \quad \forall\, \xi, \eta \in \mathcal{H}.$$

As an example, we consider $\mathcal{H} = L^2(0, 1)$ and the operator $Q : L^2(0, 1) \to L^2(0, 1)$, such that $(Qf)(x) := xf(x)$, for every $f \in L^2(0, 1)$ and $x \in (0, 1)$. It is readily seen that Q is bounded and selfadjoint.

For any selfadjoint operator T and any $\xi \in \mathcal{H}$, the number $\langle T\xi|\xi\rangle$ is real. A selfadjoint operator T is said to be *positive*, and we write $T \geq 0$, if $\langle T\xi|\xi\rangle \geq 0$, for every $\xi \in \mathcal{H}$.

B.3.2 Projection Operators

A bounded operator P is called a *projection operator* if $P = P^2 = P^*$. There is one-to-one correspondence between closed subspaces of $\mathcal{H}$ and projection operators on $\mathcal{H}$. As we already remarked, if $\mathcal{M}$ is a closed subspace of $\mathcal{H}$, then $\mathcal{H} = \mathcal{M} \oplus \mathcal{M}^\perp$. If P denotes the canonical projection of $\mathcal{H}$ onto $\mathcal{M}$, then it turns out that P is a projection operator on $\mathcal{H}$. Conversely, if P is a projection operator, then the set $\mathcal{M}_P = \{\xi \in \mathcal{H} : P\xi = \xi\}$ is a closed subspace of $\mathcal{H}$ and P is the projection operator on $\mathcal{M}_P$.

B.3.3 Isometric and Unitary Operators

A linear operator $U : \mathcal{H} \to \mathcal{H}$ is called *isometric* if $\langle U\xi|U\eta\rangle = \langle \xi|\eta\rangle$, for every $\xi, \eta \in \mathcal{H}$. One obviously has $\|U\xi\| = \|\xi\|$, for every $\xi \in \mathcal{H}$; therefore, U is bounded and $\|U\| = 1$. The operator U is injective and so U^{-1} exists. If $U^{-1} = U^*$, the operator U is called *unitary*. Equivalently, U is unitary if $U^*U = UU^* = I$, where I denotes the identity operator of $\mathcal{H}$. For a fixed $t \in \mathbb{R}$, the shift operator $U : L^2(\mathbb{R}) \to L^2(\mathbb{R})$, such that $Uf(t) := f(x + t)$, $x \in \mathbb{R}$, is an example of a unitary operator.

B.4 Unbounded Operators in Hilbert Spaces

Definition B.4.1 Let $\mathcal{H}$ be a Hilbert space, $D(T)$ a dense vector subspace of $\mathcal{H}$ and $T : D(T) \to \mathcal{H}$ a linear map, i.e.,

$$T(\alpha\xi + \beta\eta) = \alpha T\xi + \beta T\eta, \quad \forall\, \xi, \eta \in D(T) \text{ and } \forall \alpha, \beta \in \mathbb{C}.$$

Then, the pair $(T, D(T))$ is called a *linear operator* in $\mathcal{H}$. We refer to $D(T)$ as to the *domain* of T.

The set

$$\operatorname{Ran}(T) = \{\eta \in \mathcal{H} : \eta = T\xi, \quad \text{for some } \xi \in D(T)\}$$

will be called the *range* of T.

We will often speak of the *operator* T, without writing explicitly its domain. The specification of the domain is, however, essential for an appropriate definition of the operator.

Definition B.4.2 An operator $(T', D(T'))$ is called an *extension* of the operator $(T, D(T))$ and we write $T \subset T'$ if $D(T) \subset D(T')$ and $T\eta = T'\eta$, for every $\eta \in D(T)$.

The algebraic operations between linear operators T and S are defined as follows (provided that the sets indicated below as domains are dense in $\mathcal{H}$):

(i) Addition $T + S$:

$$\begin{cases} D(T+S) = D(T) \cap D(S), \\ (T+S)\xi = T\xi + S\xi, \quad \xi \in D(T+S). \end{cases}$$

(ii) The multiplication λT of T by scalars $\lambda \in \mathbb{C}$:

If $\lambda = 0$, then $\lambda T \equiv 0$, otherwise

$$\begin{cases} D(\lambda T) = D(T), \\ (\lambda T)\xi = \lambda(T\xi), \quad \xi \in D(T). \end{cases}$$

(iii) The multiplication TS:

$$\begin{cases} D(TS) = \{\xi \in D(S) : S\xi \in D(T)\}, \\ (TS)\xi = T(S\xi), \quad \xi \in D(TS). \end{cases}$$

(iv) The inverse T^{-1}:

If T is injective, then

$$\begin{cases} D(T^{-1}) = R(T), \\ T^{-1}(T\xi) = \xi, \quad \xi \in D(T). \end{cases}$$

The usual associative laws hold for the addition and multiplication. That is, given the operators T, S, Q in $\mathcal{H}$, one has

$$(T+S) + Q = T + (S+Q) \text{ and } (TS)Q = T(SQ).$$

The distributive law holds too, but one has the inclusion $T(S+Q) \supset TS + TQ$ instead of equality.

Let T be a linear operator in $\mathcal{H}$. The set

$$\mathcal{G}(T) \equiv \{(\xi, T\xi) : \xi \in D(T)\}$$

is a subspace of the direct sum $\mathcal{H} \oplus \mathcal{H}$, called the *graph* of T. It is clear that $T = S$, if and only if, $\mathcal{G}(T) = \mathcal{G}(S)$, and $T \subset S$, if and only if, $\mathcal{G}(T) \subset \mathcal{G}(S)$.

Definition B.4.3 A linear operator T in $\mathcal{H}$ is said to be *closed* if its graph $\mathcal{G}(A)$ is closed in $\mathcal{H} \oplus \mathcal{H}$, that is, if $\{\xi_n\} \subset D(T)$ is such that $\xi_n \to \xi$ and $T\xi_n \to \eta$, then $\xi \in D(T)$ and $\eta = T\xi$.

Let T be a closed operator in $\mathcal{H}$. Then, $D(T)$ can be made into a Hilbert space under the inner product

$$\langle \xi | \eta \rangle_T := \langle \xi | \eta \rangle + \langle T\xi | T\eta \rangle, \quad \xi, \eta \in D(T).$$

This Hilbert space is denoted by $\mathcal{H}_T$ and the norm $\| \cdot \|_T$ is called the *graph norm* associated to T.

Definition B.4.4 A linear operator T in $\mathcal{H}$ is said to be *closable* if it has a closed extension.

Every closable operator T has a minimal closed extension, called its *closure* and denoted by $\overline{T}$. Then, we have the following

Proposition B.4.5 *Let T be a linear operator in $\mathcal{H}$. The following statements are equivalent:*

(i) *T is closable;*
(ii) *$\overline{\mathcal{G}(T)}$ is the graph of a linear operator in $\mathcal{H}$;*
(iii) *If $\{\xi_n\} \subset D(T)$, such that $\xi_n \to 0$ and $T\xi_n \to \eta$, then $\eta = 0$.*

If one of the preceding conditions is true, then $\overline{T}$ is given by

$$\begin{cases} D(\overline{T}) = \{\xi \in \mathcal{H} : \exists \{\xi_n\} \subset D(T), \text{ such that } \xi_n \to \xi \text{ and } T\xi_n \to \eta\}, \\ \overline{T}\xi = \eta, \quad \xi \in D(\overline{T}). \end{cases}$$

If T is closed and T^{-1} exists, then T^{-1} is closed.

Next we define the *adjoint* T^* of a densely defined linear operator T in $\mathcal{H}$:

$$\begin{cases} D(T^*) = \{\eta \in \mathcal{H} : \exists \zeta \in \mathcal{H}, \text{ such that } \langle T\xi | \eta \rangle = \langle \xi | \zeta \rangle, \text{ for all } \xi \in D(T)\}, \\ T^*\xi = \zeta, \quad \xi \in D(T^*). \end{cases}$$

Since $D(T)$ is dense in $\mathcal{H}$, T^* is a well-defined linear operator in $\mathcal{H}$ but, unlike the case of bounded operators, $D(T^*)$ may not be dense in $\mathcal{H}$. For example, let f_0 be a bounded measurable function on $\mathbb{R}$ with $f_0 \notin L^2(\mathbb{R})$ and $g_0 \in L^2(\mathbb{R})$. Then, the densely defined linear operator T in $L^2(\mathbb{R})$ given by

$$\begin{cases} D(T) = \{h \in L^2(\mathbb{R}) : \int_{\mathbb{R}} |h(t) f_0(t)| dt < \infty\}, \\ Th = \langle h | f_0 \rangle g_0, \quad h \in D(T) \end{cases}$$

has $D(T^*) = \{0\}$.

The following is immediate.

Proposition B.4.6 *Let T and S be densely defined linear operators in $\mathcal{H}$. Then, the following statements hold:*

(1) *if $T \subset S$, then $S^* \subset T^*$;*
(2) $(\lambda T)^* = \overline{\lambda} T^*$, $\lambda \in \mathbb{C}$;
(3) *if $D(T+S)$ is dense in $\mathcal{H}$, then*

$$(T+S)^* \supset T^* + S^*.$$

In particular, if T or S is bounded, then

$$(T+S)^* = T^* + S^*;$$

(4) *if $D(TS)$ is dense in $\mathcal{H}$, then*

$$(TS)^* \supset S^* T^*.$$

In particular, if T is bounded, then

$$(TS)^* = S^* T^*.$$

The notions of adjoint and closure are intimately related.

Theorem B.4.7 *Let T be a densely defined linear operator in $\mathcal{H}$. The following statements hold:*

(1) *T^* is closed and $\mathcal{G}(T^*) = V(\mathcal{G}(T)^\perp)$, where V is the unitary operator on $\mathcal{H} \oplus \mathcal{H}$ defined by $V(\xi, \eta) = \langle -\eta, \xi \rangle$, $\xi, \eta \in \mathcal{H}$;*
(2) *T is closable, if and only if, $D(T^*)$ is dense in $\mathcal{H}$. If this is true, then $\overline{T} = T^{**}$ $(\equiv (T^*)^*)$;*
(3) *if T is closable, then $\overline{T}^* = T^*$.*

We consider now the particular case of bounded operators.

Theorem B.4.8 *If $(T, D(T))$ is bounded, i.e.,* $\sup\left\{\|T\xi\| : \xi \in D(T), \|\xi\| \leq 1\right\} < \infty$, *then*

(i) *T is closable and $\overline{T} \in \mathcal{B}(\mathcal{H})$;*
(ii) *T^* is everywhere defined in $\mathcal{H}$ and bounded; i.e., $T^* \in \mathcal{B}(\mathcal{H})$.*

Proof To prove (i) we show that (iii) of Proposition B.4.5 holds. Let $\{\xi_n\} \subset D(T)$ be a sequence, such that $\xi_n \to 0$ and $T\xi_n \to \eta$. Since

$$\|T\xi_n\| \leq \|T\| \, \|\xi_n\|, \quad \forall\, n \in \mathbb{N},$$

it follows that $(T\xi_n)$ converges to 0. Thus, $\eta = 0$ and T is closable. If $\xi \in \mathcal{H}$, then there exists a sequence (ξ_n) in $D(T)$, which converges to ξ. Since T is bounded, the sequence $(T\xi_n)$ is Cauchy and thus it converges in $\mathcal{H}$. By the definition of closure itself, it follows then that $\xi \in D(\overline{T})$ and $\overline{T}\xi = \lim_{n\to\infty} T\xi_n$.

The statement (ii) follows immediately from (i) and from Theorem B.3.1. □

Conversely, if T is a closed linear operator *on* $\mathcal{H}$ (that is, defined everywhere), then $T \in \mathcal{B}(\mathcal{H})$, according to the *closed graph theorem* given below (Theorem B.4.10). For the proof we need the following

Lemma B.4.9 *Let $\{\xi_n\}$ be a sequence of vectors in a Hilbert space $\mathcal{H}$. Then, the sequence of numbers $\{\langle\xi_n|\eta\rangle\}$ is bounded, for every fixed $\eta \in \mathcal{H}$, if and only if, the sequence of norms $\{\|\xi_n\|\}$ is bounded.*

Theorem B.4.10 *Every closed linear operator T on $\mathcal{H}$* (i.e., defined everywhere) *is bounded.*

Proof First we show that T^* is continuous. Indeed, take any sequence $\{\eta_n\}$ in $D(T^*)$, which converges to 0. Define $\eta_n' = \|\eta_n\|^{-1/2}\eta_n$, $n \in \mathbb{N}$, so that $\|\eta_n'\| \to 0$, as well. Then, for any $g \in \mathcal{H}$, we have $|\langle T^*\eta_n'|g\rangle| = |\langle\eta_n'|Tg\rangle \leq \|\eta_n'\|\,\|Tg\|$. By Lemma B.4.9, $\{T^*\eta_n'\}$ is a norm-bounded sequence, which implies that

$$\lim_{n\to\infty} T^*\eta_n = \lim_{n\to\infty} \|\eta_n\|^{1/2} T^*\eta_n' = 0.$$

Thus, T^* is continuous. It follows from Theorem B.4.7 that $D(T^{**})$ is dense in $\mathcal{H}$ and T^* is closed, which implies that $D(T^*) = \mathcal{H}$ and T^* is bounded. Hence, $T = T^{**}$ is also bounded. □

B.5 Symmetric and Selfadjoint Operators

Definition B.5.1 Let $(T, D(T))$ be a linear operator in $\mathcal{H}$. We say that T is *symmetric* if

$$\langle T\xi|\eta\rangle = \langle\xi|T\eta\rangle, \quad \forall\, \xi, \eta \in D(T).$$

The definition implies that $T \subset T^*$ and therefore *any symmetric operator is closable. Its closure T^{**}* (see Theorem B.4.7) *is also a symmetric operator.*

Definition B.5.2 A symmetric operator $(T, D(T))$ is said to be *selfadjoint* if $T = T^*$.

Clearly, if T is everywhere defined and symmetric, it is bounded and selfadjoint.

Example B.5.3 We produce some examples showing that the notions of symmetric and selfadjoint operators are really different. We define:

$$\begin{cases} D(S) \equiv \left\{f \in C[0,1] : f(t) - f(0) = \int_0^t f_1(s)\,ds,\, s \in [0,1], \text{ for some } f_1 \in \right. \\ \left. L^2[0,1]\right\},\ Sf = -if_1, \quad f \in D(S) \end{cases}$$

(S is often denoted by $-i\frac{d}{dt}$). We define some operators in $L^2[0,1]$, as follows:

$$\begin{cases} D(T) = \{f \in D(S) : f(1) = f(0) = 0\}, \\ T = S \upharpoonright D(T), \end{cases}$$

and, for any $\alpha \in \mathbb{C}$,

$$\begin{cases} D(T_\alpha) = \{f \in D(S) : f(1) = \alpha f(0)\}, \\ T_\alpha = S \upharpoonright D(T_\alpha). \end{cases}$$

Then, the following facts hold:

1. T is a symmetric operator, which is not selfadjoint and in fact $T^* = S$.
2. For any $\alpha \in \mathbb{C}$, with $|\alpha| = 1$, the operator

$$\begin{cases} D(T_\alpha^*) = \{f \in AC[0,1] : f(0) = \overline{\alpha} f(1)\}, \\ T_\alpha^* = S \upharpoonright D(T_\alpha^*) \end{cases}$$

is a selfadjoint extension of T. Note that $AC[0,1]$ is the algebra of all absolutely continuous $\mathbb{C}$-valued functions on $[0,1]$.

The basic criterion of selfadjointness is provided by the following

Theorem B.5.4 *A closed symmetric operator T is selfadjoint, if and only if,* $\mathrm{Ker}(T^* \pm iI) = \{0\}$.

Bibliography

Books and Theses

1. N.I. Akhiezer, I.M. Glazman, *Theory of Linear Operators in Hilbert space* (M. Dekker, New York, 1993)
2. J.-P. Antoine, A. Inoue, C. Trapani, *Partial *-Algebras and Their Operator Realizations*. Mathematics and Its Applications, vol. 553 (Kluwer, Dordrecht, 2002).
3. J.-P. Antoine, C. Trapani, *Partial Inner Product Spaces*. Theory and Aplications, Lecture Notes in Mathematics, vol. 1986 (Springer, Berlin, 2009)
4. F.F. Bonsall, J. Duncan, *Complete Normed Algebras* (Springer, Berlin, 1973)
5. N. Bourbaki, *Espaces Vectoriels Topologiques* (Hermann, Paris, 1966)
6. O. Bratteli, D.W. Robinson, *Operator Algebras and Quantum Statistical Mechanics I, II* (Springer, Berlin, 1979)
7. H. G. Dales, *Banach Algebras and Automatic Continuity* (Oxford Science Publications, Oxford, 2000)
8. R.S. Doran, V.A. Belfi, *Characterizations of C*-Algebras* (M. Dekker, New York, 1986)
9. M. Fragoulopoulou, *Topological Algebras with Involution* (North-Holland, Amsterdam, 2005)
10. M. Fragoulopoulou, A. Inoue, M. Weigt, I. Zarakas, *Genaralized B^*-Algebras. Applications* (in preparation)
11. I.M. Gel'fand, N.Ya. Vilenkin, *Generalized Functions*, vol. IV (Academic Press, New York, 1964)
12. A. Inoue, *Tomita–Takesaki Theory in Algebras of Unbounded Operators*. Lecture Notes in Mathematics, vol. 1699 (Springer, Berlin, 1998)
13. J.-P. Jurzak, *Unbounded Noncommutative Integration* (D. Reidel, Dordrecht, 1986)
14. R.V. Kadison, J.R. Ringrose, *Fundamentals of the Theory of Operator Algebras*, vol. I (Academic Press, New York, 1983)
15. R.V. Kadison, J.R. Ringrose, *Fundamentals of the Theory of Operator Algebras*, vol. II (Academic Press, New York, 1986)
16. T. Kato, *Perturbation Theory for Linear Operators* (Springer, Berlin, 1976)
17. G. Köthe, *Topological Vector Spaces*, vols. I and II (Springer, Berlin, 1969/1979)
18. T.W. Palmer, *Banach Algebras and the General Theory of *-Algebras*, vol. I (Cambridge University Press, Cambridge, 1995)
19. T.W. Palmer, *Banach Algebras and the General Theory of *-Algebras*, vol. II (Cambridge University Press, Cambridge, 2001)

M. Fragoulopoulou, C. Trapani, *Locally Convex Quasi *-Algebras and their Representations*, Lecture Notes in Mathematics 2257,
https://doi.org/10.1007/978-3-030-37705-2

20. M. Reed, B. Simon, *Methods of Modern Mathematical Physics. I. Functional Analysis* (Academic Press, New York, 1972, 1980)
21. C.E. Rickart, *General Theory of Banach Algebras* (Van Nostrand Reinhold, 1974)
22. S. Sakai, *C*-Algebras and W*-Algebras* (Springer, Berlin, 1971)
23. K. Schmüdgen, *Unbounded Operator Algebras and Representation Theory* (Akademie, Berlin, 1990)
24. S. Strătilă, *Modular Theory in Operator Algebras* (Abacus Press, Tunbridge Wells, 1981)
25. S. Strătilă, L. Zsidó, *Lectures on von Neumann Algebras* (Editura Academiei, Bucharest and Abacus Press, Tunbridge Wells, Kent, 1979)
26. M. Takesaki, *Tomita's Theory of Modular Hilbert Algebras and Its Applications*. Lecture Notes in Mathmatics, vol. 128 (Springer, Berlin, 1970)
27. M. Takesaki, *Theory of Operator Algebras*, vols. I, II (Springer, New York, 1979)
28. F. Treves, *Topological Vector Spaces, Distributions and Kernels* (Academic Press, New York, 1967)
29. H. Triebel, *Interpolation Theory, Function Spaces, Differential Operators* (Wiley, New York, 1998)
30. A.C. Zaanen, *Integration*, Chap. 15, 2nd ed. (North-Holland, Amsterdam,1961)

Articles

31. M.S. Adamo, C. Trapani, Representable and continuous functionals on Banach quasi *-algebras. Mediterr. J. Math. **14**, 157–181 (2017)
32. G.R. Allan, A spectral theory for locally convex algebras. Proc. Lond. Math. Soc. **15**, 399–421 (1965)
33. G.R. Allan, On a class of locally convex algebras. Proc. Lond. Math. Soc. **17**, 91–114 (1967)
34. J.-P. Antoine, A. Inoue, C. Trapani, Partial *-algebras of closable operators. I. The basic theory and the abelian case. Publ. RIMS, Kyoto Univ. **26**, 359–395 (1990)
35. J.-P. Antoine, A. Inoue, C. Trapani, Partial *-algebras of closable operators. II. States and representations of partial *-algebras. Publ. RIMS, Kyoto Univ. **27**, 399–430 (1991)
36. J.-P. Antoine, W. Karwowski, Partial *-algebras of closed linear operators in Hilbert space. Publ. RIMS, Kyoto Univ. **21**, 205–236 (1985). Add./Err. ibid. **22**, 507–511 (1986)
37. J.-P. Antoine, F. Mathot, Partial *-algebras of closed operators and their commutants. I. General structure. Ann. Inst. H. Poincaré **46**, 299–324 (1987)
38. J.-P. Antoine, F. Mathot, C. Trapani, Partial *-algebras of closed operators and their commutants. II. Commutants and bicommutants. Ann. Inst. H. Poincaré **46**, 325–351 (1987)
39. J.-P. Antoine, C. Trapani, F. Tschinke, Bounded elements in certain topological partial *-algebras. Stud. Math. **203**, 222–251 (2011)
40. H. Araki, E.J. Woods, Representations of the canonical commutation relations describing a nonrelativistic infinite free Bose gas. J. Math. Phys. **4**, 637–662 (1963)
41. R. Arens, The space L^{ω} and convex topological rings. Bull. Am. Math. Soc. **52**, 931–935 (1946)
42. R. Ascoli, G. Epifanio, A. Restivo, On the mathematical decription of quantized fields. Commun. Math. Phys. **18**, 291–300 (1970); *-algebrae of unbounded operators in scalar-product spaces. Riv. Mat. Univ. Parma **3**, 21–32 (1974)
43. J. Arhippainen, J. Kauppi, Nachbin spaces as CQ*-algebras. J. Math. Anal. Appl. **400**, 568–574 (2013)
44. F. Bagarello, M. Fragoulopoulou, A. Inoue, C. Trapani, The completion of a C*-algebra with a locally convex topology. J. Operator Theory **56**, 357–376 (2006)
45. F. Bagarello, M. Fragoulopoulou, A. Inoue, C. Trapani, Locally convex quasi C*-normed algebras. J. Math. Anal. Appl. **366**, 593–606 (2010)
46. F. Bagarello, A. Inoue, C. Trapani, Unbounded C*-seminorms and *-representations of partial *-algebras. Z. Anal. Anw. **20**, 295–314 (2001)

47. F. Bagarello, A. Inoue, C. Trapani, Some classes of topological quasi *-algebras. Proc. Am. Math. Soc. **129**, 2973–2980 (2001)
48. F. Bagarello, C. Trapani, States and representations of CQ*-algebras. Ann. Inst. H. Poincaré **61**, 103–133 (1994)
49. F. Bagarello, C. Trapani, CQ*-algebras: Structure properties. Publ. RIMS, Kyoto Univ. **32**, 85–116 (1996)
50. F. Bagarello, C. Trapani, L^p-spaces as quasi *-algebras. J. Math. Anal. Appl. **197**, 810–824 (1996)
51. F. Bagarello, C. Trapani, Morphisms of certain Banach C*-modules. Publ. RIMS, Kyoto Univ. **36**, 681–705 (2000)
52. F. Bagarello, C. Trapani, S. Triolo, Quasi *-algebras of measurable operators. Stud. Math. **172**, 289–305 (2006)
53. F. Bagarello, A. Inoue, C. Trapani, Representable linear functionals on partial *-algebras. Mediterr. J. Math. **9**, 153–163 (2012)
54. S.J. Bhatt, M. Fragoulopoulou, A. Inoue, Existence of well-behaved $*$-representations of locally convex $*$-algebras. Math. Nachr. **279**, 86–100 (2006)
55. S.J. Bhatt, A. Inoue, H. Ogi, Unbounded C*-seminorms and unbounded C*-spectral algebras. J. Operator Theory **45**, 53–80 (2001)
56. H.J. Borchers, Decomposition of families of unbounded operators. RCP 25 (Strasbourg) **22**, 26–53 (1975); also in *Quantum Dynamics: Models and Mathematics*, ed. by L. Streit. Acta Physics Australasian Supplementary, vol. XVI (1976), pp. 15–46
57. P.G. Dixon, Generalized B*-algebras. Proc. Lond. Math. Soc. **21**, 693–715 (1970)
58. G. Epifanio, C. Trapani, Quasi *-algebras valued quantized fields. Ann. Inst. H. Poincaré **46**, 175–185 (1987)
59. M. Fragoulopoulou, A. Inoue, K.-D. Kürsten, On the completion of a C*-normed algebra under a locally convex algebra topology. Contemp. Math. **427**, 89–95 (2007)
60. I.M. Gelfand, M.A. Naimark, On the embedding of normed rings into the ring of operators in Hilbert space. Mat. Sbornik **12**, 197–213 (1943)
61. R. Haag, D. Kastler, An algebraic approach to quantum field theorem. J. Math. Phys. **5**, 848–861 (1964)
62. S.S. Horuzhy, A.V. Voronin, Field algebras do not leave field domains invariant. Commun. Math. Phys. **102**, 687–692 (1988)
63. A. Inoue, K.-D. Kürsten, On C*-like locally convex *-algebras. Math. Nachr. **235**, 51–58 (2002)
64. R.V. Kadison, Algebras of unbounded functions and operators. Expo. Math. **4**, 3–33 (1986)
65. G. Lassner, Topological algebras of operators. Rep. Math. Phys. **3**, 279–293 (1972)
66. G. Lassner, Topologien auf Op*-algebren. Wiss. Z. KMU Leipzig, Math. Naturwiss. R. **24**, 465–471 (1975)
67. G. Lassner, Topological algebras and their applications in Quantum Statistics. Wiss. Z. KMU Leipzig, Math. Naturwiss. R. **30**, 572–595 (1981)
68. G. Lassner, Algebras of unbounded operators and quantum dynamics. Physica **A 124**, 471–480 (1984)
69. G. Lassner, G.A. Lassner, C. Trapani, Canonical commutation relation on the interval. J. Math. Phys. **28**, 174–177 (1987)
70. E. Nelson, Analytic vectors. Ann. Math. **70**, 572–615 (1959)
71. E. Nelson, Note on non-commutative integration. J. Funct. Anal. **15**, 103–116 (1974)
72. R.T. Powers, Self-adjoint algebras of unbounded operators. II. Trans. Am. Math. Soc. **187**, 261–293 (1974)
73. A. Russo, C. Trapani, Quasi *-algebras and multiplication of distributions. J. Math. Anal. Appl. **215**, 423–442 (1997)
74. K. Schmüdgen, On well-behaved unbounded representations of *-algebras. J. Operator Theory **48**, 487–502 (2002)
75. K. Schmüdgen, A strict Positivstellensatz for the Weyl algebra. Math. Ann. **331**, 779–794 (2005)

76. I.E. Segal, A noncommutative extension of abstract integration. Ann. Math. **57**, 401–457 (1953)
77. C. Trapani, Some seminorms on quasi *-algebras. Studia Math.. **158**, 99–115 (2003); Erratum/Addendum. Stud. Math. **160**, 101–101 (2004)
78. C. Trapani, C*-seminorms on partial *-algebras: an overview, in *Proceedings of the conference on Topological Algebras and Related Topics*, Bedlewo, 2003. Polish Academy of Sciences, vol. 67 (Banach Center Publications, Warsaw, 2005), pp. 369–384.
79. C. Trapani, Unbounded C*-seminorms, biweights and *-representations of partial *-algebras: a review. Int. J. Math. Math. Sci. **2006**, 79268 (2006)
80. C. Trapani, Bounded elements and spectrum in Banach quasi *-algebras. Stud. Math. **172**, 249–273 (2006)
81. C. Trapani, Bounded and strongly bounded elements of Banach quasi *-algebras, in *Proceedings of the 5th International Conference on Topological Algebras and Applications*, Athens, 2005. Contemporary Mathematics, vol. 427 (2007), pp. 417–424
82. C. Trapani, *-Representations, seminorms and structure properties of normed quasi *-algebras. Stud. Math. **186**, 47–75 (2008)
83. C. Trapani, Locally convex quasi *-algebras of operators, in *Proceedings of the Conference FAO 2010*, Krakow. Complex Analysis and Operator Theory, vol 6 (2012), pp. 719–728
84. C. Trapani, M. Fragoulopoulou, Locally convex quasi *-algebras: an overview, in *Proceedings of the International Conference on Topological Algebras and Their Applications, ICTAA 2018*. Mathematical Studies. Estonia Mathematics Society, vol. 7 (Tartu, 2018), pp. 121–136
85. C. Trapani, S. Triolo, Auxiliary seminorms and the structure of a CQ*-algebra, in *Proceedings of the 5th International Conference on Functional Analysis and Approximation Theory*. Rendiconti del Circolo Matematico di Palermo, vol. 76 (2005), pp. 587–601
86. C. Trapani, S. Triolo, Representations of modules over a $*$-algebra and related seminorms. Stud. Math. **184**, 133–148 (2008)
87. C. Trapani, F. Tschinke, Unbounded C*-seminorms and biweights on partial *-algebras. Mediterr. J. Math. **2**, 301–313 (2005)
88. C. Trapani, F. Tschinke, Partial *-algebras of distributions. Publ. RIMS, Kyoto Univ. **41**, 259–279 (2005)
89. S. Triolo, Possible extensions of the noncommutative integral. Rend. Circ. Mat. Palermo **60**(2), 409–416 (2011)
90. A. Van Daele, A new approach to the Tomita–Takesaki theory of generalized Hilbert algebras. J. Funct. Anal. **15**, 378–393 (1974)
91. A.N. Vasil'ev, Theory of representations of a topological (non–Banach) involutory algebra. Theor. Math. Phys. **2**, 113–123 (1970)
92. I. Vidav, On some *-regular rings. Acad. Serbe Sci. Publ. Inst. Math. **13**, 73–80 (1959)
93. L. Waelbroeck, Algèbres commutatives: Éléments réguliers. Bull. Soc. Math. Belg. **9**, 42–49 (1957)
94. S. Warner, Inductive limits of normed algebras. Trans. Amer. Math. Soc. **82**, 190–216 (1956)
95. B. Yood, C*-seminorms. Stud. Math. **118**, 19–26 (1996)

Suggested Readings

J. Dixmier, *Les algèbres d'opérateurs dans l'espace hilbertien (Algèbres de von Neumann)* (Gauthier-Villars, Paris, 1957)

A. Mallios, *Topological Algebras. Selected Topics* (North-Holland, Amsterdam, 1986)

E. Michael, *Locally Multiplicatively Convex Topological Algebras*, vol. 11 (Memoirs of the American Mathematical Society, 1952)

M.S. Adamo, C. Trapani, Unbounded derivations and *-automorphisms groups of Banach quasi *-algebras. Ann. Mat. Pura Appl. **198**(5), 1711–1729 (2019)

G.R. Allan, A note on B*-algebras. Math. Proc. Camb. Philos. Soc. **61**, 29–32 (1965)

J.-P. Antoine, G. Bellomonte, C. Trapani, Fully representable and *-semisimple topological partial *-algebras. Stud. Math. **208**, 167–194 (2012)

J.-P. Antoine, W. Karwowski, Partial *-algebras of closed operators, in *Quantum Theory of Particles and Fields*, ed. by B. Jancewicz, J. Lukierski (World Scientific, Singapore, 1983), pp. 13–30

J.-P. Antoine, C. Trapani, F. Tschinke, Continuous *-homomorphisms of Banach partial *-algebras. Mediterr. J. Math. **4**, 357–373 (2007)

J.-P. Antoine, C. Trapani, F. Tschinke, Spectral properties of partial *-algebras. Mediterr. J. Math. **7**, 123–142 (2010)

F. Bagarello, Algebras of unbounded operators and physical applications: a survey. Rev. Math. Phys. **19**(3), 231–272 (2007)

G. Bellomonte, C. Trapani, Quasi *-algebras and generalized inductive limits of C*-algebras. Stud. Math. **202**, 165–190 (2011)

G. Bellomonte, S. Di Bella, C. Trapani, Bounded elements of C*-inductive spaces. Ann. Mat. Pura Appl. **195**, 343–356 (2016)

S.J. Bhatt, Representability of positive functionals on abstract *-algebras without identity with applications to locally convex *-algebras. Yokohama Math. J. **29**, 7–16 (1981)

S.J. Bhatt, A. Inoue, K.-D. Kürsten, Well-behaved unbounded operator representations and unbounded C*-seminorms. J. Math. Soc. Jpn. **56**, 417–445 (2004)

S.J. Bhatt, A. Inoue, H. Ogi, On C*-spectral algebras. Rend. Circ. Math. Palermo **56**, 207–213 (1998)

S.J. Bhatt, A. Inoue, H. Ogi, Admissibility of weights on non-normed *-algebras. Trans. Am. Math. Soc. **351**, 183–208 (1999)

S.P. Gudder, W. Scruggs, Unbounded representations of *-algebras. Pac. J. Math. **70**, 369–382 (1977)

I. Ikeda, On unbounded *-representations of *-algebras. Nihonkai Math. J. **5**, 43–59 (1994)

M. Fragoulopoulou, C. Trapani, *Locally Convex Quasi *-Algebras and their Representations*, Lecture Notes in Mathematics 2257,
https://doi.org/10.1007/978-3-030-37705-2

I. Ikeda, A. Inoue, Invariant subspaces of closed *-representations. Proc. Am. Math. Soc. **116**, 737–745 (1992)

I. Ikeda, A. Inoue, On types of positive linear functionals of *-algebras. J. Math. Anal. Appl. **173**, 276–288 (1993)

I. Ikeda, A. Inoue, M. Takakura, Unitary equivalence of unbounded *-representations of *-algebras. Math. Proc. Camb. Philos. Soc. **122**, 269–279 (1997)

A. Inoue, Positive linear functionals on topological *-algebras. J. Math. Anal. Appl. **68**, 17–24 (1979)

K.-D. Kürsten, The completion of the maximal O_p*-algebra on a Fréchet domain. Publ. RIMS Kyoto Univ. **22**, 151–175 (1986)

T.W. Palmer, Spectral algebras. Rocky Mountain J. Math. **22**, 293–328 (1992)

N.C. Phillips, Inverse limits of C*-algebras. J. Operator Theory **19**, 159–195 (1988)

J.D. Powell, Representations of locally convex *-algebras. Proc. Am. Math. Soc. **44**, 341–346 (1974)

R.T. Powers, Self-adjoint algebras of unbounded operators. I. Commun. Math. Phys. **21**, 85–124 (1971)

R.T. Powers, Algebras of unbounded operators. Proc. Sympos. Pure Math. **38**, 389–406 (1982)

K. Schmüdgen, Uniform topologies and strong operator topologies on polynomial algebras and on the algebra of CCR. Rep. Math. Phys. **10**, 369–384 (1976)

C. Trapani, Quasi *-algebras of operators and their applications. Rev. Math. Phys. **7**, 1303–1332 (1995)

C. Trapani, CQ*-algebras of operators and application to Quantum Models, in *Proceedings of the 2nd ISAAC Conference*, Fukuoka, vol. 1 (Kluwer, Dordecht, 2000), pp. 679–685

C. Trapani, Sesquilinear forms on certain Banach C*-modules, in *Proceedings of the IV International Conference on Functional Analysis and Approximation Theory*. Rendiconti del Circolo Matematico di Palermo, vol. 68 (2002), pp. 855–864

C. Trapani, CQ*-algebras of operators: density properties, in *Operator Theory and Banach Algebras*, ed. by M. Chidami, R. Curto, M. Mbekhta, F.H. Vasilescu, J. Zemanek (Theta, Bucharest, 2003)

C. Trapani, F. Tschinke, Faithfully representable topological *-algebras: some spectral properties, in *Topological Algebras and Their Applications*. De Gruyter Proceedings in Mathematics (De Gruyter, Berlin, 2018), pp. 233–249

Index

M. Fragoulopoulou, C. Trapani, *Locally Convex Quasi *-Algebras and their Representations*, Lecture Notes in Mathematics 2257,
https://doi.org/10.1007/978-3-030-37705-2

LECTURE NOTES IN MATHEMATICS

Springer

Editors in Chief: J.-M. Morel, B. Teissier;

Editorial Policy

1. Lecture Notes aim to report new developments in all areas of mathematics and their applications – quickly, informally and at a high level. Mathematical texts analysing new developments in modelling and numerical simulation are welcome.

   Manuscripts should be reasonably self-contained and rounded off. Thus they may, and often will, present not only results of the author but also related work by other people. They may be based on specialised lecture courses. Furthermore, the manuscripts should provide sufficient motivation, examples and applications. This clearly distinguishes Lecture Notes from journal articles or technical reports which normally are very concise. Articles intended for a journal but too long to be accepted by most journals, usually do not have this "lecture notes" character. For similar reasons it is unusual for doctoral theses to be accepted for the Lecture Notes series, though habilitation theses may be appropriate.

2. Besides monographs, multi-author manuscripts resulting from SUMMER SCHOOLS or similar INTENSIVE COURSES are welcome, provided their objective was held to present an active mathematical topic to an audience at the beginning or intermediate graduate level (a list of participants should be provided).

   The resulting manuscript should not be just a collection of course notes, but should require advance planning and coordination among the main lecturers. The subject matter should dictate the structure of the book. This structure should be motivated and explained in a scientific introduction, and the notation, references, index and formulation of results should be, if possible, unified by the editors. Each contribution should have an abstract and an introduction referring to the other contributions. In other words, more preparatory work must go into a multi-authored volume than simply assembling a disparate collection of papers, communicated at the event.

3. Manuscripts should be submitted either online at www.editorialmanager.com/lnm to Springer's mathematics editorial in Heidelberg, or electronically to one of the series editors. Authors should be aware that incomplete or insufficiently close-to-final manuscripts almost always result in longer refereeing times and nevertheless unclear referees' recommendations, making further refereeing of a final draft necessary. The strict minimum amount of material that will be considered should include a detailed outline describing the planned contents of each chapter, a bibliography and several sample chapters. Parallel submission of a manuscript to another publisher while under consideration for LNM is not acceptable and can lead to rejection.

4. In general, **monographs** will be sent out to at least 2 external referees for evaluation.

   A final decision to publish can be made only on the basis of the complete manuscript, however a refereeing process leading to a preliminary decision can be based on a pre-final or incomplete manuscript.

   Volume Editors of **multi-author works** are expected to arrange for the refereeing, to the usual scientific standards, of the individual contributions. If the resulting reports can be forwarded to the LNM Editorial Board, this is very helpful. If no reports are forwarded or if other questions remain unclear in respect of homogeneity etc, the series editors may wish to consult external referees for an overall evaluation of the volume.

5. Manuscripts should in general be submitted in English. Final manuscripts should contain at least 100 pages of mathematical text and should always include

   – a table of contents;
   – an informative introduction, with adequate motivation and perhaps some historical remarks: it should be accessible to a reader not intimately familiar with the topic treated;
   – a subject index: as a rule this is genuinely helpful for the reader.
   – For evaluation purposes, manuscripts should be submitted as pdf files.

6. Careful preparation of the manuscripts will help keep production time short besides ensuring satisfactory appearance of the finished book in print and online. After acceptance of the manuscript authors will be asked to prepare the final LaTeX source files (see LaTeX templates online: https://www.springer.com/gb/authors-editors/book-authors-editors/manuscriptpreparation/5636) plus the corresponding pdf- or zipped ps-file. The LaTeX source files are essential for producing the full-text online version of the book, see http://link.springer.com/bookseries/304 for the existing online volumes of LNM). The technical production of a Lecture Notes volume takes approximately 12 weeks. Additional instructions, if necessary, are available on request from lnm@springer.com.

7. Authors receive a total of 30 free copies of their volume and free access to their book on SpringerLink, but no royalties. They are entitled to a discount of 33.3 % on the price of Springer books purchased for their personal use, if ordering directly from Springer.

8. Commitment to publish is made by a *Publishing Agreement*; contributing authors of multiauthor books are requested to sign a *Consent to Publish form*. Springer-Verlag registers the copyright for each volume. Authors are free to reuse material contained in their LNM volumes in later publications: a brief written (or e-mail) request for formal permission is sufficient.

Addresses:
Professor Jean-Michel Morel, CMLA, École Normale Supérieure de Cachan, France
E-mail: moreljeanmichel@gmail.com

Professor Bernard Teissier, Equipe Géométrie et Dynamique,
Institut de Mathématiques de Jussieu – Paris Rive Gauche, Paris, France
E-mail: bernard.teissier@imj-prg.fr

Springer: Ute McCrory, Mathematics, Heidelberg, Germany,
E-mail: lnm@springer.com

Printed in the United States
By Bookmasters